HARCOURT BRACE COLLEGE OUTLINE SERIES

GEOMETRY
Plane and Practical

Bruce Stephan

Webb Institute of Naval Architecture
Glen Cove, New York

Harcourt Brace College Publishers
Fort Worth   Philadelphia   San Diego
New York   Orlando   Austin   San Antonio
Toronto   Montreal   London   Sydney   Tokyo

Requests for permission to make copies of any part of the work should be mailed to:
Permissions Department
Harcourt Brace & Company
6277 Sea Harbor Drive
Orlando, Florida 32887

Printed in the United States of America

Library of Congress Cataloging-in-Publication Data

Stephan, Bruce
    Geometry: plane and practical / Bruce Stephan
        p. cm --(Harcourt Brace & Company college outline series)
    (Books for professionals)
    Includes index.
    ISBN 0-15-601664-8
    1. Geometry, Plane.   I. Title.   II. Series.   III. Series: Books for professionals.
    QA474.S83   1991
    516.22---de20

ISBN: 0-15-601664-8   91-20178
                                                                     CIP

First Edition

    1 2 3 4 5 6 145   10 9 8 7 6 5

PREFACE

Although I've been teaching mathematics in one form or another for twenty-nine years, I've never found an individual who couldn't understand the basic principles of arithmetic, algebra, and geometry if those principles were explained patiently, in a slow, deliberate fashion. The fact is, mathematics is not a set of rules carved in stone to crush the masses; nor is it a special, secret language shared by the technical elite. Instead, it is an attempt by ordinary, everyday human beings to describe the world around them. And, as a model of reality, mathematics works pretty well. Although it cannot tell us what "reality" *really* is, mathematics does let us build a descriptive system based on what appear to be "self-evident" facts. This system, then, can be developed to include other, not-so-evident facts that we can test in the real world. Thus, because everyone must deal with reality, mathematics should be for everyone.

Geometry is all about space. Space is the most intimate thing we experience. We live in it daily, and—if the physicists and chemists are right—we're even made up of it! Euclidean geometry is just one model of our everyday space. Although there are other models that are equally good at explaining space, this one has been around for a long time, and it's been very successful. Besides, we have to start someplace, and the Euclidean model provides a good beginning.

Euclidean geometry starts with the ideas of a point, a line, and a plane. These ideas seem innocuous enough—everybody knows what we mean by them. Then Euclidean geometry takes off with the idea of two lines that can go on forever without touching—parallel lines. We can take these ideas as our jumping-off place. And once we've jumped off into Euclidean space, we can go on to create a logical structure which leads, step-by-step, from one consequence to another.

In this Outline, I've tried to follow a strong line of intuitive reasoning, aimed at developing a practical understanding that can leave readers with the ability to use geometry in their daily lives. Given this practical aim, I've done some unorthodox things at times. I've emphasized constructions, I present all triangles as a sum or difference of right triangles, and I use the congruence of right triangles to establish the standard forms of congruence for other triangles. Also, I introduce simple trigonometric concepts early, and end the book with a more formal introduction to trigonometry. And since coordinate geometry is so useful in describing space, I give a brief introduction to the Cartesian coordinate system—and then shamelessly use it if it offers an easily comprehensible explanation of what's happening. So, the Pythagorean Theorem is present, but so are the more general Laws of Cosines and Sines. (These laws are used to find various parts and areas of triangles from a bare minimum of information.)

I believe that your brain is a little like a file cabinet: you can cram only so many facts into it. This is why I've written this Outline the way I have, with the emphasis on understanding, not memorization. It's just plain easier to remember a minimum of facts, along with the reasoning process that lets you derive other facts, than it is to retain a whole collection of axioms, postulates, and theorems. It's also why I've concentrated on how right triangles make up other triangles, how other figures can be decomposed into triangles for their properties, and how circles can be viewed as the limit of regular polygons.

Finally, this book has problems—lots of them. This is because mathematics is not a spectator sport. Read the material, look at the Example problems, then try the Solved Problems and Review Exercises at the end of each chapter. Then do it all over again. You'll get out of this book what you put into it. And, whatever you do, remember that mathematics is something anyone can do. . . it just takes a little practice.

* *

I'd like to dedicate this book to Johns Hopkins University, without whose inspiration—in the form of tuition bills for my children—I would never have put my ideas into print.

Piermont, New York   BRUCE STEPHAN

CONTENTS

FUNDAMENTAL CONCEPTS

THIS CHAPTER IS ABOUT

☑ **Points, Lines, and Planes**
☑ **Distance**
☑ **Circles**
☑ **Angles**
☑ **Angular Measure**
☑ **Parallel Lines**
☑ **Analytic Methods: Coordinate Representation**

Geometry is a branch of mathematics that allows us to describe the space that surrounds us. In **Euclidean plane geometry** we build a model of this space by reasoning logically from a set of "self-evident" basic assumptions called **postulates** or **axioms**. The postulates of geometry are statements about points, lines, and planes.

1-1. Points, Lines, and Planes

A. Points

A **point**, which was described by Euclid as "that which has no part," is best taken as an **undefined term**.

note: Calling a term "undefined" is a way of saying a concept is so fundamental that we can't define it in terms of anything more basic. But we *can* describe the characteristics or properties associated with undefined terms.

Intuitively, you can think of a point as a "dot" on a flat piece of paper or as a "spot" in our three-dimensional world. A point is "something" that has no length or width or depth. In short, a point is not a thing at all, but a location in space.

B. Lines, segments, and rays

A **line**, like a point, is an undefined term. Lines have no width or depth, but a line may be described as a set of points.

- For every point *A* and every other point *B*, there is a unique (one and only one) line *L* that passes through both *A* and *B*.

So any two distinct points determine a line!

We can represent a line by making two dots on a piece of paper and then placing a ruler through those dots, as in Figure 1-1. When we trace the edge of the ruler with a pencil, we have drawn part of a line. But, while the ruler tracing ends, the concept of a line does not.

- A line continues infinitely in both directions.

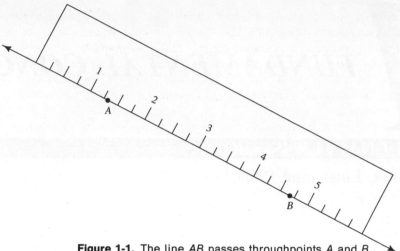

Figure 1-1. The line *AB* passes throughpoints *A* and *B*.

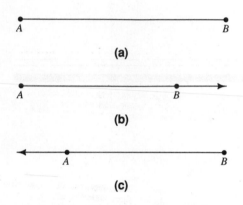

Figure 1-2. (a) Line segment *AB*. **(b)** Ray *AB*. **(c)** Ray *AB*.

To emphasize this infinite extension, we use the notation $\overleftrightarrow{AB}$ for the line determined by points *A* and *B*, where the arrowheads indicate that the line extends infinitely beyond the two points.

The *part* of a line that falls between point *A* and another point *B* is called a **line segment**, which is denoted simply by *AB*. Figure 1-2a represents a line segment.

Similarly, the part of the line that extends from *A* through *B* and on infinitely beyond *B* (Figure 1-2b) is called a **ray** and denoted by $\overrightarrow{AB}$, while ray $\overleftarrow{AB}$ extends from *B* through *A* (Figure 1-2c). The direction of the arrow indicates the direction of the ray's infinite extension.

C. Planes

In two-dimensional Euclidean plane geometry, a **plane**, another undefined term, is simply the set of all points and lines under consideration. Intuitively, a plane is a flat surface—like this page—that is extended infinitely in all directions. If we extend each line segment that can be drawn on the page's surface to a full line, the result of all these extensions is our plane!

EXAMPLE 1-1: For points *A*, *B*, *C*, and *D* of Figure 1-3a, draw (1) $\overleftrightarrow{AD}$, (2) $\overrightarrow{AC}$, (3) *AB*, and (4) $\overleftarrow{BD}$.

Solution: Lay a ruler along each pair of points and trace it from point to point.

(1) Since the arrowheads for *AD* are over both *A* and *D*, *AD* is a full line. Extend the tracing past *A* and *D* and draw an arrowhead on each end.
(2) For $\overrightarrow{AC}$, start in the same way; but since the arrowhead is over *C* only, $\overrightarrow{AC}$ is a ray. Extend the tracing past *C* and place an arrowhead on that extension's end, but don't extend your tracing past *A*.
(3) *AB* is a line segment. Your ruler tracing of *AB* should end at both *A* and *B*.
(4) $\overleftarrow{BD}$ is also a ray and your tracing of it should extend past *B* but not past *D*.

Your drawing should look like Figure 1-3b.

1-2. Distance
A. Properties of distance

Everybody knows what "distance" is—that is, we all have some idea of how far one thing is from another. And we all have some sense of what it means to speak of the "distance between two points *A* and *B*." This distance, which we'll initially denote $d(A, B)$, is a real number that conveys a sense of how "close" A is to B or, equivalently, how "close" B is to A. But all our intuition about and experience of distance do not make a *definition*. (Notice that we've put the word "close" in quotation marks. This is because it is, in effect, a "weasel" word—if we can say what "close" is, we must have a meaning for "distance.") How, then, can we define distance and be sure that our definition is a good one?

We know that a good definition is one that includes all the essential properties of the thing we are trying to define. Therefore, before we attempt to define distance, let's list all of the important properties that must be covered by any definition of distance. Then we can say that *any* way of assigning real numbers to a pair of points (A, B) that satisfies the properties listed is a "distance" between A and B.

$d(A, B) \geq 0$ — The number that represents the distance between two points A and B must not be negative.

$d(A, A) = 0$ — The distance from A to itself is the smallest possible distance—zero. Or, we can say that the closest any point B can get to point A is the distance between A and B when B *coincides with* (is in the same place as) A.

$d(A, B) = d(B, A)$ — The distance from A to B is the same as the distance from B to A. (That is, distance doesn't depend on where we start measuring.)

$d(A, B) \leq d(A, C) + d(C, B)$ — The shortest distance between two points A and B is the straight line segment AB.

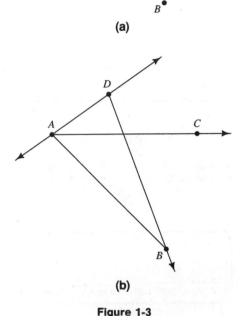

(a)

(b)

Figure 1-3

note: The property $d(A, B) \leq d(A, C) + d(C, B)$, which is sometimes called the **triangle inequality**, is easier to see if you draw it. Start with a line segment AB, as in Figure 1-4a. Then add any point C to the drawing, as in Figure 1-4b. Now, measure $d(A, B)$, $d(A, C)$, and $d(C, B)$. You'll see that the distance between A and B is never greater than the distance between A and C plus the distance between B and C, no matter where you put C.

Enough gobbledygook! Let's get down to something concrete!

B. Measuring distance

To measure the distance between any two points A and B in a plane, we use a device, such as a ruler, marked with standard units of measurement, such as feet or meters. These units are measures of *length*. We place the ruler on the plane so that it passes through the given points.

- The **distance** between two points A and B may then be defined as the number of scribe marks on the ruler between A and B.

If the points A and B do not fall exactly on the scribe marks, we must refine the resolution of the ruler by choosing smaller standard units. In the metric system of measurement, for instance, the standard unit of length is one *meter* (1 m). We refine the resolution of a metric ruler by dividing each meter into 10 equal *decimeters* (1 m = 10 dm). If points A and B do not fall on the decimeter scribe marks, we can refine the ruler's resolution again by dividing each decimeter into 10 equal *centimeters* (1 dm = 10 cm). And if

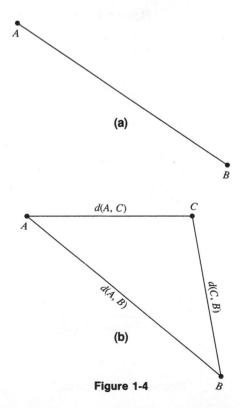

(a)

(b)

Figure 1-4

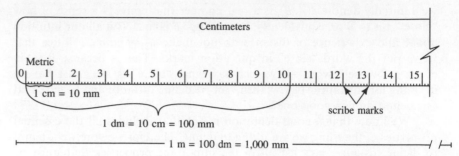

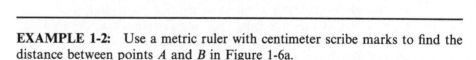

Figure 1-5. A metric ruler with centimeter and millimeter scribe marks.

A and *B* do not fall on the centimeter scribe marks, we can repeat the refining process by dividing each centimeter into 10 equal *millimeters* (1 cm = 10 mm; see Figure 1-5). In fact, this refining process may be repeated as many times as is necessary to get an exact measurement of the distance between *A* and *B*.

note: You may be wondering where these "standard units of measurement" come from. In fact, the meter is an internationally agreed-upon length, now defined as the distance light travels in 1/299,792,458 seconds.

The ruler we choose to measure the distance from *A* to *B* need not even extend from *A* to *B*. If our ruler isn't long enough, we can stretch a taut piece of string end to end over our standard ruler and then tally up the scribe marks. This tally is now our distance.

Having seen that we can measure distance by comparing it to a standard device inscribed with known units of length measurement, we have a way to define distance itself:

- The **distance** $d(A, B)$ is the *length* of the line segment AB between the endpoints of the segment.

EXAMPLE 1-2: Use a metric ruler with centimeter scribe marks to find the distance between points *A* and *B* in Figure 1-6a.

Solution: Place the ruler across *A* and *B* as in Figure 1-6b with scribe mark 0 on point *A*. The scribe mark labeled 4 falls over *B*. The distance (or *length* of line segment *AB*) is 4 centimeters.

1-3. Circles

Now that we have distance, we can have *circles*.

- A **circle** is a set of all points in a plane that are a fixed distance from one chosen point 0. The point 0 is called the **center** of the circle and the fixed distance is called the **radius**.

This definition is a recipe for drawing a circle in the plane. Once we know the center of the circle, we may place the point of a compass at that location, as illustrated in Figure 1-7. (A compass is a necessary tool to draw circles and arcs, bisect lines, and do many other useful things. If you don't have one yet, you should get one *immediately*, so you can do the rest of this chapter's problems. With the spread of the compass's legs, you can lay out equal distances many times without repeatedly measuring with a ruler.) We spread the two legs of the compass until the distance between them is equal to the desired radius. Then, by rotating the free leg of the compass about the center, we trace out a circle.

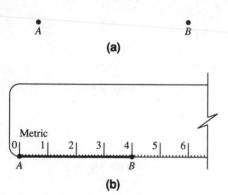

(a)

(b)

Figure 1-6

Figure 1-7. A compass.

EXAMPLE 1-3: Draw a circle whose center is point P and whose radius is 3 centimeters.

Solution

Step 1: Measure 3 centimeters with a compass: That is, place the sharp point at scribe mark 0 on a centimeter ruler, then open the compass until the pencil point is at scribe mark 3.

Step 2: Place the sharp point on your paper and call that mark P.

Step 3: Rotate the compass to draw the desired circle, as in Figure 1-8.

EXAMPLE 1-4: Finding the Midpoint of a Line Segment
Use a ruler and compass to locate the *midpoint M* of line segment AB in Figure 1-9a.

Solution: You *could* use the scribe marks of the ruler to carefully measure the segment's length, divide by 2, and then measure to the midpoint. But the resolution of the ruler may not be good enough to locate M accurately. Luckily, you can do better with a ruler and compass:

Step 1: Open the legs of your compass to an amount larger than a rough estimate of half the length of line segment AB.

Step 2: Place the point of the compass at A and draw a part of the circle (an **arc**) that intersects AB. Then, without changing the radial distance, move your compass point to B and draw another arc intersecting AB. The arcs you draw should be large enough so that they intersect each other twice—once above AB and once below, as in Figure 1-9b. Label these intersection points P_1 and P_2.

Step 3: Draw a line through P_1 and P_2. The midpoint M is the spot where the line drawn through P_1 and P_2 intersects AB (see Figure 1-9b).

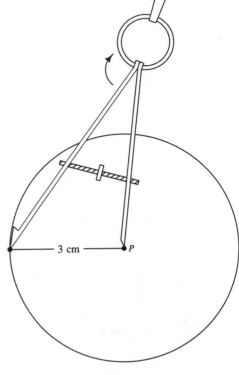

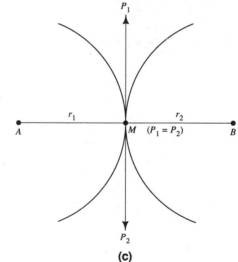

Figure 1-8

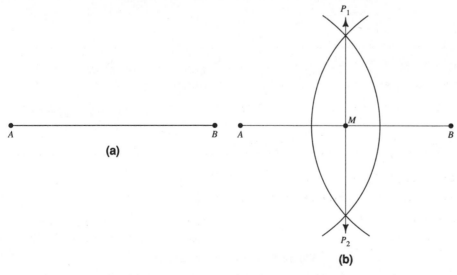

Figure 1-9

In order to understand why you can be sure that the intersection of AB and $\overleftrightarrow{P_1P_2}$ is the midpoint M, think of yourself as repeating this drawing over and over again, steadily decreasing the compass's opening, until P_1 and P_2 coincide. That is, eventually, you will end up with two circles of equal radius, and the radius of each of these circles will be equal to half the length of AB (see Figure 1-9c). Also notice that, whatever their radius, the arcs of the circles drawn in this manner always intersect on the line $\overleftrightarrow{P_1P_2}$.

1-4. Angles

An angle is defined by two rays:

- If two rays $\overrightarrow{AB}$ and $\overrightarrow{AC}$ emanate from the same point A, then these rays define an **angle**.

Intuitively, an angle, as shown in Figure 1-10, is the opening between these rays. Point A is called the **vertex** of the angle, and the set of points between the rays is called the **interior** of the angle. The common notations we use for the angle formed by rays AB and $\overrightarrow{AC}$ are $\angle A$, $\angle BAC$, and $\angle CAB$.

note: We are mainly interested in the case when $\overrightarrow{AB}$ and $\overrightarrow{AC}$ are not the same ray, that is, when point C does not lie on AB. If, however, $\overrightarrow{AB}$ and $\overrightarrow{AC}$ are the same ray, we say that the interior of the angle is *empty*.

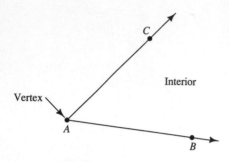

Figure 1-10. An angle.

A. Congruent angles

It's helpful to think of angles as rigid elements that can be picked up and slid around the plane. Suppose we move $\angle ABC$, as in Figure 1-11, so that vertex B is on vertex F and ray $\overrightarrow{BA}$ is on ray $\overrightarrow{FE}$. If ray $\overrightarrow{BC}$ then falls on ray $\overrightarrow{FG}$, we say that these angles are congruent angles. **Congruent angles** are angles that have the same openings and therefore are, in effect, merely repositionings of each other; that is, congruent angles are equal. The notation for indicating that two angles are congruent is $\angle ABC \cong \angle EFG$.

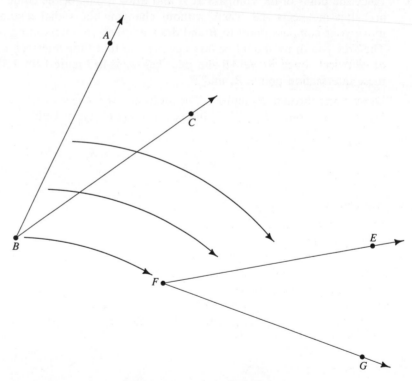

Figure 1-11. Congruent angles.

B. Supplementary angles

- If ray $\overrightarrow{BA}$ and ray $\overrightarrow{BC}$ lie on a line $\overleftrightarrow{AC}$, as in Figure 1-12a. Then $\angle ABC$ gets the special name **straight angle**.
- If two angles, $\angle ACB$ and $\angle BCD$, share a common vertex C and a ray $\overrightarrow{CB}$, as in Figure 1-12b, such that the angle formed by rays $\overrightarrow{CA}$ and $\overrightarrow{CD}$ is a straight angle, then the two angles, $\angle ACB$ and $\angle BCD$, are called **supplementary angles**.

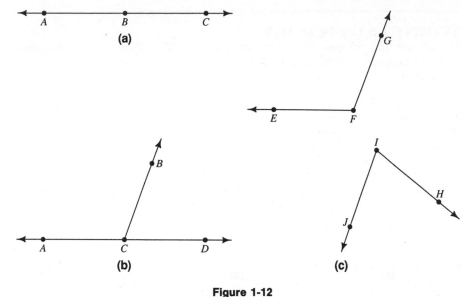

Figure 1-12

- If any two angles, ∠*EFG* and ∠*HIJ*, are congruent to a pair of supplementary angles, ∠*ACB* and ∠*BCD*, as in Figure 1-12c, then ∠*EFG* and ∠*HIJ* are also referred to as a *supplementary pair of angles.*

C. Right angles

Right angles are a special and most important class of angles:

- All **right angles** are congruent to each other and any two of them form a supplementary pair of angles.

  When we draw right angles, we usually distinguish them from other angles by placing a little square box in the interior of the angle, as in Figure 1-13a. If we extend the rays of a right angle to make two intersecting lines, as in Figure 1-13b, these two lines are *perpendicular.*

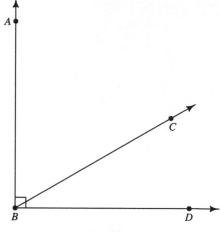

Figure 1-13

D. Complementary angles

Just as two angles that can be combined to make a straight angle are called *supplementary angles,*
- Any two angles, ∠*ABC* and ∠*CBD*, with a common vertex *B* and ray $\overrightarrow{BC}$ that can be combined to make a right angle (as in Figure 1-14) are called **complementary angles**.
- Any other two angles, ∠*EFG* and ∠*HIJ*, that are congruent to a pair of complementary angles, ∠*ABC* and ∠*CBD*, are also referred to as a *complementary pair of angles.*

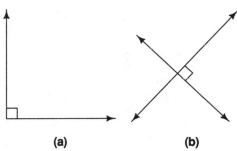

Figure 1-14

E. Acute and obtuse angles

If the opening of ∠*ABC* is less than that of a right angle, meaning that there is some other angle, ∠*DEF*, complementary to ∠*ABC*, then ∠*ABC* is called an **acute angle**. Both ∠*ABC* and ∠*CBD* in Figure 1-14 are acute angles that are complementary. If an angle, ∠*BAC*, is neither an acute nor a right angle, that is, if its opening is larger than that of a right angle but smaller than that of a straight angle, then ∠*BAC* is an **obtuse angle**. There are no angles that are complementary to obtuse angles, but each obtuse angle has an acute angle that is supplementary to it. In Figure 1-15, for instance, ∠*BAC* is obtuse, while ∠*CAD* is the acute angle supplementary to ∠*BAC*.

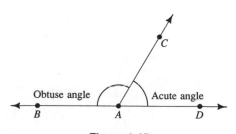

Figure 1-15

EXAMPLE 1-5: Copying an Angle

Use a ruler and compass to make a copy of $\angle ABC$ in Figure 1-16a. In other words, construct $\angle HFG$ congruent to $\angle ABC$.

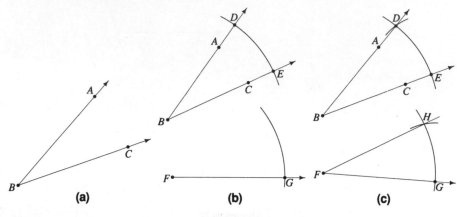

Figure 1-16

Solution

Step 1: The opening of an angle can be measured by a compass. First open a compass and place the point at the vertex B of $\angle ABC$. Then swing an arc with the pencil tip until the arc cuts both rays. Label these points D and E, as in Figure 1-16b.

Step 2: Without changing the compass opening, place the point of the compass at some point F, which you choose to be the vertex of the new angle. Swing an arc of approximately the same length as the first. Then draw a ray from F that intersects the arc. Label the intersection G as in Figure 1-16b.

Step 3: Next, place the point of the compass at point E on $\angle ABC$ and open the compass until the pencil touches point D. Now, without changing the compass opening, place the compass point at G on $\overrightarrow{FG}$ and sweep another arc that intersects the previous arc. Label this intersection point H, as in Figure 1-16c.

Step 4: Finally, draw ray FH. $\angle HFG$ is congruent to $\angle ABC$.

EXAMPLE 1-6: Bisecting an Angle

Find a point D in Figure 1-17a so that ray BD splits $\angle ABC$ into two congruent angles, $\angle ABD$ and $\angle DBC$. (Ray BD is called the **bisector** of $\angle ABC$.)

Solution

Step 1: As in Example 1-5, place a compass point at vertex B and sweep an arc that intersects the angle's rays. Label these intersection points E and F. (See Figure 1-17b.)

Step 2: Find the midpoint of the arc EF by the reasoning you used to find the midpoint of a line segment in Example 1-4. Place your compass point first at E and then at F and sweep two arcs of equal radius that intersect each other at a point in the interior of the angle. Label this intersection point D, as in Figure 1-17b.

Step 3: Draw ray BD.

Ray BD bisects EF, so it must also bisect $\angle ABC$. Therefore, $\angle ABD$ and $\angle DBC$ are congruent.

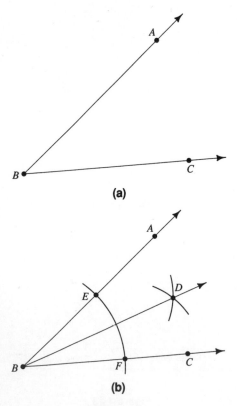

Figure 1-17

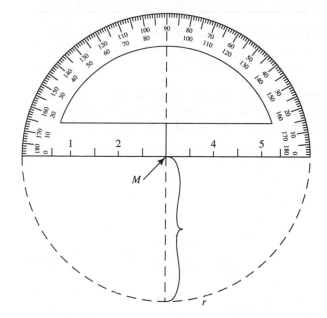

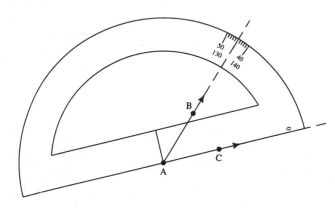

Figure 1-18

1-5. Angular Measure

Just as we measure distance by comparing a length to standard units marked on a ruler, we measure the opening of an angle by comparing the arc that its rays cut from a circle centered at the angle's vertex to a standard unit of opening called the degree. A **degree** (denoted by the symbol °) is 1/360th of any circle and can be used as a standard to measure any part of a circle, such as an arc.

The device we use to measure angles is called a **protractor**. It consists of a straight ruler and a half-circle arc, as illustrated in Figure 1-18a. The midpoint of the ruler's edge is the center of a circle whose radius is exactly one-half the ruler's length, and the top of the protractor is exactly one-half of that circle. Degrees are indicated by scribe marks placed along the semicircle's edge. In a half circle there are 180 standard degrees, denoted as 180°.

To measure an angle, first place the edge of the protractor along one ray, so that the ruler's center is at the vertex. You should select the ray so that the other ray (possibly extended) lies under the circular part of the protractor, as in Figure 1-18b. The measure of this angle is then the number of scribe marks that fall between the two rays. If you place the protractor so that one ray crosses the reference circle at the 0° mark, then you may read the angle's measurement from where the second ray crosses the circle.

note: A degree is *not* a unit of length; it is a part of a circle. A protractor measures the *opening* of an angle, and the measurement is not affected by the size of the protractor's radius.

The special angles discussed in Section 1-4 may now be described in terms of degree measurements.

(a) *Congruent angles* have the same degree measure.
(b) A *straight angle* measures 180°.
(c) A *right angle* measures 90°.
(d) The sum of the degree measures of a *supplementary pair of angles* is 180°.
(e) The sum of the degree measures of a *complementary pair of angles* is 90°.
(f) An *acute angle* measures less than 90°.
(g) An *obtuse angle* measures between 90° and 180°.

note: Another standard unit of angular measurement, called the *radian*, is also in common use. We'll examine it in detail in Chapters 7 and 12.

EXAMPLE 1-7: Use a protractor to measure the number of degrees in each of the angles in Figure 1-19a.

Solution

Step 1: Place the center mark of the protractor at the vertex of each angle. Rotate the protractor until the straight side falls on one ray of the angle and that ray crosses the circular part at the scribe mark at 0°.

Step 2: If the other ray does not intersect the circular part of the protractor, extend the ray as in ∠1 of Figure 1-19b.

Step 3: Read off the scribe mark where the other ray crosses the circular part. You should find that ∠1 = 30° and that ∠2 = 135°.

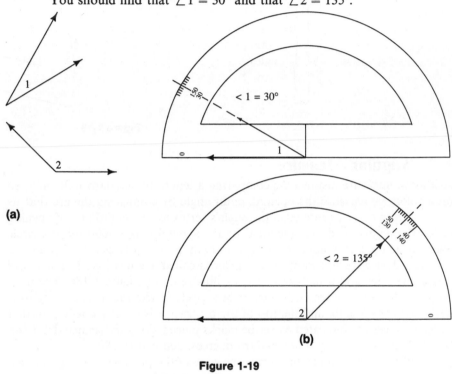

(a)

(b)

Figure 1-19

1-6. Parallel Lines

Two lines in a plane are said to be **parallel** if they do not intersect. In other words, if there is no point that lies on both lines. The fundamental postulate that characterizes Euclidean plane geometry is called the **parallel postulate**:

- For any line L and any point P not lying on L, there is one and only one line that passes through P and is parallel to L.

  note: There is no proof for the parallel postulate. In fact, this assumption has been controversial for centuries. Moreover, if we do *not* accept this postulate, we can develop *other* models of the space surrounding us that are perfectly consistent and equally good at explaining what we perceive. In these models, measuring the distance between two points A and B is slightly more involved than the simple ruler method of Section 1-2, and the measurement of angles is more involved than the method shown in Section 1-4. But it is the acceptance of this postulate that characterizes Euclidean geometry, which to date has been successful in describing our everyday world. We define the **Euclidean plane** as a plane where the parallel postulate holds. **Euclidean geometry** is then the geometry that this postulate forces on us through the measurement of distances and angles as discussed in Sections 1-2 and 1-4.

Since we can draw only line segments, not complete lines, there is really no way to decide from the definition if two lines are parallel. There are, however, other properties equivalent to the parallel postulate that we can use to prove that two lines are parallel. Before stating the other properties, however, we'll need some new vocabulary, which deals with intersecting lines.

- A **transversal** is a line that intersects two or more lines. For example, line $\overleftrightarrow{EF}$ in Figure 1-20 is a transversal of $\overleftrightarrow{AB}$ and $\overleftrightarrow{CD}$.
- When any two lines cross, four angles are formed, such that two pairs of the angles may be seen as *adjacent* (next to each other), and two pairs may be seen as *opposite* each other. The pairs of nonadjacent angles are called **vertical angles**. In Figure 1-20, $\angle 1$ and $\angle 5$ are a pair of vertical angles, and $\angle 6$ and $\angle 4$ are also a pair of vertical angles.
- When two or more lines are crossed by a transversal, the angles that have one ray along the segment of the transversal between the two lines and their other ray on each of the two lines, but on opposite sides of the transversal, are called **alternate interior angles**. In Figure 1-20, $\angle 1$ and $\angle 3$ are a pair of alternate interior angles, as are $\angle 4$ and $\angle 2$.
- When any two lines are crossed by a transversal, the angles whose rays lie on the same side of the transversal and on the same side of their respective lines are called **corresponding angles**. In Figure 1-20, $\angle 5$ and $\angle 3$ are a pair of corresponding angles, as are $\angle 6$ and $\angle 2$.

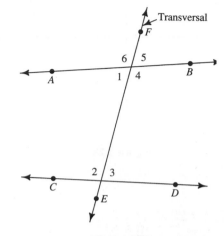

Figure 1-20

Now, with the vocabulary of intersecting lines in mind, we can state the principles that are equivalent to the parallel postulate:

(a) Two lines that are cut by a transversal have congruent alternate interior angles if and only if the lines are parallel.
(b) Two lines that are cut by a transversal have congruent corresponding angles if and only if the lines are parallel.
(c) Two lines are parallel if and only if they are perpendicular to a common transversal.
(d) Two lines that are parallel to a third line are parallel to each other.

EXAMPLE 1-8: Constructing a Line Parallel to a Given Line

Construct a line L_2 through point P that is parallel to line L_1, as in Figure 1-21a.

Solution: Use principle (**a**).

Step 1: Draw a line (any line) through P until it intersects L_1, as in Figure 1-21b. Label the intersection point A. Line $\overleftrightarrow{AP}$ will be a transversal of the two lines L_1, which is given, and L_2, which will be drawn parallel to L_1.

Step 2: Notice that when $\overleftrightarrow{AP}$ is drawn, two angles are formed at A—an acute angle and an obtuse angle, which are supplementary. Each of these angles will be one of a pair of alternate interior angles, which can be copied at P. Use the procedure described in Example 1-5 to copy the acute angle on the opposite side of $\overleftrightarrow{AP}$ (see Figure 1-21b):

- Place your compass point at A and swing arc 1, which intersects $\overleftrightarrow{AP}$ at C and L_1 at B.
- Without changing the compass opening, place the compass point at P and swing arc 2, which intersects $\overleftrightarrow{AP}$ at D.
- Set your compass opening to equal the distance between B and C. Then place the compass point at D and swing arc 3, which intersects arc 2 at E.

Step 3: Draw PE, which is a line segment of L_2. Since the acute angle $\angle P$ on the left side of $\overleftrightarrow{AP}$ is congruent to the acute angle $\angle A$ on the right side of $\overleftrightarrow{AP}$, line L_2 must be parallel to L_1 by principle (**a**).

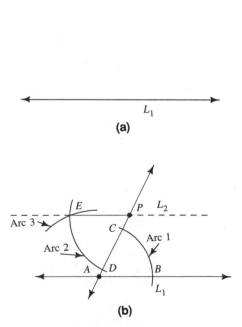

Figure 1-21

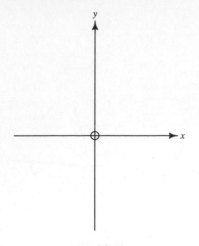

Figure 1.22

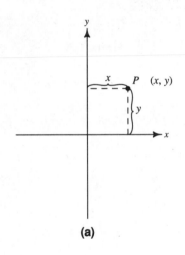

(a)

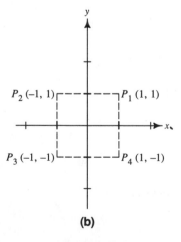

(b)

Figure 1-23

1-7. Analytic Methods: Coordinate Representation

Analytic geometry offers us a way to take the *geometric* things we've already examined, such as points, lines, and angles, and express them as *analytic* things, such as numbers and equations. Sometimes, the analytic approach enables us to solve difficult geometric problems by simpler manipulations of numbers or equations. Similarly, involved analytic ideas can sometimes take a simpler geometric form. Because it gives us the ability to move back and forth between two forms of representation, analytic geometry is a powerful tool of modern science. It allows us, in effect, to pick the ground on which we do battle with nature.

The fundamental idea in analytic geometry is **coordinate representation**, which is a way of describing the points on the flat, two-dimensional Euclidean plane. To do this, we begin with the Euclidean plane and select a point of reference. This reference point is called the **origin**. Then we draw two perpendicular lines—a horizontal line usually called the **x axis** and a vertical line usually called the **y axis**. (See Figure 1-22.)

A. Expressing points as a pair of coordinates

We can describe any point P in the plane by its distance from each of the two axes. First, we draw a line through P parallel to the x axis; then we draw another line through P parallel to the y axis. The length of the line segment from P to the y axis (P's distance from the y axis) is denoted by x, and the length of the segment from P to the x axis (P's distance from the x axis) is denoted by y, as in Figure 1-23a.

These distances appear to specify P, a point, in terms of (x, y), a pair of numbers. Thus, $(1, 1)$ corresponds to a point that falls one unit of distance from each axis. But after a moment's thought you should notice that there are *four* points ($P_1, P_2, P_3,$ and P_4) that fall one unit of distance from each of the two axes, as shown in Figure 1-23b. The association of numbers with points is therefore not one-to-one, because points can fall either above or below the x axis and to the left or the right of the y axis and still satisfy our definition of distance. But we can eliminate this ambiguity if we agree to tag the first number, the **x coordinate** or **abscissa** with a minus sign when the point falls to the left of the y axis. Similarly, we'll tag the second number, the **y coordinate** or **ordinate**, with a minus sign when the point falls below the x axis. With this modification, we can distinguish points $P_1, P_2, P_3,$ and P_4 by designating them $(1, 1)$, $(-1, 1)$, $(-1, -1)$, and $(1, -1)$, respectively.

The plane in which we represent points by this method is called the **Cartesian plane**. We've represented the geometric idea of points by the analytic idea of a pair of numbers (x, y), referred to as the **coordinates** of the point. To remind us of the sign convention, the axes are usually drawn with arrowheads pointing in the direction associated with their positive value.

B. Distance in the Cartesian plane

When we use the Cartesian plane to represent points by coordinates, we can replace the crude ruler scheme of distance measurement by a simple equation. If P_1 and P_2 are points with coordinates (x_1, y_1) and (x_2, y_2), respectively, then the distance equation

DISTANCE BETWEEN TWO POINTS IN THE CARTESIAN PLANE	

$$d(P_1, P_2) = \sqrt{(x_1 - x_2)^2 + (y_1 - y_2)^2} \quad \textbf{(1-1)}$$

meets all the requirements of a distance as described in Section 1-2. Furthermore, the number that results from Eq. (1-1) is the same as the ruler measurement of the distance between P_1 and P_2.

EXAMPLE 1-9: Plot the points whose coordinates are $(2, 3)$, $(-1, 3)$, $(0, 2)$, $(2, -1)$, and $(-2, -1)$ on the Cartesian plane.

Solution: See Figure 1-24.

EXAMPLE 1-10: Find the distance between the points whose coordinates are $(2, 3)$ and $(-1, 2)$.

Solution: Use Eq. (1-1):

$$d = \sqrt{(x_1 - x_2)^2 + (y_1 - y_2)^2} = \sqrt{(2 - (-1))^2 + (3 - 2)^2}$$
$$= \sqrt{3^2 + 1^2}$$
$$= \sqrt{9 + 1} = \sqrt{10}$$

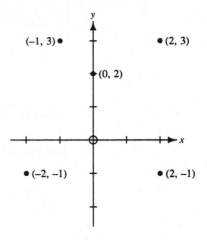

Figure 1-24

SUMMARY

1. Geometry is a mathematical model of the space that surrounds us.
2. A *point* is an undefined term that may be thought of as a location.
3. A *line* is determined by two distinct points.
4. Lines extend forever.
5. A *ray* is a piece of a line with one endpoint.
6. A *line segment* is a piece of a line with two endpoints.
7. A *distance* between two points is a number that tells how far apart the points are.
8. Euclidean distance is measured with a ruler marked in standard unit lengths.
9. A *circle* is a set of points that fall a specific distance, the *radius*, from a fixed point, the *center*.
10. An *angle* is the opening between two rays.
11. An angle is measured with a *protractor*.
12. A *straight angle* is the same as a line and measures 180°.
13. A *right angle* is half a straight angle, or 90°.
14. An *acute angle* is less than 90°.
15. An *obtuse angle* is greater than 90°.
16. Two *supplementary angles* add to 180°.
17. Two *complementary angles* add to 90°.
18. *Parallel* lines do not intersect.
19. The Euclidean model of geometry is based on the assumption that one and only one line can be constructed through a given point parallel to another line.
20. When a *transversal* cuts two parallel lines, the *alternate interior angles* formed are congruent.
21. Points can be represented by coordinate pairs of real numbers.
22. In the Euclidean model, the distance between points (x_1, y_1) and (x_2, y_2) is

$$d = \sqrt{(x_1 - x_2)^2 + (y_1 - y_2)^2}$$

RAISE YOUR GRADES
Can you . . . ?

☑ draw a circle, given a radius and a center point
☑ measure an angle in degrees
☑ find the coordinates of a point with respect to a set of coordinate axes
☑ copy an angle with a ruler and compass
☑ bisect an angle with a ruler and compass
☑ bisect a line segment with a ruler and compass

☑ construct a line through a point perpendicular to a given line
☑ construct a line through a point parallel to a given line
☑ find the distance between two points
☑ find the measure of an angle that is either complementary or supplementary to a given angle

SOLVED PROBLEMS

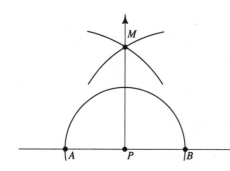

Figure 1-25

PROBLEM 1-1   Given two points A and B, draw a circle that passes through A and B and whose center lies on the line segment AB.

Solution

Step 1:   Placing your compass point first at A and then at B, swing intersecting arcs, as in Example 1-4, to locate the midpoint M of the line segment AB.

Step 2:   Point M is the center of the circle, so set your compass point at M. The radius is the distance either from A to M or from M to B.

Step 3:   Set the compass opening to this radius and draw the circle as in Figure 1-25.

PROBLEM 1-2   Use a ruler and compass to construct a right angle with vertex at a given point, point P.

Solution

Step 1:   Use a ruler to draw a line segment that includes point P somewhere near the middle.

Step 2:   Place the point of the compass at P and sweep out an arc that crosses the line segment at two points. Label the points A and B, as in Figure 1-26.

Step 3: $\angle APB$ is a straight angle, and a right angle is half of a right angle. So, placing your compass point first at A and then at B, swing intersecting arcs, as in Example 1-6, to construct point M; then draw ray $\overrightarrow{PM}$, which bisects $\angle APB$. Both angles, $\angle APM$ and $\angle MPB$, shown in Figure 1-26 are right angles.

PROBLEM 1-3   Without using a protractor, draw an angle complementary to $\angle ABC$ of Figure 1-27a.

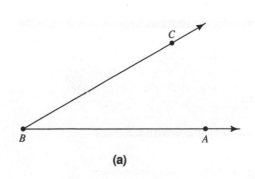

(a)

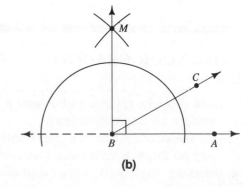

(b)

Figure 1-27

Solution

Step 1: Extend ray $\overrightarrow{BA}$ past B to make line $\overleftrightarrow{BA}$.

Step 2: As in Problem 1-2, construct a right angle with vertex at B so that $\overrightarrow{BA}$ *is one of its defining rays* and $\overrightarrow{BM}$ is the other defining ray. As in Figure 1-27b ray $\overrightarrow{BC}$ falls within $\angle MBA$.

Step 3: Since $\angle MBA$ is a right angle, and is the sum of $\angle MBC$ and $\angle CBA$, $\angle MBC$ is complementary to $\angle CBA$.

PROBLEM 1-4 Classify each of the angles in Figure 1-28 as either an acute or obtuse angle.

Solution Find the degree measure of each angle by using a protractor.

Angle 1 measures $45°$, and $45 < 90$, so $\angle 1$ is an acute angle.
Angle 2 measures $135°$, and $135 > 90$, so $\angle 2$ is an obtuse angle.
Angle 3 measures $30°$, and $30 < 90$, so $\angle 3$ is acute also.

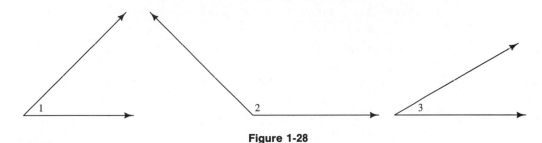

Figure 1-28

PROBLEM 1-5 An angle measures $37°$. Find **(a)** the measure of its complementary angle and **(b)** the measure of its supplementary angle.

Solution

(a) *Step 1:* Let x be the measure of the complement.

Step 2: Express the relation between the angle and x by an equation:

$$x + 37° = 90°$$

Step 3: Solve for x:

$$x = 90° - 37 = 53°$$

(b) *Step 1:* Let y be the measure of the supplement.

Step 2: $$y + 37° = 180°$$

Step 3: Solve for y:

$$y = 180° - 37° = 143°$$

PROBLEM 1-6 Given point P and line $\overleftrightarrow{AB}$, use a ruler and compass to construct a line through P that is perpendicular to $\overleftrightarrow{AB}$.

Solution Point P can lie either on $\overleftrightarrow{AB}$ or off $\overleftrightarrow{AB}$, so this problem is really asking two questions. If P is *on* line $\overleftrightarrow{AB}$, then the problem is equivalent to constructing a right angle at P by bisecting straight angle APB, which you've already done in Problem 1-2. So let's turn our attention to the case when P is *not* on $\overleftrightarrow{AB}$.

Step 1: Place the compass point at P and sweep an arc that cuts $\overleftrightarrow{AB}$ at points C and D, as in Figure 1-29.

Step 2: Now place the compass point at C and sweep an arc of radius equal to the distance between C and P on the other side of $\overleftrightarrow{AB}$ from P.

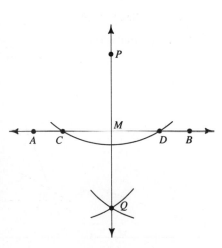

Figure 1-29

Step 3: Place the compass point at D and repeat Step 2. Label the intersection point of these arcs Q.

Step 4: Draw line $\overrightarrow{PQ}$, and label its intersection with line segment CD as point M.

Step 5: Review the steps you've taken: You've just bisected $\angle CMD$. Since $\angle CMD$ is a straight angle, $\angle PMC$ and $\angle PMD$ are right angles. Thus, $\overleftrightarrow{PQ}$ is perpendicular to $\overleftrightarrow{AB}$.

PROBLEM 1-7   Points P_1, P_2, and P_3 have coordinates respectively, $(-1, 2)$, $(1, 4)$, and $(3, -5)$, in the Cartesian plane. Find the Euclidean distance between each pair of points.

Solution: Use Eq. (1-1):

$$d(P_1, P_2) = \sqrt{(x_1 - x_2)^2 + (y_1 - y_2)^2}$$
$$= \sqrt{(-1 - 1)^2 + (2 - 4)^2} = \sqrt{(-2)^2 + (-2)^2}$$
$$= \sqrt{4 + 4}$$
$$= 2\sqrt{2}$$

$$d(P_1, P_3) = \sqrt{(x_1 - x_3)^2 + (y_1 - y_3)^2}$$
$$= \sqrt{(-1 - 3)^2 + (2 - (-5))^2}$$
$$= \sqrt{(-4)^2 + (7)^2}$$
$$= \sqrt{16 + 49}$$
$$= \sqrt{65}$$

$$d(P_2, P_3) = \sqrt{(x_2 - x_3)^2 + (y_2 - y_3)^2}$$
$$= \sqrt{(1 - 3)^2 + (4 - (-5))^2}$$
$$= \sqrt{(-2)^2 + 9^2}$$
$$= \sqrt{4 + 81}$$
$$= \sqrt{85}$$

Review Exercises

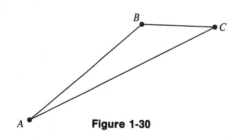

Figure 1-30

EXERCISE 1-1   Draw any two points A and B. Then draw (**a**) line $\overleftrightarrow{AB}$, (**b**) ray $\overrightarrow{AB}$, (**c**) ray $\overleftarrow{AB}$, and (**d**) line segment AB.

EXERCISE 1-2   Measure the distances between the three points A, B, C in Figure 1-30 by using an appropriate ruler. Verify property (**d**) in Section 1-2.

EXERCISE 1-3   Draw three circles with the same center and radii of 1 inch, 3 centimeters, and $1\frac{1}{4}$ inches. (note: When two or more circles have the same center, the circles are said to be *concentric*.)

EXERCISE 1-4   Measure each of the four angles in Figure 1-31 and express your result in degrees.

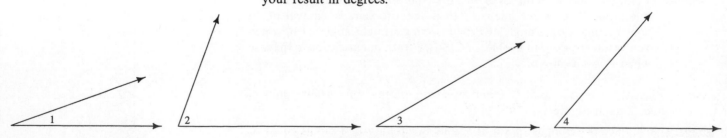

Figure 1-31

EXERCISE 1-5   Give the number of degrees in the angles supplementary to each of the angles in Figure 1-31.

EXERCISE 1-6   Give the number of degrees in the angles complementary to each of the angles in Figure 1-31.

EXERCISE 1-7   Use a ruler and compass to construct the angle complementary to the angle in Figure 1-32.

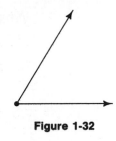

Figure 1-32

EXERCISE 1-8   Given the three points *A*, *B*, and *C* of Figure 1-33, construct a line (**a**) through *C* parallel to *AB* and (**b**) through *C* perpendicular to *AB*.

EXERCISE 1-9   Are the lines *AB* and *CD* in Figure 1-34 parallel or not? Why?

EXERCISE 1-10   Identify the angles in Figure 1-35 as acute, obtuse, right, or straight.

Figure 1-33

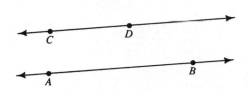

Figure 1-34

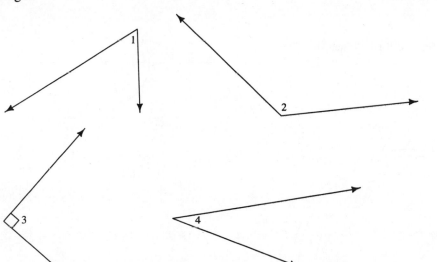

Figure 1-35

EXERCISE 1-11   Give the Cartesian coordinates of points *A*, *B*, *C*, and *D* shown in Figure 1-36.

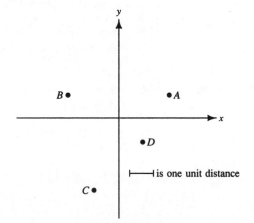

Figure 1-36

EXERCISE 1-12   Find the distance between the points whose Cartesian coordinates are (**a**) $(1, 2)$ and $(5, 5)$; (**b**) $(-1, 2)$ and $(11, -3)$.

Answers to Review Exercises

1-1 See Figure 1-37.

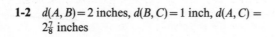

(a)

(b)

(c)

(d)

Figure 1-37

1-2 $d(A, B) = 2$ inches, $d(B, C) = 1$ inch, $d(A, C) = 2\frac{7}{8}$ inches

1-3 See Figure 1-38.

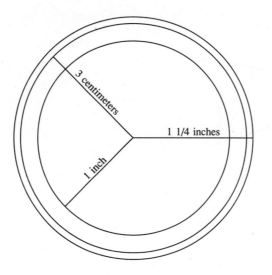

3 centimeters

1 1/4 inches

1 inch

Figure 1-38

1-4 $\angle 1 = 20°$, $\angle 2 = 70°$, $\angle 3 = 30°$, $\angle 4 = 50°$

1-5 160°, 110°, 150°, 130°

1-6 70°, 20°, 60°, 40°

1-7 See Figure 1-39.

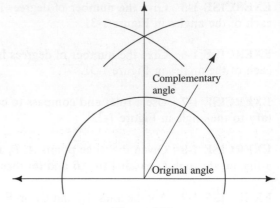

Complementary angle

Original angle

Figure 1-39

1-8 See Figure 1-40.

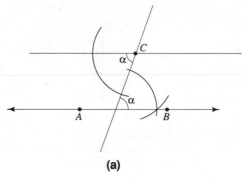

α

α

A B

(a)

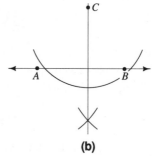

C

A B

(b)

Figure 1-40

1-9 Draw a transversal and measure the alternate interior angles. They are equal, so the lines are parallel.

1-10 $\angle 1$ and $\angle 4$ are acute, $\angle 2$ is obtuse, $\angle 3$ is a right angle.

1-11 A is $(2, 1)$, B is $(-2, 1)$, C is $(-1, -3)$, and D is $(1, -1)$.

1-12 (a) 5   (b) 13

2 MATHEMATICAL LOGIC

THIS CHAPTER IS ABOUT

☑ **Inductive and Deductive Reasoning**
☑ **Symbolic Logic**
☑ **Correct and Incorrect Arguments**
☑ **Methods of Proof**
☑ **Strategies of Proving Things**

The ability to *reason* (think) correctly is fundamental to all mathematics. It is especially crucial in Euclidean geometry, where we construct a model of our surrounding space based on a set of self-evident axioms that we assume to be true. Although logical reasoning is a separate subject far beyond the scope of this outline, this chapter will arm you with the logic you'll need to do two important things in geometry:

(1) Use a correct argument to prove a geometric fact.
(2) Recognize an incorrect argument in a proof.

2-1. Inductive and Deductive Reasoning

Reasoning is the process of drawing a **conclusion**, a new "fact," from a first "fact" that we believe is true. This first fact is called the **premise** or **hypothesis**. The reasoning process is usually classified into one of two types—inductive or deductive.

A. Inductive reasoning

Every human being is able to look around and observe facts. You know that the sun will rise tomorrow morning. You know that you'd experience pain if you stuck yourself with a pin. These are facts that you know well! But *how* do you know them?

A reasonable answer would be that you've *observed* the sun rise every morning of your life, and so has everyone else in all recorded history. Similarly, when you've stuck yourself with a sharp object, it has hurt! You know these facts because you've seen them and felt them: They are conclusions you've drawn based on numerous observations and past experiences.

- The drawing of conclusions based on experience and observation is called **inductive reasoning**.

There are two important characteristics of *inductive* reasoning that distinguish it from another kind of reasoning that we'll discuss in a moment.

- We can't be absolutely certain that the conclusion we draw through inductive reasoning is indeed true.

All we can do is treat the conclusion as something that will occur with a high probability. If we find an experiment where the expected outcome does not occur, then we must either discard or revise our conclusion.

• Conclusions we draw through inductive reasoning contain new information.

That is, we can learn new facts, make advances in science and technology, and so on. And, although inductive reasoning can't guarantee the absolute truth of a conclusion, we can increase the probability that the conclusion is true with further observation and experimentation.

EXAMPLE 2-1:   Consider the digits that make up the numbers 15, 21, 27, 57, 144, 252, and 3135. The sum of each number's digits is divisible by 3. So is the number. Draw an inductive conclusion based on this observation.

Number	Sum of Digits	Factored Sum	Factored Number
15	$1 + 5 = 6$	$6 = 3 \times 2$	$15 = 3 \times 5$
21	$2 + 1 = 3$	$3 = 3 \times 1$	$21 = 3 \times 7$
27	$2 + 7 = 9$	$9 = 3 \times 3$	$27 = 3 \times 9$
57	$5 + 7 = 12$	$12 = 3 \times 4$	$57 = 3 \times 19$
144	$1 + 4 + 4 = 9$	$9 = 3 \times 3$	$144 = 3 \times 48$
252	$2 + 5 + 2 = 9$	$9 = 3 \times 3$	$252 = 3 \times 84$
3135	$3 + 1 + 3 + 5 = 12$	$12 = 3 \times 4$	$3135 = 3 \times 1045$

Solution: One reasonable inductive conclusion is that any number whose digits add up to a number divisible by 3 must also be divisible by 3 itself. A second reasonable inductive conclusion would be that if a number is divisible by 3, then the sum of its digits must also be divisible by 3.

B. Deductive reasoning

Deductive reasoning takes a different approach—one that is more abstract. In **deductive reasoning** we reason *from* a premise, that is, we develop a logical argument *from* the premise *to* the conclusion. If the premise is true, then the deductive process demands that the conclusion also be true. There is no uncertainty. If fact one is true, then fact two is also true. This is the nature of deductive reasoning.

• If we use deductive reasoning correctly to draw a conclusion from a *true* premise or hypothesis, then that conclusion is also true.

If a premise is true in a specific example, and the conclusion of the deductive argument is seen to be false, then either the argument that led to this conclusion must be incorrect or the premise must have been false after all!

• Conclusions we draw through deductive reasoning contain no new information.

note: You're probably asking yourself now, "What's the purpose of deductive reasoning if it gives us no new information?" Consider the premises

"Tom is older than Sue"

"Sue is older than Alice"

and the conclusion

"Tom is older than Alice"

The conclusion gives us no new information. That is, everything in the conclusion is stated—perhaps implicitly—in the premises. What *is* new, however, is the *arrangement* or *combination* of the original facts. It could be that we had the information at our fingertips, but we'd never thought of it in that particular way before.

Deductive reasoning goes hand in hand with inductive reasoning. It is the other part of the core of scientific investigation. Through inductive reasoning we observe what is happening around us and draw conclusions. These conclusions become the premises for new deductive arguments. Using deductive reasoning, we can discover (*deduce*) other conclusions that are not obvious. Since these conclusions must be true if the premises (hypotheses) were true, the scientific method then requires that we perform more experiments to see if the predicted conclusions are indeed occurring.

We have used inductive reasoning to begin to build our Euclidean model for the geometry of our surrounding space. When we developed the undefined terms in Chapter 1, such as a point or a line, we used our experience with space to say that "This is what a point is" and "This is what a line is." From now on, we'll use deductive reasoning to discover other properties, or facts, about space that must be true if the fundamental concepts discussed in Chapter 1 fit the world we live in.

The parallel postulate is the major inductively derived fact distinguishing Euclidean geometry from other models of plane geometry. If this postulate is valid in our world, then so are all the other facts (*theorems*) we deduce from it. As long as we don't meet up with an example in which our space defies one of the deduced conclusions, then Euclidean geometry is a valid model for the space around us.

EXAMPLE 2-2: An integer m is divisible by 3 if there is another integer k such that $m = 3k$. The decimal representation of the $n + 1$ digit number $a_n \cdots a_2 a_1 a_0$ stands for

$$a_0 + a_1(10) + a_2(10)^2 + \cdots + a_n(10)^n$$

For example, $237 = 7 + 3(10) + 2(100)$. Reason deductively that if the number $a_n \cdots a_2 a_1 a_0$ has $a_n + \cdots + a_2 + a_1 + a_0 = 3k$ for some integer k—that is, the sum of the digits is divisible by 3—then the original number must also be divisible by 3.

Solution

Step 1: Assume $a_0 + a_1 + a_2 + \cdots + a_n = 3k$.

Step 2: Notice that $10 = 9 + 1$, $(10)^2 = 100 = 99 + 1$, $(10)^3 = 1000 = 999 + 1$, $\ldots$, and $(10)^n = 100 \cdots 00 = 99 \cdots 99 + 1$.

Step 3: Thus $a_0 + a_1(10) + a_2(10)^2 + \cdots + a_n(10)^n$
$$= a_0 + (9a_1 + a_1) + (99a_2 + a_2) + \cdots + (99 \cdots 99a_n + a_n)$$
$$= (a_0 + a_1 + a_2 + \cdots + a_n) + 9a_1 + 99a_2 + \cdots + 99 \cdots 99a_n$$

Step 4: Replace $a_0 + a_1 + a_2 + \cdots + a_n$ with $3k$, so that

$$3k + 9a_1 + 99a_2 + \cdots + 99 \cdots 99a_n = 3k + 9(a_1 + 11a_2 + \cdots + 11 \cdots 11a_n)$$
$$= 3[k + 3(a_1 + 11a_2 + \cdots + 11 \cdots 11a_n)]$$
$$= 3 \times \text{(some integer)}$$

Thus $a_n \cdots a_2 a_1 a_0$ is divisible by 3. You have used deductive reasoning to confirm the inductive conclusion of Example 2-1.

2-2. Symbolic Logic

The process of deductive reasoning is synonymous with developing a correct logical argument. This process can become involved, but we can simplify it by using **symbolic logic**.

A. Definitions, notation, and operations

• A **simple statement** is a sentence or phrase declaring some fact.

For example, the two sentences

"It is snowing outside"

and

"I am cold"

are both simple statements.

• A **compound** (or **composite**) **statement** is a sentence or phrase made by *joining* two or more statements (simple or compound) with the word AND or OR.

For example, the two sentences

"It is snowing outside AND I am cold"

and

"It is snowing outside OR it is raining outside"

are both compound statements.

We can perform operations with statements in symbolic logic just as we perform operations (such as + or −) with numbers in arithmetic and algebra. Specifically, there are four fundamental operations in symbolic logic: conjunction, disjunction, negation, and implication. When performing these operations, we use lowercase letters to represent simple statements, such as p for premise and q for conclusion, and special symbols to represent the operations. Thus—

if **p** = "It is snowing outside" and **q** = "I am cold," then:

• The **negation** of one of the statements, either "It is NOT snowing outside" or "I am NOT cold," is written $\sim$**p** or $\sim$**q**, respectively.
• The **conjunction** of the two simple statements "It is cold outside AND I am cold" is written **p** $\wedge$ **q**.
• The **disjunction** of the two statements "It is snowing outside OR I am cold" is written **p** $\vee$ **q**.
• The two statements may also be joined by writing them as "IF it is snowing outside, THEN I am cold." This compound statement is an **implication**, which we can write as **p** $\Rightarrow$ **q**.

note: We read the symbolic statements as follows:

negation:	$\sim$p	"NOT p"
conjunction:	p $\wedge$ q	"p AND q"
disjunction:	p $\vee$ q	"p OR q"
implication:	p $\Rightarrow$ q	"p IMPLIES q" or "IF p, THEN q"

EXAMPLE 2-3: Use the notation of symbolic logic to express the following compound statements:

(a) Mary likes John AND John likes Sue.
(b) Sue likes Sam OR Sam likes Mary.
(c) John does NOT like Sam AND Mary does NOT like Sue.
(d) IF John likes Sue, then Mary does NOT like Sue.

Solution

Step 1: Represent the simple statements with letters:

Let p = "Mary likes John"
     q = "John likes Sue"
     r = "Sue likes Sam"
     s = "Sam likes Mary"
     t = "John likes Sam"
     u = "Mary likes Sue"

Step 2: Use the letters and operational symbols to represent the statements:

(a) $p \wedge q$
(b) $r \vee s$
(c) $(\sim t) \wedge (\sim u)$
(d) $q \Rightarrow (\sim u)$

Notice the use of parentheses here to indicate which operation should be performed first. Just as in algebra, you have to simplify everything in parentheses before performing any other operations.

B. Truth tables

- Every statement has a truth value; that is, a statement can be either true or false, but not both.

For example, if p = "It is snowing outside," and all we can see for miles is blue sky and sunshine, then the truth value of p would be false.

If all statements have truth values, then the compound statements $p \wedge q$, $p \vee q$, and $p \Rightarrow q$, and the negation $\sim p$ must also have truth values. But the truth values of compound statements depend on the truth values of simple statements. For example, consider the conjunction $p \wedge q$ = "It is snowing outside AND I am cold." It's easy to see that $p \wedge q$ is true only when *both* p and q are true; in all other cases (i.e., when both p and q are false, when p is true and q is false, and when p is false and q is true), $p \wedge q$ is false. Similarly, it makes sense that if p is true, then $\sim p$ is false; but if p is false, then $\sim p$ is true.

Using words to describe the truth values of compound statements can get complicated, so we use symbols laid out in a truth table. A **truth table** is a device that allows us to describe the truth values of compound (and more complex) statements and to show how the truth values of compound statements depend on the truth values of the simple statements being combined.

A truth table is made up of columns and rows that intersect to form squares. There is one column for each of the simple and compound statements involved in a given statement, and there are 2^n rows, where *n* is the number of simple statements. Once the empty table has been constructed, we fill the squares formed by the rows and columns with symbols—T for true and F for false—representing all the possible truth values of each of the statements we have, so that the last column corresponds to the truth value of the complete statement.

Let's construct the truth values for negation $\sim p$, conjunction $p \wedge q$, disjunction $p \vee q$, and implication $p \Rightarrow q$, where

p = "It is snowing"
q = "I am cold"

1. Negation: ~p

Step 1: We have two statements, p and ~p, so we make two columns. We have only one simple statement, p, so we make $2^n = 2^1 = 2$ rows:

	col. 1	col. 2
	p	~p
row 1		
row 2		

Step 2: We enter the possible truth values (T for true, F for false) for the simple statement p in the first column:

p	~p
T	
F	

Step 3: We fill in the last column with the possible truth values of ~p. If p is T, then ~p is F; and if p is F, then ~p is T:

TABLE 2-1: Negation, or NOT

p	~p	
T	F	If it's true that it is snowing, then it's false that it is not snowing.
F	T	If it's false that it is snowing, then it's true that it is not snowing.

Now we have the completed truth table for negation, or NOT. Notice that the last column is a succinct, and easy-to-remember, representation of the logical results of the ~ operation:

- In negation, ~p is true only when p is false.

2. Conjunction: p ∧ q

Step 1: We have three statements in all: p, q, and p ∧ q. So we make three columns. Of these statements, two are simple statements, so we make $2^n = 2^2 = 4$ rows:

p	q	p∧q

Step 2: We enter all the possible combinations for the truth values of the simple statements p and q.

p	q	p∧q
T	T	
T	F	
F	T	
F	F	

Step 3: We fill in the last column with the truth values resulting from each combination:

TABLE 2-2: Conjunction, or AND

p	q	p∧q	
T	T	T	If "It is snowing" is true and "I am cold" is true, then "It is snowing AND I am cold" is true.
T	F	F	If "It is snowing" is true but "I am cold" is false, then "It is snowing AND I am cold" is false.
F	T	F	If "It is snowing" is false but "I am cold" is true, then "It is snowing AND I am cold" is false.
F	F	F	If "It is snowing" is false and "I am cold" is false, then "It is snowing AND I am cold" is false.

- In conjunction, p ∧ q is true *only* when both p *and* q are true.

3. Disjunction: p ∨ q

The first two steps in constructing the truth table for disjunction are the same as those we used for conjunction, so let's skip right to Step 3:

TABLE 2-3: Disjunction, or OR

p	q	p∨q	
T	T	T	If "It is snowing" is true and "I am cold" is true, then "It is snowing OR I am cold" is true.
T	F	T	If "It is snowing" is true but "I am cold" is false, then "It is snowing OR I am cold" is true.
F	T	T	If "It is snowing" is false but "I am cold" is true, then "It is snowing OR I am cold" is true.
F	F	F	If "It is snowing" is false and "I am cold" is false, then "It is snowing OR I am cold" is false.

- In disjunction, p ∨ q is true whenever *either* of the simple statements p *or* q is true.

4. Implication: p ⇒ q

The truth table for p ⇒ q is shown in Table 2-4.

TABLE 2-4: Implication, or "IF . . . , THEN . . .**"**

p	q	p⇒q
T	T	T
T	F	F
F	T	T
F	F	T

Table 2-4 shows us that

- The implication p ⇒ q is true for every pair of p and q values *except* when p is true and q is false.

It means that in a true implication, a true statement cannot imply a false one. The compound statement

"If I see a polar bear, THEN $1 \times 2 = 7$"

is false if indeed I do see a polar bear (since 1×2 does not equal 7); but if I do not see a polar bear, then the implication is logically true!

C. Using truth tables to show the interactions among operations

1. Interactions between conjunctions and disjunctions

Truth tables are invaluable in determining the truth values of more involved compound statements. Without truth tables, for example, it would be difficult to show that the following relationship between disjunctions and conjunctions holds:

INTERACTIONS BETWEEN CONJUNCTIONS AND DISJUNCTIONS		
	$(p \vee q) \wedge r = (p \wedge r) \vee (q \wedge r)$	(2-1a)
	$p \wedge (q \vee r) = (p \wedge q) \vee (p \wedge r)$	(2-1b)

note: By saying that two compound statements are *equal*, we mean that they return the same truth values when given the same combinations of truth values for their respective simple statements.

These two relationships show that we can manipulate statements in a manner similar to arithmetic. Just as $(a + b) \times c = a \times c + b \times c$, $(p \vee q) \wedge r = (p \wedge r) \vee (q \wedge r)$. It seems that $\vee$ (OR) plays a role in logic similar to $+$ (addition) in arithmetic, while $\wedge$ plays the role of $\times$ (multiplication).

EXAMPLE 2-4:   Use a truth table to show that rule (2-1a) is true.

Solution

Step 1: Set up truth tables for $(p \vee q) \wedge r$ and $(p \wedge r) \vee (q \wedge r)$. (Notice that there are $2^3 = 8$ rows: 3 simple statements with 2 choices each.) Fill in all the possible combinations of T and F for p, q, and r.

$(p \vee q) \wedge r$

p	q	r	$p \vee q$	$(p \vee q) \wedge r$
T	T	T		
T	T	F		
T	F	T		
T	F	F		
F	T	T		
F	T	F		
F	F	T		
F	F	F		

Note the starting pattern for the simple statements.

(p ∧ r) ∨ (q ∧ r)

p	q	r	p ∧ r	q ∧ r	(p ∧ r) ∨ (q ∧ r)
T	T	T			
T	T	F			
T	F	T			
T	F	F			
F	T	T			
F	T	F			
F	F	T			
F	F	F			

Step 2: Fill in the column for p ∨ q in the first table and the columns for p ∧ r and q ∧ r in the second table. (Follow the rules for the conjunctions shown in Table 2-2 and disjunctions shown in Table 2-3.)

(p ∨ q) ∧ r

p	q	r	p ∨ q	(p ∨ q) ∧ r
T	T	T	T	
T	T	F	T	
T	F	T	T	
T	F	F	T	
F	T	T	T	
F	T	F	T	
F	F	T	F	
F	F	F	F	

(p ∧ r) ∨ (q ∧ r)

p	q	r	p ∧ r	q ∧ r	(p ∧ r) ∨ (q ∧ r)
T	T	T	T	T	
T	T	F	F	F	
T	F	T	T	F	
T	F	F	F	F	
F	T	T	F	T	
F	T	F	F	F	
F	F	T	F	F	
F	F	F	F	F	

Step 3: Again following the rules for conjunction and disjunction, fill in the last columns of each table shown in Step 2.

(p ∨ q) ∧ r

p	q	r	p ∨ q	(p ∨ q) ∧ r
T	T	T	T	T
T	T	F	T	F
T	F	T	T	T
T	F	F	T	F
F	T	T	T	T
F	T	F	T	F
F	F	T	F	F
F	F	F	F	F

(p ∧ r) ∨ (q ∧ r)

p	q	r	p ∧ r	q ∧ r	(p ∧ r) ∨ (q ∧ r)
T	T	T	T	T	T
T	T	F	F	F	F
T	F	T	T	F	T
T	F	F	F	F	F
F	T	T	F	T	T
F	T	F	F	F	F
F	F	T	F	F	F
F	F	F	F	F	F

Step 4: Since the last column is identical in both tables shown in Step 3, the compound statements are equal. Thus rule (2-1a) is true.

2. DeMorgan's laws

There are two very useful rules that relate how conjunctions and disjunctions interact with negation. These are called **DeMorgan's Laws** and are expressed as

DEMORGAN'S LAWS

$$\sim(p \wedge q) = (\sim p) \vee (\sim q) \tag{2-2}$$

$$\sim(p \vee q) = (\sim p) \wedge (\sim q) \tag{2-3}$$

EXAMPLE 2-5:   Construct a truth table to show that law (2-2) is true.

Solution

Step 1: The table for the left side is

∼(p ∧ q)

p	q	p ∧ q	∼(p ∧ q)
T	T	T	F
T	F	F	T
F	T	F	T
F	F	F	T

Note that ∼ (NOT) reverses the truth values for conjunction.

Step 2: The table for the right side is

$(\sim p)\wedge(\sim q)$

p	q	~p	~q	$(\sim p)\vee(\sim q)$
T	T	F	F	F
T	F	F	T	T
F	T	T	F	T
F	F	T	T	T

Step 3: Thus $\sim(p\wedge q)=(\sim p)\vee(\sim q)$. So law (2-2) is true.

> *note:* An easy way to remember DeMorgan's Laws is to think of distributing the $\sim$ over each item in the parentheses. Whenever you encounter $\sim\wedge$, replace it with $\vee$, and whenever you encounter $\sim\vee$, replace it with $\wedge$.

EXAMPLE 2-6: Use DeMorgan's Laws to rewrite $\sim((p\wedge q)\vee(\sim r))$ with negations on the simple statements p, q, and r only.

Solution

Step 1: Distribute the negation $\sim$ over the items in the parentheses; that is, think of $\sim\wedge$ as $\vee$ and $\sim\vee$ as $\wedge$:

$$\sim((p\wedge q)\vee(\sim r))=\sim(p\wedge q)``\sim\vee"(\sim(\sim r))\quad\text{or}\quad\sim(p\wedge q)\wedge(\sim(\sim r))$$

Step 2: But $\sim(\sim r)=r$ (see Exercise 2-5), and

$$\sim(p\wedge q)=(\sim p)``\sim\wedge"(\sim q)\quad\text{or}\quad(\sim p)\vee(\sim q)$$

Step 3: Combine the results:

$$\sim((p\wedge q)\vee(\sim r))=((\sim p)\vee(\sim q))\wedge r$$

2-3. Correct and Incorrect Arguments

• Deductive arguments are basically *true implications* (p $\Rightarrow$ q), where p is the premise and q is the conclusion.

Sometimes in reasoning, however, we may want (or need) to use a variation of the implication p $\Rightarrow$ q in order to prove that our conclusion is true. We must then make sure that the variation we use is *equivalent* to—has the same truth values as—the original implication.

There are three implications that are commonly used as substitutes for p $\Rightarrow$ q in deductive arguments. Two of these are not equivalent to p $\Rightarrow$ q and thus give us incorrect conclusions, while the other is equivalent and thus gives us a *correct* (true, valid) conclusion.

A. Converse

• The **converse** of the implication, p $\Rightarrow$ q is the implication **q $\Rightarrow$ p.**

The converse q $\Rightarrow$ p is *not* equivalent to the original implication p $\Rightarrow$ q! (See Problem 2-4.) In incorrect arguments we frequently assume that, since the original implication holds (is in a true state), the converse must also hold. Or we try to prove that the converse of an implication holds, and then claim that the original implication also holds. For example, if q is the statement "I am a human being" and p is the statement "I am a man," then the implication p $\Rightarrow$ q "IF I am a man, THEN I am a human being" is correct. (That is, p $\Rightarrow$ q is in a true state since a true p cannot imply that q is false. All men *are* human beings.) The converse of this implication is q $\Rightarrow$ p, or "IF I am a human being, THEN I am a man." This is clearly not correct, as any woman will quickly—and rightly—point out!

The error that we make when we argue from the converse results from the fact that the implication p ⇒ q could be in a true state (a correct implication) when p = F and q = T. With these values for p and q, the implication q ⇒ p would *not* be in a true state.

- Arguments based on the converse are not correct.

EXAMPLE 2-7: If p = "I am taking a shower" and q = "I am wet," show that p ⇒ q and its converse q ⇒ p are not equivalent.

Solution

p ⇒ q: IF I am indeed taking a shower, which involves being wet by definition, THEN I am indeed wet.

q ⇒ p: IF I am wet, I may be in the shower, but I may also be standing in the rain at a bus stop or swimming in a lake or playing in a fountain. Therefore, this converse is not equivalent to the implication.

B. Inverse

- The **inverse** of the implication p ⇒ q is the implication ∼**p** ⇒ ∼**q**.

The inverse is also *not* equivalent to the original implication. If, for example, we use the statement "I am a cat" as p and "I am NOT a dog" as q, then p ⇒ q ("IF I am a cat, THEN I am not a dog") is true if said by a cat or a monkey (or anybody!). The inverse ∼p ⇒ ∼q would be "IF I am not a cat, THEN I am a dog," which is no longer true when said by anyone other than a cat or a dog! The two implications are not the same.

- Arguments based on the inverse rather than the original implication are not correct.

note: The error that we make when we argue from the converse or the inverse is that the implication p ⇒ q could be in a true state without p's being true. Think about it.

EXAMPLE 2-8: Construct truth tables for the implication p ⇒ q and its inverse ∼p ⇒ ∼q. Show that the two implications are not the same—that is, that they have different values for the same combination of values for p and q.

Solution

Step 1: The truth table for p ⇒ q is

p	q	p ⇒ q
T	T	T
T	F	F
F	T	T
F	F	T

Step 2: The table for the inverse is

p	q	~p	~q	~p ⇒ ~q
T	T	F	F	T
T	F	F	T	T
F	T	T	F	F
F	F	T	T	T

Step 3: The last columns in rows 2 and 3 of the tables do not agree, so the inverse is not equivalent to the original implication.

C. Contrapositive

- The **contrapositive** of the implication p ⇒ q is the implication **~q ⇒ ~p.**

The contrapositive *is* equivalent to the original implication p ⇒ q. Let's check the tricky case where p is T and q is F. If the contrapositive holds, then if q is false (i.e., ~q is true), ~p must unquestionably be true (i.e., p must be false). It works! A quick construction of a truth table for each implication (Exercise 2-4) will show that the original implication and the contrapositive are indeed equivalent implications.

- Arguments based on the contrapositive are correct.

EXAMPLE 2-9: Write the contrapositive of the implication "IF I am a cat, THEN I am NOT a dog."

Solution: Reverse the premise "I am a cat" and the conclusion "I am NOT a dog" and then negate each of them. Thus you should have "IF I am a dog, THEN I am NOT a cat."

note: The double negation "I am NOT NOT a dog" is the same as "I am a dog."

D. Summary

An **incorrect argument** is one in which the premise p is true, the implication p ⇒ q holds, and the conclusion q is false. A **correct argument** is one in which the premise p is true, the implication p ⇒ q holds, and the conclusion is true.

The *converse* and *inverse* of an implication may be in a true state under conditions that permit the original implication to be in a false state. Thus, even though p is true, q *could* be false. We cannot assume that the original implication is true when the converse or inverse is true. Conclusions based on converses or inverses may not be true!

The *contrapositive* of the implication p ⇒ q can be in a true state with a false conclusion, q, for the original implication only when the original premise, p, is also false. Thus, if the contrapositive is in a true state and the original premise p is true, then q must also be true. Conclusions based on contrapositives will always be true when the premise is true!

2-4. Methods of Proof
A. Direct proof

- A **direct proof** follows a series of implications from the hypothesis (or premise), p, to the conclusion, q.

In a direct proof, we wish to show that when p is true, then q must also be true and hence p ⇒ q is true. (In other words: A true p cannot imply a false q.)

Suppose we already know that the implications p ⇒ r, r ⇒ s, and s ⇒ q are true. Then a true p forces r to be true, which in turn forces s to be true, which finally forces q to be true. We follow a chain of implications from the starting hypothesis, p, steadily moving through intermediate facts until we reach the one that forces q to be true. This process of steadily working from fact to fact to the conclusion is a direct proof. Example 2-2 is a direct proof.

B. Indirect proof

- An **indirect proof**—also called **proof by contradiction** or *reductio ad absurdum* [reduction to the absurd]—follows from the premise, p, to the conclusion, q.

An indirect proof is based on the contrapositive. Instead of showing that when p is true, q must be true and hence p ⇒ q is true, an indirect proof shows that a false conclusion q can only come from a false premise p.

We first assume that the opposite (negation) of the conclusion (~q) is true. Next, we show that the truth of the negation forces (~p) to be true also. It follows that the contrapositive ~q ⇒ ~p is true. And, as we have already established, the contrapositive is true precisely when the original implication is true. This means that a true premise p forces a true conclusion q, so p ⇒ q is true!

Let's look at indirect proof again, this time in plain English. What we really do is to assume the opposite of what we want to prove. Then we show that if that opposite were the case, we'd wind up in a situation where the hypothesis (which we're already treating as true) can't *possibly* be true. Thus we must have done something wrong. If we've reasoned correctly, only the assumption that the hoped-for conclusion was false could be erroneous. The conclusion must then be true. QED!

note: QED stands for the Latin phrase *quod erat demonstrandum*, which means "which was to be proved." This is a phrase that harks back to the days when most mathematical stuff was done in Latin. Some people still use this abbreviation at the end of proofs to say, in effect, "I did it!"

EXAMPLE 2-10:   Prove that $\sqrt{2}$ is not a rational number. (A rational number is one that can be written as the ratio a/b of two integers.)

Solution: Use indirect proof.

Step 1:   Assume that $\sqrt{2}$ *is* a rational number. Thus $\sqrt{2} = a/b$, where a and b are integers with no common factors. (We've canceled out any common factors.)

Step 2:   Then, square both sides:

$$2 = \frac{a^2}{b^2}$$

so that

$$a^2 = 2b^2$$

Thus, a^2 is an even integer.

note: An even integer has the form $2n$, where n is another integer. An odd integer has the form $2n + 1$, where n is some integer. The square of an even integer is even and the square of an odd integer is odd.

Step 3: Now since a^2 is even, a must be even, so $a = 2k$ for some integer k. Thus,

$$a^2 = 2b^2$$

$$(2k)^2 = 2b^2$$

$$4k^2 = 2b^2$$

Step 4: If $4k^2 = 2b^2$, then

$$2k^2 = b^2$$

This shows that b^2 is even, so b must be even and have the form $b = 2j$ for some integer j.

Step 5: But wait! This is absurd, for then a has a factor 2 and b also has a factor 2. But a and b can have no common factors! This is a contradiction.

Step 6: Thus $\sqrt{2}$ is not rational. QED

2-5. Strategies of Proving Things

In most of this chapter, we've been concerned with a stiff formalism of logic and ways of reasoning. The proofs in the rest of the outline concern geometric properties. Let's stop at this point and get down to some basic helps and hints about how you can actually develop such proofs, or show that an apparent fact is false.

A. Diagrams

Many geometric proofs are direct proofs through a series or chain of implications. A diagram of the conditions stated in the premise of an argument is often very helpful in building the chain that leads to the ultimate conclusion.

- Always draw a diagram wherever possible.

Be careful, however, *not* to draw special cases. There is a great temptation to use a property of the special case in an argument when that property isn't really there. For example,

(1) If the hypothesis states something about two lines, don't draw two lines that look parallel, as in Figure 2-1a. Rather, draw the two lines clearly *not* parallel, as in Figure 2-1b. If you drew the lines as in Figure 2-1a, you might easily forget later that the lines are not indeed parallel, and then use some fact about parallel lines.

(2) If the hypothesis states something about intersecting lines, don't draw the lines as perpendicular if that's not part of the hypothesis. You might look at the figure later and assume you have a right angle.

(3) If the problem calls for a **quadrilateral** (a four-sided figure), don't draw the special cases in which the figure has perpendicular sides (Figure 2-2a), or opposite sides parallel (Figure 2-2b), or opposite sides equal (Figure 2-2c). All these quadrilaterals have special properties that the general quadrilateral (Figure 2-2d) does not have.

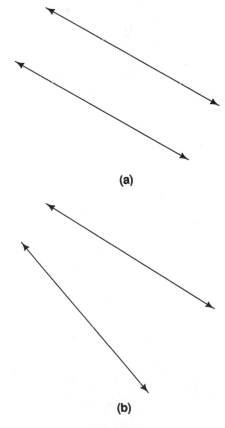

(a)

(b)

Figure 2-1

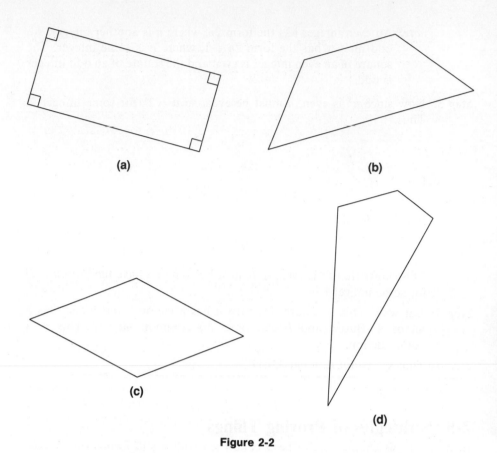

Figure 2-2

(4) If the hypothesis speaks of angles and says nothing about acute or obtuse angles, make two diagrams, one with each type of angle. If you can prove the desired conclusion for one diagram, can you do the same for the other? If not, ask yourself if you have used a special property for one of the figures.

(5) **Triangles** (three-sided figures) have special properties if one angle is 90°. Draw such a triangle only when a 90° angle is specifically mentioned in the hypothesis. In general, try to draw figures with three unequal sides and three unequal angles, as in Figure 2-3a—not the special cases of two equal sides, as in Figure 2-3b.

B. Chain of reason

After you draw your diagram, write the hypothesis down clearly. Label with letter names any points, angles, or lines mentioned in the hypothesis. List any facts that you're given about these labeled items. If two angles are equal, indicate that equality on the diagram.

At this point, go down the list of known facts. Look at the diagram. Think about what facts (theorems) have already been developed about the figures you've drawn. You'll usually see that some small additional fact, which was not listed in the hypothesis, is also true. List this fact along with the information you've been given, and also clearly state or list the reason for this new item on the side. Look at the diagram again with this new information and repeat the process!

• This process of drawing, looking, and listing is called a **chain of reason**, in which one fact leads to another, to another, to another . . . .

If you can't develop a chain of implications that leads you to the desired conclusion, try an indirect proof. Assume the conclusion is the opposite of what you desire. Ask youself: What would happen if this is so? Does this

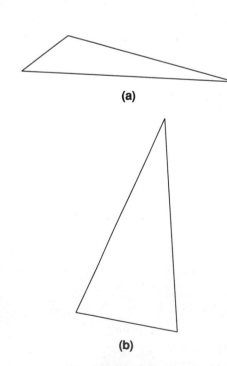

Figure 2-3

assumption develop a chain of reasoning to something that contradicts one of the facts in the hypothesis? If it does, then your assumption of an opposite conclusion is wrong and the original implication is correct.

C. Counterexamples

If neither direct nor indirect proof leads you to a true conclusion, maybe the implication is not true! If the implication is true, the conclusion *must* be true in *all* cases when the hypothesis is true. Possibly, the hypothesis does not imply the conclusion. Think: Can you construct an example in which the hypothesis is true and yet the conclusion in your example is false? If you can, then you have developed a **counterexample**.

note: Since it usually takes a fair effort to prove something, it is usually a good idea to begin by asking yourself if the desired implication might have a counterexample. Counterexamples are not that hard to find. The best way to do this is to redraw a diagram in an extreme situation. If, for instance, two line segments are not supposed to be equal, draw then *really* not equal—make one small and one huge.

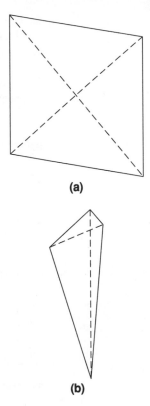

(a)

(b)

Figure 2-4

EXAMPLE 2-11: Do the line segments drawn from opposite points of a quadrilateral as in Figure 2-4a bisect each other?

Solution: Before trying to prove this hypothesis, redraw the diagram in the extreme form, as in Figure 2-4b. In this distorted figure, the line segments are certainly not halved. Figure 2-4b is a counterexample to the claim that the lines are indeed bisected. So this hypothesis is not true.

SUMMARY

1. There are two types of reasoning: inductive and deductive reasoning.

   (a) Inductive reasoning draws conclusions from observations, and inductive conclusions are not certain.
   (b) Deductive reasoning draws conclusions from a premise, and deductive conclusions are true if the premise is true.

2. Symbolic logic aids deductive reasoning.
3. Symbolic logic uses letters for statements and symbols for operations:

   (a) $p \wedge q$ represents the conjunction p AND q; conjunctions are true when *both* p and q are true.
   (b) $p \vee q$ represents the disjunction p OR q; disjunctions are true when *either* p or q is true.
   (c) $\sim p$ represents the negation NOT p; negations are true when p is false.
   (d) $p \Rightarrow q$ represents the implication IF p, THEN q; an implication is true when a true p forces a true q.

4. Logical operations interact with each other in the following ways:

$$\sim(\sim p) = p$$
$$(p \vee q) \wedge r = (p \wedge r) \vee (q \wedge r) \quad \text{and} \quad p \wedge (q \vee r) = (p \wedge q) \vee (p \wedge r)$$
$$\sim(p \vee q) = (\sim p) \wedge (\sim q) \quad \text{and}$$
$$\sim(p \wedge q) = (\sim p) \vee (\sim q) \qquad \text{DeMorgan's Laws}$$

5. Proofs in which we assert that a true premise p forces the conclusion q to be true are flawed if we reason through the converse or inverse:

   (a) The converse $q \Rightarrow p$ of an implication $p \Rightarrow q$ is not equivalent to the original implication.
   (b) The inverse $\sim p \Rightarrow \sim q$ of an implication $p \Rightarrow q$ is not equivalent to the original implication.

6. The contrapositive $\sim q \Rightarrow \sim p$ of the implication $p \Rightarrow q$ is equivalent to the original implication.

7. In direct proofs we argue from the premise through a chain of reason (or chain of implications) to the conclusion.

8. In indirect proofs we argue from the fact that a false conclusion can come only from a false premise.

RAISE YOUR GRADES

Can you . . . ?

☑ distinguish between inductive and deductive reasoning
☑ write a statement in the notation of symbolic logic
☑ construct a truth table for compound logical statements
☑ write DeMorgan's laws
☑ simplify a compound statement with conjunctions, disjunctions, and negations
☑ write the contrapositive of an implication
☑ write the inverse of an implication
☑ show that the contrapositive of the converse is the inverse
☑ construct an indirect proof

SOLVED PROBLEMS

Inductive and Deductive Reasoning

PROBLEM 2-1   Observe the following pattern for the odd integers $n = 3, 5, 7, 9, \ldots$

Integer	Integer Squared	Integer Squared -1	Factored Form of Integer Squared -1
3	9	8	4×2
5	25	24	$4 \times 2 \times 3$
7	49	48	$4 \times 2 \times 6$
9	81	80	$4 \times 2 \times 10$
11	121	120	$4 \times 2 \times 15$
13	169	160	$4 \times 2 \times 20$

Draw an inductive conclusion about odd integers based upon these observations.

Solution   Several common factors appear. Notice, for example, that 5 seems to factor $n^2 - 1$ for any odd number larger than 7. But the most obvious conclusion seems to be that $4 \times 2 \, (=8)$ must factor $n^2 - 1$ for any odd number greater than 1!

   note: $15^2 - 1 = 225 - 1 = 224 = 4 \times 2 \times 28$. Thus the possible conclusion that 5 is a common factor for $n^2 - 1$ for odd n greater than 7 is not true!

PROBLEM 2-2   Use deductive reasoning to show that the square of an odd integer is odd.

Solution

Step 1: By definition any odd integer n must be of the form $2k + 1$ for some integer k.

Step 2: If n is odd, then $n^2 = (2k + 1)^2$.

Step 3: But

$$(2k + 1)^2 = (2k + 1)(2k + 1) = 4k^2 + 4k + 1$$
$$= 2k(2k + 2) + 1$$

Step 4: Set $m = k(2k + 2)$. Then

$$(2k + 1)^2 = 2m + 1$$

note: $m = k(2k + 2)$ is a perfectly good integer.

Step 5: Since $(2k + 1)^2$ can be written as $2m + 1$, where m is an integer, the square of an odd integer is odd.

Symbolic Logic

PROBLEM 2-3 Construct a truth table for each side of DeMorgan's Law (2-3): $\sim(p \vee q) = (\sim p) \wedge (\sim q)$. Show that this law is valid.

Solution The left-hand side has 2 simple statements, p and q, and 2 compound statements, $p \vee q$ and $\sim(p \vee q)$, so the table has 4 columns (one for each statement) and $2^2 = 4$ rows (2^n rows, where n is the number of simple statements).

Step 1: Construct the empty table for the left-hand side.

Step 2: Fill in the columns with truth values (T or F) for the simple statements, following the usual pattern (TT, TF, FT, FF).

Step 3: Fill in the column for the statement $p \vee q$, following the rules for disjunction. (See Table 2-3; remember that a disjunction is false only if both p and q are false.)

Step 4: Fill in the column for $\sim(p \vee q)$, following the rules for negation on column $p \vee q$ (reversing the truth values because of the "not").

p	q	$p \vee q$	$\sim(p \vee q)$
T	T	T	F
T	F	T	F
F	T	T	F
F	F	F	T

The right-hand side has 5 statements in all, only 2 of which are simple: p and q. So its table has 4 rows and 5 columns.

Step 1: Construct the empty table with 4 rows and 5 columns; the columns are headed p, q, $\sim p$, $\sim q$, and $(\sim p) \wedge (\sim q)$.

Step 2: Fill in the columns for p and q.

Step 3: Fill in the columns for $\sim p$ and $\sim q$, following the rule for negation.

Step 4: Fill in the column for $(\sim p) \wedge (\sim q)$, following the rules for conjunction (see Table 2-2) on columns $\sim p$ and $\sim q$.

p	q	$\sim p$	$\sim q$	$(\sim p) \wedge (\sim q)$
T	T	F	F	F
T	F	F	T	F
F	T	T	F	F
F	F	T	T	T

Since the last columns agree for all the corresponding values of p and q, the two expressions are equivalent. So, DeMorgan's Law (2-3) is valid.

Correct and Incorrect Arguments

PROBLEM 2-4 Use truth tables to show that an implication $p \Rightarrow q$ and its converse $q \Rightarrow p$ are not equivalent.

Solution Follow the steps for constructing truth tables; then use the rules for implication to fill in the truth values for the last columns. (Remember that an implication $p \Rightarrow q$ is true for every pair of p and q except when p is true and q is false; see Table 2-4.)

p	q	$p \Rightarrow q$
T	T	T
T	F	F
F	T	T
F	F	T

p	q	$q \Rightarrow p$
T	T	T
T	F	T
F	T	F
F	F	T

Since the second and third rows differ in the last column, the two implications $p \Rightarrow q$ and $q \Rightarrow p$ are not equivalent.

PROBLEM 2-5 The following argument illustrates a typical misunderstanding in deductive reasoning or the use of an implication.

If there is a mysterious force destroying ships in the Bermuda Triangle, then ships would disappear there. Ships disappear in the Bermuda Triangle. Thus, there must be a mysterious force destroying ships in the Bermuda Triangle.

Use symbolic logic to show the error in this argument.

Solution Let

p = "There is a mysterious force destroying ships
        in the Bermuda Triangle"

q = "Ships disappear in the Bermuda Triangle"

Now we have the implication $p \Rightarrow q$.

The argument claims that $p \Rightarrow q$ is true and that q is also true. Since $p \Rightarrow q$ is true, we are dealing with row 1, 3, or 4 of the truth table (2-4) for an implication. The fact that q is also true further restricts us to rows 1 and 3. But since row 3 would still permit p to be false, we cannot claim that p is true.

This argument is really an attempt to claim that the converse $q \Rightarrow p$ is the equivalent of the implication $p \Rightarrow q$. The truth table of Problem 2-4 shows that this is not the case. The states of true and false for $p \Rightarrow q$ do not all correspond to the true or false states for $q \Rightarrow p$.

PROBLEM 2-6 An argument we hear is "Since you can't prove that p is false, p must be true." For instance, "You can't prove that alien astronauts didn't visit the earth; thus alien astronauts must have been here in ancient times." Refute this argument and use symbolic logic to point out its error.

Solution Basically, a statement can be either true or false. Not knowing which value the statement has does not leave you free to pick whichever choice you'd like.

You could formally proceed as follows.

Let p = "I can prove that alien astronauts didn't visit the earth" and q = "Alien astronauts didn't visit the earth." The arguer is taking $p \Rightarrow q$ as true and $\sim p$ as true. The false conclusion reached is that $\sim q$ is true. This is just the common pitfall of thinking that the inverse of an implication is true when the implication is true.

Methods of Proof

PROBLEM 2-7   Give a direct proof that the product ab of two even integers a and b is even. [*Hint:* An even number has the form $2k$, where k is an integer.]

Solution

Step 1:   Let $a = 2m$ and $b = 2n$ for some integers m and n, respectively.

Step 2:   Then $a \times b = (2m) \times (2n) = 2(m \times 2n)$.

Step 3:   Let k be the integer $m \times 2n$.

Step 4:   Then $2(m \times 2n) = 2k$. And since $a \times b = 2k$, the product ab is even.

PROBLEM 2-8   Prove or disprove that when the product $a \times b$ of two integers a and b is odd, then both a and b must be odd.

Solution   It's best to use an indirect proof here. Assume that both a and b are *not* odd. That means at least one of them, say a, must be even. Let $a = 2n$. Thus $a \times b = (2n) \times b = 2(n \times b)$. Since $a \times b$ is $2 \times$ (some integer), $a \times b$ is even. But this is ridiculous, since you already know that $a \times b$ is odd. Thus a cannot be even.

Strategies of Proving Things

PROBLEM 2-9   Prove or disprove that when the product $a \times b$ of two integers a and b is even, then both a and b are even.

Solution   You probably know from experience that the implication (even $a \times b$) $\Rightarrow$ (even a) and (even b) is false. If it were true, it would hold for *every* a and b. So, it's fairly easy to disprove this implication by counterexample. All you need to do is find one pair a and b where the premise is true ($a \times b$ is even) and the conclusion is false. If $a = 2$ (even) and $b = 3$ (odd), $a \times b = 6$, which is an even number.

PROBLEM 2-10   Triangle 1 has sides of length a, b, and c and triangle 2 has sides of length e, f, and g. Prove or disprove the claim that if $a = e$, $b = f$, and if the measure of the angle formed by sides of length b and c equals the measure of the angle formed by sides of length f and g, then c must equal g.

Solution   Let's draw some triangles.

Step 1:   First draw a line segment of length b.

Step 2:   Next draw a dotted line from the right end of this segment as in Figure 2-5a. The side of length c will lie along this dotted line.

Step 3:   Open your compass to a length a and place the point on the left end of the line segment.

Step 4:   Swing the compass and let the pencil trace out an arc. Draw this as generally as possible. The arc should cross the dotted line in two places as in Figure 2-5b.

Step 5:   Join the left end of the line segment to one of these crossing points. There are two possible triangles as in Figure 2-5c. Either of these shapes can be triangle 1. One triangle is obtuse; the other one is acute.

Step 6:   Repeat the process and construct the two possible triangles for triangle 2 as in Figure 2-5d with $f = b$ and $e = a$. Be sure to make the angles between the dotted line and the line segments the same.

Step 7:   Look at the diagrams. If you had selected both triangles 1 and 2 as acute, or both triangles 1 and 2 as obtuse, the claim would appear reasonable. However, if triangle 1 is the acute version of Figure 2-5c while triangle 2 is the obtuse version of Figure 2-5d, then $g > c$ and the claim is false.

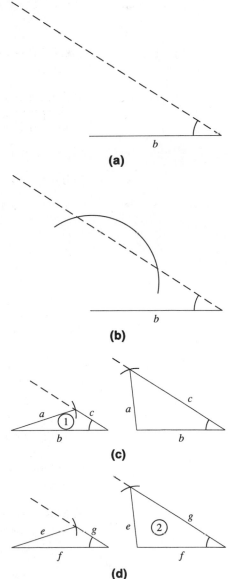

(a)

(b)

(c)

(d)

Figure 2-5

Review Exercises

EXERCISE 2-1 Write the following statements in the notation of symbolic logic:

(a) If I work hard, then I am rich.
(b) If I'm rich, then I am happy.
(c) I work hard and I am not happy.
(d) I am happy or I am rich.
(e) Being rich does not make me happy. [*Hint:* Rewrite **(e)** in "If, then" language.]

EXERCISE 2-2 For each of the statements of Exercise 2-1, write the negation in **(1)** symbols and **(2)** words.

EXERCISE 2-3 For each of the *implications* in Exercise 2-1, write the

(1) converse
(2) inverse
(3) contrapositive

EXERCISE 2-4 Make a truth table to show that $p \Rightarrow q$ is equivalent to $(\sim q) \Rightarrow (\sim p)$.

EXERCISE 2-5 Make a truth table to show that $\sim(\sim p) = p$.

EXERCISE 2-6 Make a truth table to show that $p \wedge (q \vee r) = (p \wedge q) \vee (p \wedge r)$ [see rule (2-2)] is true.

EXERCISE 2-7 If $p = T$, $q = F$, and $r = T$, find the values of

(a) $(\sim p \wedge r) \vee q$
(b) $(p \Rightarrow q) \wedge (r \vee q)$
(c) $(p \wedge q) \wedge r$
(d) $\sim(p \Rightarrow (q \wedge r))$
(e) $\sim((\sim p) \wedge (q \vee (\sim r)))$

EXERCISE 2-8 Simplify:

(a) $\sim(r \vee (\sim q))$
(b) $\sim(p \vee (\sim q)) \vee (\sim p) \wedge r$
(c) $\sim((\sim((\sim r) \wedge (\sim p))) \vee (\sim q))$

EXERCISE 2-9 Prove that the sum of two odd integers is even.

EXERCISE 2-10 Prove that the sum of an even integer and an odd integer is odd.

EXERCISE 2-11 Comment on the correctness of the following arguments:

(a) *Given:* If I study hard, then I will pass the test.
I passed the test.
Conclusion: I studied hard.

(b) *Given:* If I study hard, then I will pass the test.
I did not pass the test.
Conclusion: I didn't study hard.

(c) *Given:* If I study hard, then I will pass the test.
I studied hard.
Conclusion: I passed the test.

(d) *Given:* If I study hard, then I will pass the test.
I didn't study hard.
Conclusion: I didn't pass the test.

EXERCISE 2-12 Fill in the missing piece needed in the following to make the arguments correct:

(a) *Given:* If I drink and then drive my car, I will have an accident.
I did not have an accident.
Conclusion: _____

(b) *Given:* If I drink and then drive my car, I will have an accident.
I drive my car
_____
Conclusion: I had an accident.

(c) *Given:* _____
I am an honest man.
Conclusion: I have no worries.

Answers to Review Exercises

2-1 Let p = "I work hard"
q = "I am rich"
r = "I am happy"

(a) p ⇒ q   (d) r ∨ q
(b) q ⇒ r   (e) ~(q ⇒ r)
(c) p ∧ (~r)

2-2 (1: a) ~(p ⇒ q)
b) ~(q ⇒ r)
c) ~(p ∧ (~r)) or (~p) ∨ r
d) ~(r ∨ q) or (~r) ∧ (~q)
e) ~(~(q ⇒ r)) or q ⇒ r

(2: a) Working hard does not make me rich.
b) Being rich does not make me happy.
c) I do not work hard or I am happy.
d) I am not happy and I am not rich.
e) If I am rich, then I am happy.

2-3 (1: a) If I am rich, then I work hard.
b) If I am happy, then I am rich.
(2: a) If I do not work hard, then I do not get rich.
b) If I am not rich, then I am not happy.
(3: a) If I am not rich, then I do not work hard.
b) If I am not happy, then I am not rich.

2-4

p	q	(~q)	(~p)	(~q) ⇒ (~p)
T	T	F	F	T
T	F	T	F	F
F	T	F	T	T
F	F	T	T	T

2-5

p	(~p)	~(~p)
T	F	T
F	T	F

2-6

p	q	r	(q ∨ r)	p ∧ (q ∨ r)
T	T	T	T	T
T	T	F	T	T
T	F	T	T	T
T	F	F	F	F
F	T	T	T	F
F	T	F	T	F
F	F	T	T	F
F	F	F	F	F

p	q	r	p ∧ q	p ∧ r	(p ∧ q) ∨ (p ∧ r)
T	T	T	T	T	T
T	T	F	T	F	T
T	F	T	F	T	T
T	F	F	F	F	F
F	T	T	F	F	F
F	T	F	F	F	F
F	F	T	F	F	F
F	F	F	F	F	F

2-7   (a) F   (d) T
        (b) F   (e) T
        (c) F

2-8   (a) $(\sim r) \wedge q$
        (b) $(\sim p) \wedge (q \vee r)$
        (c) $((\sim r) \wedge (\sim p)) \wedge q$ or $(\sim (r \vee p)) \wedge q$

2-9   Let $a = 2n + 1$ and $b = 2m + 1$. Then

$$
\begin{aligned}
a + b &= (2n + 1) + (2m + 1) \\
&= (2n + 2m) + (1 + 1) \\
&= 2(n + m) + 2 = 2(n + m + 1)
\end{aligned}
$$

2-10   Let $a = 2n$ and $b = 2m + 1$. Then

$$
a + b = 2n + 2m + 1 = 2(n + m) + 1
$$

2-11   (a) Incorrect. It uses the converse.
        (b) Correct. It uses the contrapositive.
        (c) Correct. The implication and the premise of the implication are true, which forces the conclusion of the implication to be true.
        (d) Incorrect. It uses the inverse.

2-12   (a) I did not drink or I did not drive my car.
        (b) I drank.
        (c) If I am an honest man, then I have no worries.

3 TRIANGLES

THIS CHAPTER IS ABOUT

☑ **Triangle Parts**
☑ **Special Triangles**
☑ **Properties of Right Triangles**
☑ **Congruent General Triangles**
☑ **Similar Triangles**

3-1. Triangle Parts

A. Sides and vertices

A **triangle** is a geometric figure formed by joining three line segments together at their endpoints. The triangle formed from the line segments AB, BC, and CA is shown in Figure 3-1. The line segments AB, BC, and CA are the **sides of the triangle**, and the points A, B, and C where the line segments meet are the **vertices of the triangle**.

note: The singular of "vertices" is "vertex" as in "Point A is a vertex of the triangle." We usually refer to the triangle itself as $\triangle ABC$.

caution: Figure 3-1 fits our usual intuitive image of a triangle—the sides enclose space and no vertex lies on the opposite side. But if our definition of a triangle is not limited to this image unless we add a requirement that the three potential vertices not lie on the same line, i.e., be *noncollinear*. If we do not add this noncollinearity requirement, the definition includes a figure like that shown in Figure 3-2, which is formed from the line segments AB, AC, and CB. This figure looks like a line segment, but it does meet our original definition of a triangle. We sometimes refer to such a figure as a **degenerate triangle**, and we can view it as the result of steadily moving vertex C to side AB in Figure 3-1. In this outline we'll usually assume that triangles have noncollinear vertices like those in Figure 3-1. We may, however, need to draw on such extreme cases as Figure 3-2 to make points from time to time.

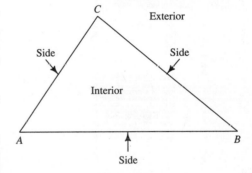

Figure 3-1

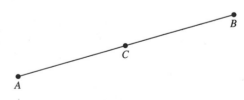

Figure 3-2

B. Angles

Angles $\angle CAB$, $\angle ABC$, and $\angle BCA$ in $\triangle ABC$ are called the **interior angles of the triangle** (see Figure 3-3). When no ambiguity can occur, we may use the shorter representation $\angle A$, $\angle B$, and $\angle C$ for $\angle CAB$, $\angle ABC$, and $\angle BCA$, respectively.

If we extend one of the sides of $\triangle ABC$ past a vertex, as in Figure 3-3, we see that the interior angle is supplementary to the angle formed by the extended ray and the nonextended side. This new angle is called an **exterior angle of the triangle**. For example, $\angle EAC$ and $\angle BCD$ of Figure 3-3 are both exterior angles for $\triangle ABC$.

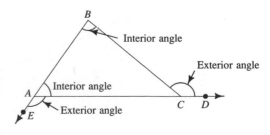

Figure 3-3

C. Interior, exterior, and perimeter

The **interior** of the triangle is the set of points that lie inside all three interior angles of a triangle. The **exterior** of a triangle is the set of points that neither lie in the interior of the triangle (inside the triangle) nor belong to one of the sides (lie on the triangle). The **perimeter** of the triangle is the set of points that actually falls on the sides of the triangle. The perimeter separates the interior and the exterior points of the triangle.

caution: Some people use the term "perimeter" to mean the set of actual points of the sides of the triangle, while others use the term to mean the sum of the lengths of the three sides (distance around the triangle). Still others use the term "perimeter" to mean both things at once. The intended meaning is usually clear from the context. If the meaning isn't clear, it's a good idea to check the beginning of whatever book you're reading for a clear statement of what the author means! In this outline I will try to be consistent in using "perimeter" for the actual points and "length of the perimeter" for the distance. If I should happen to lapse into the ambiguity too, I apologize in advance and beg your indulgence.

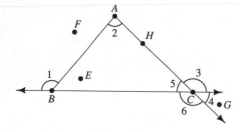

Figure 3-4

EXAMPLE 3-1:   **(a)** Classify the points E, F, G, and H in Figure 3-4 as interior, exterior, or perimeter points. **(b)** Classify the angles $\angle 1$, $\angle 2$, $\angle 3$, $\angle 4$, $\angle 5$, and $\angle 6$ as interior or exterior angles where possible.

Solution

(a) E is an interior point. H is a perimeter point. F and G are exterior points.
(b) Angles $\angle 1$, $\angle 3$, and $\angle 6$ are exterior angles. Angles $\angle 2$ and $\angle 5$ are interior angles. Angle $\angle 4$ is neither an interior nor an exterior angle.

3-2. Special Triangles

A. Scalene, acute, and obtuse triangles

The term **scalene triangle** is the general name we use for an ordinary, run-of-the-mill triangle with three unequal sides and three unequal interior angles. If all the interior angles of a scalene triangle are acute, we may call the triangle an **acute triangle**. If one of the interior angles of a scalene triangle is obtuse, we may call the triangle an **obtuse triangle**.

B. Isosceles triangles

An **isosceles triangle** is the special case in which a triangle has two sides of equal length. Although this term implies nothing about the length of the third side, we usually reserve the term "isosceles" for the triangle whose third side does not have the same length as the other two sides. The angle between the two equal sides is sometimes called the **apex angle** of the isosceles triangle, and the side opposite this angle is called the **base** of the isosceles triangle.

Figure 3-5 shows an isosceles triangle, where the matching slash marks indicate the equal sides AB and AC. BC is the base and $\angle A$ is the apex angle for the isosceles triangle in Figure 3-5.

Figure 3-5

C. Equilateral triangles

If all three sides of a triangle are of equal length, as in Figure 3-6, this triangle is called an **equilateral triangle**. An equilateral triangle is a special case of isosceles triangles. Even though equilateral triangles are isosceles triangles, the terms "base" and "apex," which are used with isosceles triangles, are rarely used with equilateral triangles.

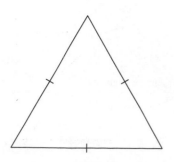

Figure 3-6

D. Right triangles

A **right triangle** is a triangle in which one of the interior angles is 90°, so that the two sides that meet at this angle are perpendicular. The two perpendicular sides are called the **legs** of the right triangle, and the remaining side is called the **hypotenuse**. The hypotenuse is always opposite the right angle. Figure 3-7 shows a right triangle, where $\angle C$ is the right angle, sides CA and CB are the legs, and side AB is the hypotenuse.

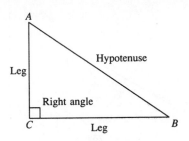

Figure 3-7

EXAMPLE 3-2: Give the terms that best describe the triangles shown in Figure 3-8.

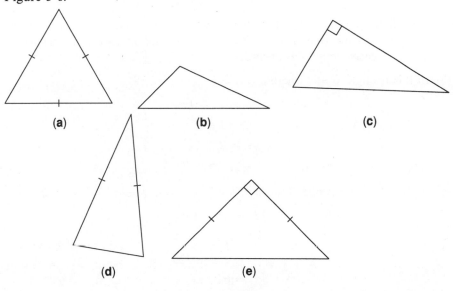

Figure 3-8

Solution (**a**) Equilateral triangle. (**b**) Obtuse scalene triangle. (**c**) Right triangle. (**d**) Isosceles triangle. (**e**) Isosceles right triangle.

3-3. Properties of Right Triangles
A. The Pythagorean Theorem

The relationship between the legs of a right triangle and its hypotenuse is one of the oldest known geometric facts and is called the **Pythagorean Theorem**:

- **The sum of the squares of the lengths of the legs equals the square of the length of the hypotenuse.**

If we draw a right triangle as in Figure 3-9, and use lowercase letters a, b, and c to represent the lengths of the sides opposite $\angle A$, $\angle B$, and $\angle C$, respectively, then the Pythagorean Theorem takes the algebraic form

PYTHAGOREAN THEOREM
$$a^2 + b^2 = c^2 \qquad \text{(3-1)}$$

If we are given information about the lengths of any two sides of a right triangle, we can use the algebraic form of the Pythagorean Theorem to find the length of the other side.

note: The Pythagorean Theorem is the basis for extending the ruler-and-string concept of distance expressed in Section 1-2 to formula (1-1), $d(P_1, P_2) = \sqrt{(x_1 - x_2)^2 + (y_1 - y_2)^2}$, which expresses the Euclidean distance between two points in the Cartesian plane. Notice that the distance d between two points P_1 and P_2 is the square root of the sum of two squares.

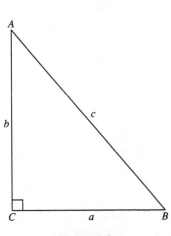

Figure 3-9

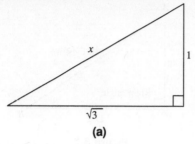

(a)

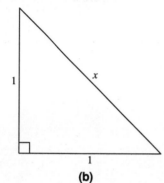

(b)

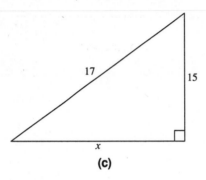

(c)

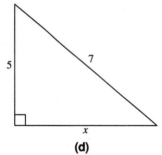

(d)

Figure 3-10

EXAMPLE 3-3: For each of the right triangles shown in Figure 3-10 find the length x of the side about which you are given no information.

Solution: You know that $a^2 + b^2 = c^2$.

(a) In this triangle, x is the hypotenuse, so $x = c$, where $a = 1$ and $b = \sqrt{3}$.

$$x^2 = 1^2 + (\sqrt{3})^2$$
$$= 1 + 3$$
$$= 4$$
$$x = \sqrt{4} = \pm 2 = 2$$

> *note:* Since you're looking for a length x, you know that x must be positive. Thus, even though $\sqrt{4} = 2$ or -2, you should choose 2, because a "negative length" is absurd.

(b) In this triangle $x = c$, where $a = b = 1$.

$$x^2 = 1^2 + 1^2$$
$$= 1 + 1$$
$$= 2$$
$$x = \sqrt{2}$$

(c) In this triangle x is one of the legs, say a, so $x = a$, where $c = 17$ and $b = 15$.

$$17^2 = x^2 + 15^2$$
$$x^2 = 17^2 - 15^2$$
$$= 289 - 225$$
$$= 64$$
$$x = \sqrt{64} = 8$$

(d) In this triangle, $x = a$ where $c = 7$ and $b = 5$.

$$7^2 = x^2 + 5^2$$
$$x^2 = 7^2 - 5^2$$
$$= 49 - 25$$
$$= 24$$
$$x = \sqrt{24} = \sqrt{4 \times 6} = 2\sqrt{6}$$

or

$$x = 4.898$$

> *note:* Most calculators have a button that computes square roots directly. Such a calculator will make your life much easier.

B. Angles of a right triangle

Consider the non–right angles $\angle A$ and $\angle B$ of right triangle $\triangle ABC$ (see Figure 3-7). These are the two angles opposite the legs of the right triangle. If leg CB (opposite $\angle A$) is very small, then $\angle A$ must be almost $0°$ and certainly acute. If leg CB is steadily lengthened, $\angle A$ increases, but it remains acute until it reaches $90°$. But if $\angle A$ were ever to reach $90°$, then *both* sides AB and CB would be perpendicular to side AC. And if that were the case, sides AB and CB would be on parallel lines. (Remember that one of the equivalences to the Parallel Postulate states that any two lines perpendicular to a third line are parallel.) But parallel lines *cannot* have point

B in common! Thus ∠*A* must always be acute. Similar reasoning shows that ∠*B* is also acute. Thus a right triangle always has one 90° angle and two acute angles. And—

- **The two acute angles of a right triangle are complementary.**

EXAMPLE 3-4: Prove that the acute angles of a right triangle are complementary.

Solution

Step 1: Draw a right triangle, as in Figure 3-11, with the right angle at vertex *C*.

Step 2: Construct a line segment *AE* perpendicular to side *AC* at vertex *A*. This line $\overleftrightarrow{AE}$ creates ∠1 with the hypotenuse *AB*.

Step 3: Line segments *AE* and *CB* are both perpendicular to *AC*, and therefore $\overleftrightarrow{AE}$ and $\overleftrightarrow{CB}$ are parallel.

Step 4: ∠*B* and ∠1 are alternate interior angles that are cut from these two parallel lines by transversal $\overleftrightarrow{AB}$. Thus ∠*B* = ∠1.

> *note:* Although ∠*B* and ∠1 are not *the same* angle, they do have equal measure. When two angles have equal measure, we sometimes say that such angles are *equal*.

Step 5: By construction (i.e., because you constructed it that way), ∠*A* + ∠1 = 90°.

Step 6: Thus, ∠*A* + ∠1 = ∠*A* + ∠*B* = 90°.

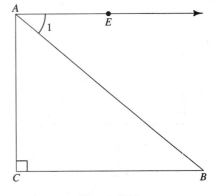

Figure 3-11

C. Sine, cosine, and tangent

If the length of the hypotenuse of any right triangle is given, then the openings of the two acute angles determine the lengths of the two legs. Similarly, if the length of one leg is given, then the openings of the two acute angles determine the lengths of the other leg and the hypotenuse. Finally, if we are given the measures of the two acute angles and the length of one of the sides, we can quickly construct the entire right triangle using a ruler, a compass, and a protractor; then, after we have drawn the triangle, we can find the lengths of the remaining sides.

By the same reasoning, if we are given the lengths of two sides of a right triangle, we can easily find the length of the third side by using the Pythagorean Theorem (just as we did in Example 3-3), as long as we know whether the two given sides are both legs or one leg and a hypotenuse. Then we can construct the entire right triangle using a ruler and compass, and we can measure the acute angles with a protractor.

The point of all this is that—

- **The acute angles of a right triangle are uniquely determined either by the ratio of the two legs or by the ratio of one leg and the hypotenuse—and vice versa.**

Let's see how this works.

Consider any two right triangles having a pair of equal acute angles ∠α, as shown in Figure 3-12. Since we know that the two acute angles of any right triangle are complementary, we know that we can subtract the measure of ∠α from 90° to find the measure of the other acute angle ∠β in each triangle. And, of course, we know that the third angle in each triangle has measure 90°. Thus, we can see that all of the corresponding angles of the two triangles shown in Figure 3-12 are equal.

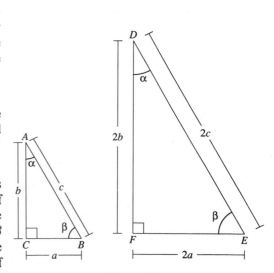

Figure 3-12

- When the acute angles of two (or more) right triangles are congruent, the triangles are called **similar right triangles.**

We can see from Figure 3-12 that similar right triangles have the same shape but not necessarily the same size. In this case, each side of $\triangle DEF$ is exactly twice as long as the corresponding side of $\triangle ABC$, so $DE = 2AB = 2c$, $EF = 2BC = 2a$, and $DF = 2AC = 2b$. Now consider the fraction, or *ratio*, obtained by dividing the length of one of the legs, say DF (the side opposite $\angle \beta$) by the length of the hypotenuse DE in the larger triangle. Since

$$\frac{DF}{DE} = \frac{2AC}{2AB} = \frac{AC}{AB} \quad \text{or} \quad \frac{DF}{DE} = \frac{2b}{2c} = \frac{b}{c}$$

the factor 2 cancels out, and we get the same ratio of the two lengths as for the smaller triangle. This is an example of the most important property of similar triangles:

- The corresponding sides of similar triangles are *proportional.*

This means that there is always some number k such that each side of one of a pair of similar triangles is k times as long as the corresponding side of the other triangle. As long as we compare ratios of corresponding sides, rather than the lengths of the sides themselves, this factor k always cancels out, so we get equal ratios. Thus, the acute angle of any *right* triangle determines the ratios of each pair of its sides, even though it does not determine how big the triangle is.

These ratios of right triangles are important enough to have names of their own. We call them the **sine**, the **cosine**, and the **tangent**, abbreviated sin, cos, and tan, respectively. For $\angle \alpha$ the ratios are defined as follows:

SINE OF $\angle \alpha$ $\sin \alpha = \dfrac{\text{opposite leg}}{\text{hypotenuse}} = \dfrac{a}{c}$

COSINE OF $\angle \alpha$ $\cos \alpha = \dfrac{\text{adjacent leg}}{\text{hypotenuse}} = \dfrac{b}{c}$

TANGENT OF $\angle \alpha$ $\tan \alpha = \dfrac{\text{opposite leg}}{\text{adjacent leg}} = \dfrac{a}{b}$

These ratios can also be defined in terms of the other acute angle $\angle \beta$, of course. Thus we see that

$$\text{Sin } \beta = \frac{\text{opposite leg}}{\text{hypotenuse}} = \frac{b}{c}$$

$$\cos \beta = \frac{\text{adjacent leg}}{\text{hypotenuse}} = \frac{a}{c}$$

$$\tan \beta = \frac{\text{opposite leg}}{\text{adjacent leg}} = \frac{b}{a}$$

And from this it's easy to see that

$$\sin \beta = \cos \alpha = \cos(90° - \beta)$$

$$\cos \beta = \sin \alpha = \sin(90° - \beta)$$

and

$$\tan \beta = \frac{1}{\tan \alpha}$$

Table 3-1 gives the values for the sine, cosine, and tangent of the angles $0°$ through $90°$ in increments of $1°$. These values will prove useful in many different types of calculations. For example, since it is difficult to measure angles to a very great degree of accuracy with a protractor, we can use these numbers, called the **trigonometric ratios**, to help us draw triangles.

TABLE 3-1: Trigonometric Ratios

Angle (degrees)	sine	cosine	tangent	Angle (degrees)	sine	cosine	tangent
0	0.0000	1.0000	0.0000	46	0.7193	0.6947	1.0355
1	0.0175	0.9998	0.0175	47	0.7314	0.6820	1.0724
2	0.0349	0.9994	0.0349	48	0.7431	0.6691	1.1106
3	0.0523	0.9986	0.0524	49	0.7547	0.6561	1.1504
4	0.0698	0.9976	0.0699	50	0.7660	0.6428	1.1918
5	0.0872	0.9962	0.0875	51	0.7771	0.6293	1.2349
6	0.1045	0.9945	0.1051	52	0.7880	0.6157	1.2799
7	0.1219	0.9925	0.1228	53	0.7986	0.6018	1.3270
8	0.1392	0.9903	0.1405	54	0.8090	0.5878	1.3764
9	0.1564	0.9877	0.1584	55	0.8192	0.5736	1.4281
10	0.1736	0.9848	0.1763	56	0.8290	0.5592	1.4826
11	0.1908	0.9816	0.1944	57	0.8387	0.5446	1.5399
12	0.2079	0.9781	0.2126	58	0.8480	0.5299	1.6003
13	0.2250	0.9744	0.2309	59	0.8572	0.5150	1.6643
14	0.2419	0.9703	0.2493	60	0.8660	0.5000	1.7321
15	0.2588	0.9659	0.2679	61	0.8746	0.4848	1.8040
16	0.2756	0.9613	0.2867	62	0.8829	0.4695	1.8807
17	0.2924	0.9563	0.3057	63	0.8910	0.4540	1.9626
18	0.3090	0.9511	0.3249	64	0.8988	0.4384	2.0503
19	0.3256	0.9455	0.3443	65	0.9063	0.4226	2.1445
20	0.3420	0.9397	0.3640	66	0.9135	0.4067	2.2460
21	0.3584	0.9336	0.3839	67	0.9205	0.3907	2.3559
22	0.3746	0.9272	0.4040	68	0.9272	0.3746	2.4751
23	0.3907	0.9205	0.4245	69	0.9336	0.3584	2.6051
24	0.4067	0.9135	0.4452	70	0.9397	0.3420	2.7475
25	0.4226	0.9063	0.4663	71	0.9455	0.3256	2.9042
26	0.4384	0.8988	0.4877	72	0.9511	0.3090	3.0777
27	0.4540	0.8910	0.5095	73	0.9563	0.2924	3.2709
28	0.4695	0.8829	0.5317	74	0.9613	0.2756	3.4874
29	0.4848	0.8746	0.5543	75	0.9659	0.2588	3.7321
30	0.5000	0.8660	0.5774	76	0.9703	0.2419	4.0108
31	0.5150	0.8572	0.6009	77	0.9744	0.2250	4.3315
32	0.5299	0.8480	0.6249	78	0.9781	0.2079	4.7046
33	0.5446	0.8387	0.6494	79	0.9816	0.1908	5.1446
34	0.5592	0.8290	0.6745	80	0.9848	0.1736	5.6713
35	0.5736	0.8192	0.7002	81	0.9877	0.1564	6.3138
36	0.5878	0.8090	0.7265	82	0.9903	0.1392	7.1154
37	0.6018	0.7986	0.7536	83	0.9925	0.1219	8.1443
38	0.6157	0.7880	0.7813	84	0.9945	0.1045	9.5144
39	0.6293	0.7771	0.8098	85	0.9962	0.0872	11.4300
40	0.6428	0.7660	0.8391	86	0.9976	0.0698	14.3007
41	0.6561	0.7547	0.8693	87	0.9986	0.0523	19.0812
42	0.6691	0.7431	0.9004	88	0.9994	0.0349	28.6364
43	0.6820	0.7314	0.9325	89	0.9998	0.0175	57.2900
44	0.6947	0.7193	0.9657	90	1.0000	0.0000	undefined
45	0.7071	0.7071	1.0000				

EXAMPLE 3-5: Construct a right triangle whose hypotenuse is 3.5 inches long and whose base angle $\angle \beta$ has a 37° measure. Find the other parts of the triangle: **(a)** the measure of the second acute angle and **(b)** the lengths of the two legs.

Solution

Step 1: Draw a line segment extending 3.5 inches from A to B. This line segment AB is the hypotenuse c, where $c = 3.5$ inches (given).

Step 2: Use a protractor to measure a 37° angle at B and draw the ray forming this angle.

Step 3: Draw a line perpendicular to the ray you've just drawn from A. Label the intersection of the ray and the perpendicular C.

Now you have the right triangle $\triangle ABC$ whose hypotenuse is 3.5 inches long and whose base angle is 37°. Your triangle should look like the triangle shown in Figure 3-13.

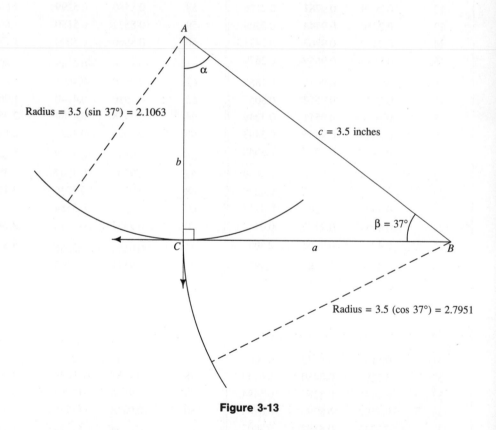

Figure 3-13

Next, you want to find the measure of the other acute angle and the lengths of the two legs.

(a) Finding the measure of the acute angle $\angle \alpha$ is easy. Since the two angles must be complementary, you just subtract 37° from 90°.

$$\angle \alpha = 90° - 37° = 53°$$

(b) To find the length of $BC = a$, you can use a combination of simple algebra and the trigonometric ratios (Table 3-1). If

$$\cos \beta = \frac{a}{c}$$

then

$$\cos 37° = \frac{a}{3.5 \text{ inches}}$$

so

$$a = (\cos 37°)(3.5 \text{ inches})$$

To find the value of $\cos 37°$, find the angle $37°$ in Table 3-1 and read across to the cosine column.

$$\cos 37° = 0.7986$$

Then

$$a = (0.7986)(3.5 \text{ inches}) = 2.7951 \text{ inches}$$

Now you can use the same reasoning to find $AC = b$:

$$\sin \beta = \frac{b}{c}$$

$$\sin 37° = \frac{b}{3.5 \text{ inches}}$$

And from the table you have $\sin 37° = 0.6018$, so

$$0.6018 = \frac{b}{3.5 \text{ inches}}$$

$$b = (0.6018)(3.5 \text{ inches})$$
$$= 2.1063 \text{ inches}$$

EXAMPLE 3-6: Find to three decimal places the sine, cosine, and tangent of the acute angles $\angle \alpha$ and $\angle \beta$ in the triangle in Example 3-5.

Solution: Either use Table 3-1 or a good scientific calculator in "degree mode."

$\sin 37° = 0.602$	$\cos 37° = 0.799$	$\tan 37° = 0.754$
$\cos 53° = 0.602$	$\sin 53° = 0.799$	$\tan 53° = 1.327$

D. Congruence of right triangles

1. Definition of congruence

It's possible to think of triangles as rigid pieces of cardboard that can be picked up and slid around (and out of) the plane. If we move one triangle, $\triangle ABC$, in space and find that $\triangle ABC$ can be placed exactly on top of another triangle, $\triangle DEF$, so that the sides and angles match precisely, we say that these two triangles are **congruent**. When triangles are congruent, each one has exactly the same size and shape as the other. This means that the lengths of their three sides are *respectively* equal, as are the measures of their three angles.

Let's consider, for example, the triangles $\triangle ABC$, $\triangle DEF$, and $\triangle GJH$ in Figure 3-14. It's easy to see that triangles $\triangle ABC$ and $\triangle DEF$ have corresponding parts that can be made to fit onto each other:

$A \leftrightarrow D$ (A corresponds to D):	$AB \leftrightarrow DE$	$\angle A \leftrightarrow \angle D$
$B \leftrightarrow E$ (B corresponds to E):	$BC \leftrightarrow EF$	$\angle B \leftrightarrow \angle E$
$C \leftrightarrow F$ (C corresponds to F):	$AC \leftrightarrow DF$	$\angle C \leftrightarrow \angle F$

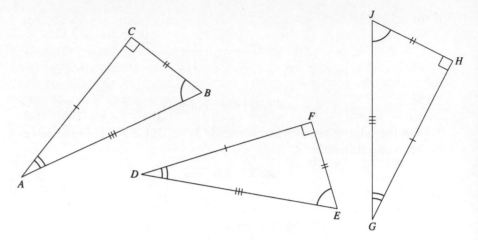

Figure 3-14

And the relationship between the members of these corresponding pairs is an *equivalence* relationship, so that the sides and angles of these triangles are congruent:

$$AB \cong DE \qquad \angle A \cong \angle D$$
$$BC \cong EF \qquad \angle B \cong \angle E$$
$$AC \cong DF \qquad \angle C \cong \angle F$$

Thus we say that the triangles $\triangle ABC$ and $\triangle DEF$ are congruent, $\triangle ABC \cong \triangle DEF$.

note: We use the symbol $\leftrightarrow$ to denote correspondence and the symbol $\cong$ to denote congruence.

Triangles $\triangle ABC$ and $\triangle DEF$ could be superimposed exactly if we were to slide one of them on the page. But what about $\triangle GJH$? We can't slide $\triangle GJH$ onto $\triangle ABC$, but we could pick it up and flip it through space to get the following correspondences:

$$A \leftrightarrow G: \qquad AB \leftrightarrow GJ \qquad \angle A \leftrightarrow \angle G$$
$$B \leftrightarrow J: \qquad BC \leftrightarrow JH \qquad \angle B \leftrightarrow \angle J$$
$$C \leftrightarrow H: \qquad AC \leftrightarrow GH \qquad \angle C \leftrightarrow \angle H$$

And since each of these correspondence relationships is also an equivalence relationship, $\triangle ABC \cong \triangle GJH$.

note: The slash marks and special angle markers indicate corresponding parts of the triangles shown in Figure 3-14. You can make use of special markings of this type to keep track of corresponding parts in diagrams and complicated proofs.

2. Recognizing congruent right triangles

We've already learned that we can get all the information we need about a right triangle when we are given much less than the lengths of all three sides and the measures of all three angles. If, for example, we are given the lengths of the two legs, we can use the Pythagorean Theorem to find the hypotenuse. Then the ratio of the lengths of the legs to that of the hypotenuse uniquely determines the measures of the acute angles. Or, if we know the measure of one acute angle and the length of one side (one of the legs or the hypotenuse) of a right triangle, we can obtain the measure of the other acute angle and the lengths of the other two sides.

We can use what we know about the determination of right triangles to show that two (or more) right triangles are congruent. Since the two acute angles of a right triangle are complementary, all we really need to show that the two right triangles are congruent is (**1**) that an acute angle of one triangle equals (is congruent to) an acute angle of the other triangle and that the opposite sides, the adjacent sides or the hypotenuses have equal lengths; or (**2**) that two of the sides of one triangle are equal to two of the sides of the other.

In short, we usually need only two pieces of information to show that two right triangles are congruent. Let's summarize these results in detail as follows.

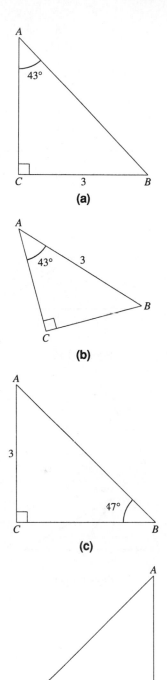

- Two right triangles are congruent if—

  (**a**) the lengths of two legs of one triangle are equal to the lengths of two legs of the other triangle.
  (**b**) the lengths of one leg and the hypotenuse of one triangle are respectively equal to the lengths of a leg and the hypotenuse of the other triangle.
  (**c**) an acute angle of one triangle is congruent to an acute angle of the other triangle and their hypotenuses are equal in length.
  (**d**) an acute angle of one triangle is congruent to an acute angle of the other triangle and their adjacent legs are equal in length.
  (**e**) an acute angle of one triangle is congruent to an acute angle of the other triangle and their opposite legs are equal in length.

  note: When we refer to these rules for determining congruence in right triangles, we can use the following short forms: (**a**) **leg–leg,** (**b**) **leg–hypotenuse,** (**c**) **angle–hypotenuse,** (**d**) **angle–adjacent leg,** and (**e**) **angle–opposite leg.**

  caution: We *cannot* show that two or more right triangles are necessarily congruent if the measures of their acute angles are equal. This is enough to establish that right triangles are *similar* (see Figure 3-12), but we need more information about the sides before we can show congruence.

EXAMPLE 3-7: Determine which of the four right triangles shown in Figure 3-15 is/are congruent to another. Explain your answer.

Solution: Triangles (**a**) and (**d**) have respectively $\angle A = 43°$ and $\angle A = 90° - 47° = 43°$. The sides opposite this angle in both triangles have length 3. Thus these triangles are congruent. Even though (**c**) and (**b**) have sides of length 3 and angle 43°, the side of length 3 is not opposite the 43° angle. So these triangles cannot be congruent to (**a**) or (**d**)! Nor can (**c**) and (**b**) be congruent to each other because the lengths of two of their sides are not equal.

Figure 3-15

3-4. Congruent General Triangles

We can use what we've learned about congruence in right triangles to show whether or not any two general triangles are congruent. As with right triangles, we don't need to know that all three corresponding sides and all three corresponding angles are congruent to show that two general triangles are congruent. What we do need to know is how to break down, or *decompose*, any general triangle into right triangle parts.

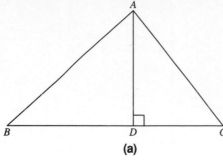

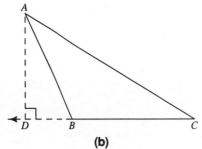

Figure 3-16

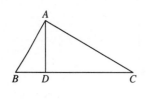

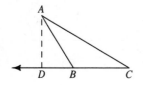

Figure 3-17

A. Decomposition into right triangles

For any general triangle $\triangle ABC$, we can pick a vertex, say A, and construct a line segment AD such that this line segment is perpendicular to the opposite side BC (as in Figure 3-16a) or to an extension of the opposite side $\overline{BC}$ (as in Figure 3-16b). This line segment AD in Figure 3-16a and b is called the **altitude** of the triangle. And the side BC, with respect to which AD is drawn, is called the **base** of the triangle. (Sometimes, the word "base" is also used for the *length* of this line segment.)

We can view $\triangle ABC$ in Figure 3-16a as the result of combining (adding) right triangles $\triangle ABD$ and $\triangle ADC$. And we can view $\triangle ABC$ in Figure 3-16b as the result of removing (subtracting) $\triangle ABD$ from $\triangle ADC$. Thus—

- When two triangles are both broken down into a sum of right triangles (as in Figure 3-16a), or a difference of right triangles (as in Figure 3-16b), the original triangles are congruent when the corresponding right triangles are congruent.

EXAMPLE 3-8: Show that the sum of the interior angles of any triangle is 180°.

Solution: Show two triangles, one acute and the other obtuse, as in Figure 3-17. Label the vertices of the acute triangles A, B, and C, and draw the altitude AD. Label the vertices of the obtuse triangle A, B, and C, and draw the altitude AD.

Acute Triangle

Step 1: Because the two acute angles of a right angle are complementary, you know that

$$\angle C + \angle DAC = 90°$$

and

$$\angle B + \angle DAB = 90°$$

Step 2: You know by construction that

$$\angle A = \angle DAC + \angle DAB$$

Step 3: Thus

$$\angle C + \angle A + \angle B = \angle C + \angle DAC + \angle DAB + \angle B$$
$$= 90° + 90° = 180°$$

Obtuse Triangle

Step 1: You know by construction that $\angle DBA$ and $\angle ABC$ are supplementary, so

$$\angle DBA + \angle ABC = 180°$$

Step 2: Because the two acute angles of a right triangle are complementary, you know that

$$\angle DAC + \angle C = 90°$$

and

$$\angle DAB + \angle DBA = 90°$$

Step 3: Thus

$$\angle DAC + \angle C = \angle DAB + \angle DBA$$

or

$$\angle DBA = (\angle DAC - \angle DAB) + \angle C$$

Step 4: By construction $\angle DAC - \angle DAB = \angle BAC$, so

$$\angle DBA = \angle BAC + \angle C$$

Step 5: Substituting this value of $\angle DBA$ into the equation in Step 1, you get

$$\angle BAC + \angle C + \angle ABC = 180° \qquad \text{QED}$$

note: Since the sum of all the interior angles in a triangle is 180°, as Example 3-8 proves, there can be only one obtuse angle in a triangle.

B. Rules for determining congruence

There are just three rules that we can use to determine if any pair of triangles are congruent. These rules are known by their acronyms—SSS, ASA, and SAS—and are discussed below.

1. SSS congruence

- **Two triangles are congruent if their three corresponding sides are congruent.**

  This rule is the **side-side-side** congruence, abbreviated **SSS**.

EXAMPLE 3-9: $\triangle ABC$ and $\triangle EFG$ in Figure 3-18a are two acute triangles having congruent sides. Show that SSS establishes congruence between these triangles.

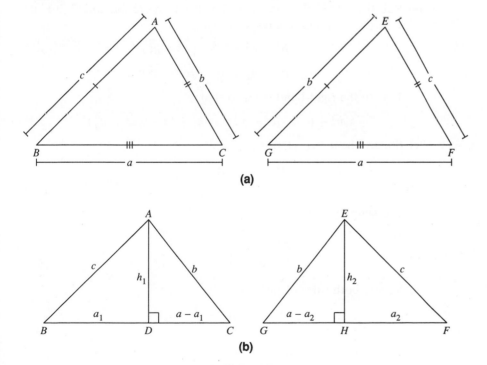

Figure 3-18

Solution

Step 1: From the biggest angles $\angle A$ and $\angle E$ in $\triangle ABC$ and $\triangle EFG$ draw altitudes AD and EH, as in Figure 3-18b. Label the lengths of AD and EH as h_1 and h_2, respectively.

note: The biggest angle of a triangle is opposite the biggest side (Problem 3-4). Thus if the triangles have corresponding sides, $\angle A$ and $\angle E$ are both opposite sides with the same length a.

Step 2: You know that the bases of $\triangle ABC$ and $\triangle EFG$ both have length a. Thus you can say that the base of the right triangle $\triangle ABD$ has length a_1 and the base of right triangle $\triangle ACD$ has length $a - a_1$, where $a_1 + a - a_1 = a$. Similarly, the base of $\triangle EFG$ also has length a, so the bases of $\triangle EFH$ and $\triangle EGH$ can be a_2 and $a - a_2$, respectively. Label these bases as in Figure 3-18b.

Step 3: Show that $h_1 = h_2$.

(a) From the Pythagorean Theorem

$$a_1 = \sqrt{c^2 - h_1^2} \tag{3-2a}$$

and

$$a_2 = \sqrt{c^2 - h_2^2} \tag{3-2b}$$

where a_1 and h_1 are the legs of right triangle $\triangle ABD$, a_2 and h_2 are the legs of the right triangle $\triangle EFH$, and c is the hypotenuse of both $\triangle ABD$ and $\triangle EFH$. Then

$$b^2 = h_1^2 + (a - a_1)^2 \tag{3-3a}$$

and

$$b^2 = h_2^2 + (a - a_2)^2 \tag{3-3b}$$

where h_1 and $a - a_1$ are the legs of right triangle $\triangle ACD$, h_2 and $a - a_2$ are the legs of right triangle $\triangle EGH$, and b is the hypotenuse of both $\triangle ACD$ and $\triangle EGH$.

(b) Substitute the values for a_1 and a_2 from Eqs. (3-2a) and (3-2b) into Eqs. (3-3a) and (3-3b), respectively.

$$b^2 = h_1^2 + (a - \sqrt{c^2 - h_1^2})^2$$
$$b^2 = h_2^2 + (a - \sqrt{c^2 - h_2^2})^2$$

Equate the right-hand sides so that

$$h_1^2 + (a - \sqrt{c^2 - h_1^2})^2 = h_2^2 + (a - \sqrt{c^2 - h_2^2})^2$$

(c) Square the $a - \sqrt{}$ terms

$$\cancel{h_1^2} + \cancel{a^2} - 2a\sqrt{c^2 - h_1^2} + \cancel{c^2} - \cancel{h_1^2}$$
$$= \cancel{h_2^2} + \cancel{a^2} - 2a\sqrt{c^2 - h_2^2} + \cancel{c^2} - \cancel{h_2^2}$$

and simplify

$$-2a\sqrt{c^2 - h_1^2} = -2a\sqrt{c^2 - h_2^2}$$
$$\sqrt{c^2 - h_1^2} = \sqrt{c^2 - h_2^2} \tag{3-4}$$

(d) Square both sides of Eq. (3-4):

$$c^2 - h_1^2 = c^2 - h_2^2$$

or

$$h_1^2 = h_2^2$$

(e) If $h_1^2 = h_2^2$, then $\sqrt{h_1^2} = \sqrt{h_2^2}$; and since h_1 and h_2 must be positive,

$$h_1 = h_2$$

Step 4: If $h_1 = h_2$, then $\triangle ABD \cong \triangle EFH$ and $\triangle ACD \cong \triangle EGH$ by the hypotenuse-leg rule. And if the component right triangles of a pair of triangles are congruent, then the original triangles are congruent.

QED

2. ASA congruence

- **Two triangles are congruent if one triangle has two angles and the side joining their vertices congruent to two angles and the connecting side of the other.**

This is called **angle-side-angle** congruence and is abbreviated by **ASA**.

EXAMPLE 3-10: Show that ASA establishes congruence between any pair of triangles.

Solution

Step 1: You already know from Section 3-3 that right triangles in this situation are congruent. Thus, draw two general triangles △*ABC* and △*EFG* as in Figure 3-19a and indicate the known corresponding parts as shown.

> *note:* The two triangles could be either both acute, as in Figure 3.19a, or both obtuse. Since the sum of the interior angles of a triangle is 180°, you know the third angles in both triangles are also equal. Let's prove the theorem for the acute case. You should be able to modify this proof for the obtuse case (see Problem 3-6).

Step 2: Drop a perpendicular (altitude) from the vertex of one of the known angles (∠*A*) to the opposite side so as to form a right triangle with the known side as the hypotenuse. In Figure 3-19b this is *AD* in △*ABC*. Do the same thing in △*EFG*, dropping the altitude *EH* from the congruent angle, ∠*E*.

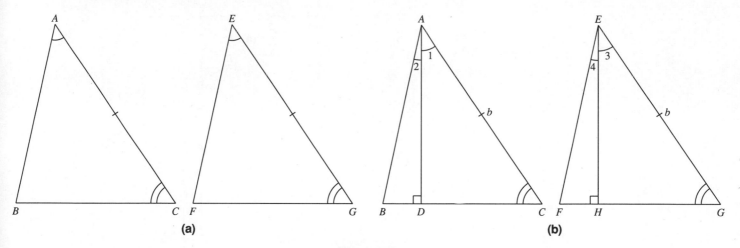

(a)   **(b)**

Figure 3-19

Step 3: The fact that *AC* ≅ *EG* is given as true, as is the fact that ∠*C* ≅ ∠*G*; so the right triangles △*ADC* and △*EHG* are congruent by hypotenuse-angle.

Step 4: *AD* ≅ *EH*, *DC* ≅ *HG*, and ∠1 ≅ ∠3 since they are corresponding parts of congruent triangles.

Step 5: You are given that ∠*A* ≅ ∠*E* and you know from construction that ∠*A* = ∠1 + ∠2 and ∠*E* = ∠3 + ∠4, so you have ∠2 ≅ ∠4.

Step 6: △*ADB* and △*EHF* are right triangles, ∠2 ≅ ∠4, and *AD* ≅ *EH*, and thus △*ADB* ≅ △*EHF* by angle–adjacent leg.

Step 7: △*ABC* ≅ △*EFG* since the component right triangles of the pair of triangles are congruent.   QED

3. SAS congruence

- **Two triangles are congruent if two sides and the angle between them in one are congruent to two sides and the contained angle in the other.**

This is called **side-angle-side** congruence and is abbreviated by SAS.

EXAMPLE 3-11: Show that SAS establishes congruence between any pair of triangles.

Solution

Step 1: You already know from Section 3-3 that right triangles are congruent in this situation. Thus draw two pairs of general triangles △*ABC* and △*EFG* as in Figure 3-20a and 3-20b.

> *note:* Again, there are two cases. Let's prove the acute triangle case of Figure 3-20a and leave the obtuse case as an exercise.

Step 2: Drop a perpendicular (altitude) from one of the angles *not* given, say *A*, to a side that *is* given. This is line *AD* in △*ABC* in Figure 3-20c. Do the same thing for △*EFG*, dropping an altitude from the corresponding angle to the congruent side.

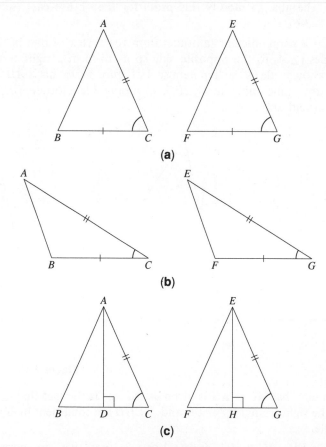

Figure 3-20

Step 3: Since △*ADC* and △*EHG* are right triangles with $AC \cong EG$ (given) and $\angle C \cong \angle G$ (given), △*ADC* ≅ △*EHG* by hypotenuse-angle.

Step 4: $AD \cong EH$ and $DC \cong HG$, since these line segments are corresponding parts of congruent triangles.

Step 5: But $BC = BD + DC$ and $FG = FH + HG$. Therefore, since $BC \cong FG$, you have $BD \cong FH$.

Step 6: Since $\triangle ABD$ and $\triangle EFH$ are right triangles with congruent legs ($AD \cong EH$, $BD \cong FH$), $\triangle ABD \cong \triangle EFH$.

Step 7: $\triangle ABC \cong \triangle EFG$ since the component right triangles of the pair of triangles are congruent. QED

caution: **Side-side-angle (two sides and the adjoining angle) SSA does NOT prove congruence!**

At this point, you might be tempted to believe that any three congruent parts are enough to prove that any two triangles are congruent. This is not so. When two sides and the adjoining angle of two triangles are congruent, the triangles may be as different as those shown in Figure 3-21, where one triangle is acute and the other is obtuse. Remember: The only sufficient criteria for proving congruence in two general triangles are SSS, SAS, and ASA.

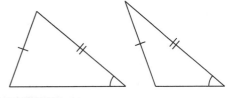

Figure 3-21

3-5. Similar Triangles

Triangles that have three congruent angles are not necessarily congruent.

- When two triangles have three congruent angles, AAA, the triangles are called **similar triangles**.

Similar triangles, like the similar right triangles shown in Figure 3-12, have the same shape, but not the same size. Instead of having congruent sides, similar triangles have *proportional* sides. That is,

- the lengths of the sides opposite congruent angles in one triangle are a common multiple of the lengths of the sides opposite the corresponding congruent angles.

If we have two similar triangles, $\triangle ABC$ and $\triangle EFG$ for example, the lengths of the sides of $\triangle ABC$ stand in common ratio to the lengths of the sides of $\triangle EFG$. That is, if $A \leftrightarrow E$, $B \leftrightarrow F$, and $C \leftrightarrow G$, then

$$\frac{AB}{EF} = \frac{AC}{EG} = \frac{BC}{FG}$$

Also, the relative ratios of two sides in one triangle are the same as the relative ratios of the other. Thus

$$\frac{AB}{AC} = \frac{EF}{EG}, \qquad \frac{BC}{AC} = \frac{FG}{EG}, \qquad \frac{AB}{BC} = \frac{EF}{FG}$$

EXAMPLE 3-12: Show that the altitude to the hypotenuse of a right triangle produces two similar triangles.

Solution

Step 1: Draw a right triangle $\triangle ABC$, as shown in Figure 3-22.

Step 2: Draw the altitude CD to the hypotenuse AB. Label the angles $\angle ACD$, $\angle A$, $\angle B$, and $\angle BCD$ as $\angle 1$, $\angle 2$, $\angle 3$, and $\angle 4$, respectively, for convenience.

Step 3: You know by construction that $\angle 1$ and $\angle 4$ are complementary (i.e., $\angle 1 + \angle 4 = 90°$). You also know that $\angle 1$ and $\angle 2$ are complementary because $\angle 1$ and $\angle 2$ are the two acute angles of the right triangle $\triangle CDA$. Thus if $\angle 4$ is complementary to $\angle 1$ and $\angle 2$ is complementary to $\angle 1$, then $\angle 2 \cong \angle 4$.

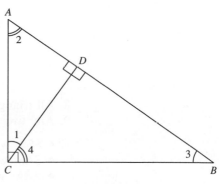

Figure 3-22

Step 4: By the same reasoning as that in Step 3, if $\angle 1$ is complementary to $\angle 4$ (by construction) and $\angle 3$ is complementary to $\angle 4$ (because $\angle 3$ and $\angle 4$ are the two acute angles in right triangle $\triangle CDB$), then $\angle 1 \cong \angle 3$.

Step 5: Angles $\angle CDA$ and $\angle CDB$ are both right angles by construction, so $\angle CDA \cong \angle CDB$.

Step 6: All three angles in $\triangle CDA$ and $\triangle CDB$ are congruent, so $\triangle CDA$ is similar to $\triangle CDB$.   QED

note: This also means that

$$\frac{AD}{CD} = \frac{CD}{BD}$$

SUMMARY

1. A triangle is a three-sided figure formed by joining three line segments together at their endpoints:

   - an isosceles triangle has two equal sides;
   - an equilateral triangle has three equal sides;
   - a right triangle has one 90° angle.

2. The two acute angles of a right triangle are complementary, that is, they sum to 90°.

3. The Pythagorean Theorem states that the square of the length of the hypotenuse of a right triangle is equal to the sum of the squares of the lengths of the two legs.

4. The interior angles of any triangle sum to 180°.

5. Congruent triangles have the same size and shape; they have three corresponding congruent sides and three corresponding congruent angles.

6. Any two right triangles can be shown to be congruent by—

   - leg-leg: when the legs of one triangle are congruent to the legs of the other;
   - leg-hypotenuse: when a leg and the hypotenuse of one triangle are respectively congruent to a leg and the hypotenuse of the other;
   - angle-hypotenuse: when an acute angle and the hypotenuse of one triangle are respectively congruent to an acute angle and the hypotenuse of the other;
   - angle–opposite leg: when an acute angle and the opposite leg of one triangle are respectively congruent to an acute angle and the opposite leg of the other;
   - angle–adjacent leg: when an acute angle and the adjacent leg of one triangle are respectively congruent to the acute angle and adjacent leg of the other.

7. For the acute angles of any right triangle the ratios of the associated legs and hypotenuse are called trigonometric ratios and are as follows:

$$\text{sine} = \frac{\text{opposite leg}}{\text{hypotenuse}}$$

$$\text{cosine} = \frac{\text{adjacent leg}}{\text{hypotenuse}}$$

$$\text{tangent} = \frac{\text{opposite leg}}{\text{adjacent leg}}$$

8. All triangles can be decomposed into two right triangles.

9. Any two triangles can be shown to be congruent by

   - SSS: when their three sides are congruent;
   - SAS: when two of their sides and the included angle are congruent;
   - ASA: when two of their angles and the side joining the vertices of these angles are congruent.

10. Similar triangles have the same shape but not the same size:
- the corresponding angles of similar triangles are congruent;
- the corresponding sides of similar triangles are proportional.

11. The altitude to the hypotenuse of a right triangle divides the triangle into two similar triangles.

RAISE YOUR GRADES
Can you . . . ?

☑ identify the names of the different line segments and angles associated with a triangle
☑ identify the type of a particular triangle
☑ find the side of a right triangle when given the other two sides
☑ find one acute angle of a right triangle when given the other
☑ find the sine, cosine, and tangent of an acute angle from the table
☑ find the sine, cosine, and tangent of an angle when given the sides of a right triangle
☑ decompose a triangle into two right triangles
☑ find the third interior angle, given two interior angles of a triangle
☑ explain how to prove that two right triangles are congruent
☑ explain how to prove that two general triangles are congruent
☑ explain how to prove that two triangles are similar

SOLVED PROBLEMS

PROBLEM 3-1 Draw a right triangle with leg 3 and hypotenuse 5 using just a straightedge and a compass.

Solution

Step 1: Open a compass 3 units. Pick a point A, place the sharp point of the compass on A, and sweep an arc. Draw the line segment AC, where C is any point on the arc.

Step 2: Open the compass a distance of 5 units, place the sharp point on A, and sweep another arc. (Notice that the arc you've just drawn is part of a circle whose center is A and whose radius is equal to the given hypotenuse of the triangle.)

Step 3: Open the compass a distance of $\sqrt{5^2 - 3^2} = \sqrt{25 - 9} = \sqrt{16} = 4$ units, place the sharp point at C, and sweep another arc that intersects the arc you drew in Step 2. Label this intersection point B. (Notice that the arc you just drew is part of a circle whose center is C and whose radius is 4—and 4 is the length of the third leg of the desired triangle as required by the Pythagorean Theorem.)

Step 4: Draw the line segments BC and AB. The points A, B, and C are the vertices of a right triangle with leg 3 and hypotenuse 5.

See Figure 3-23.

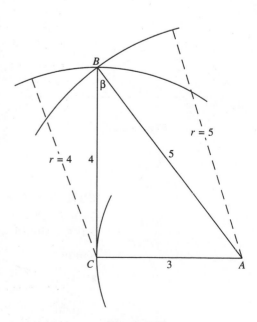

Figure 3-23

PROBLEM 3-2 Find (a) the sine, cosine, and tangent of the angles $\angle\alpha$ and $\angle\beta$ in the triangle $\triangle ABC$ from Problem 3-1. (b) Use these values and Table 3-1 to find the approximate values of the angles $\angle\alpha$ and $\angle\beta$ in degrees.

Solution

(a) Use the definitions of the trigonometric ratios:

$$\sin\alpha = \frac{\text{opposite leg}}{\text{hypotenuse}} = \frac{4}{5} = 0.800$$

$$\cos\alpha = \frac{\text{adjacent leg}}{\text{hypotenuse}} = \frac{3}{5} = 0.600$$

$$\tan\alpha = \frac{\text{opposite leg}}{\text{adjacent leg}} = \frac{4}{3} = 1.333\ldots$$

Then, because $\angle\beta$ is complementary to $\angle\alpha$,

$$\sin\beta = \cos\alpha = \frac{3}{5} = 0.600$$

$$\cos\beta = \sin\alpha = \frac{4}{5} = 0.800$$

$$\tan\beta = \frac{1}{\tan\alpha} = \frac{3}{4} = 0.750$$

(b) You can use any one of these values to find the value of either angle. If, for example, you choose $\tan\beta$, you see from Table 3-1 that the angle whose tangent is close to 0.750 is the angle that has a measure of 37°. Therefore, $\angle\beta \approx 37°$. And if $\angle\beta$ is approximately 37°, then $\angle\alpha$ is $90° - 37° = 53°$, or approximately 53°.

> *note:* You can see from Table 3-1 that the value of 0.750 in the tangent column actually falls between the rows for 36° and 37° in the angle column. In the case of $\angle\beta$ a value of 37° is appropriate. But a more accurate value for $\angle\beta$ is 36.87°.

PROBLEM 3-3 Show that the exterior angle of a triangle is equal to the sum of the two interior angles that do not have the same vertex as that exterior angle.

Solution

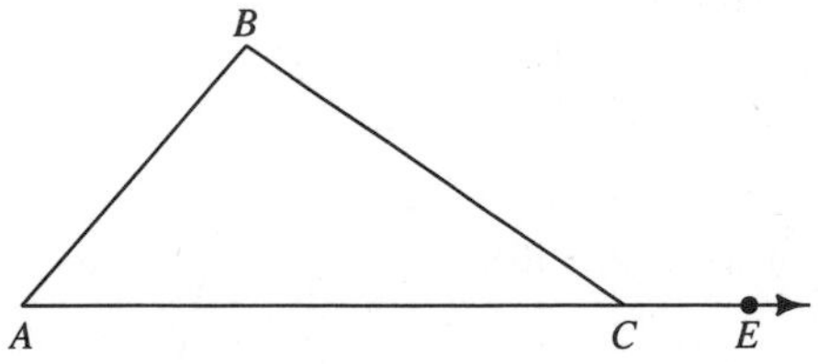

Step 1: Draw a triangle $\triangle ABC$ with the exterior angle $\angle BCE$, as in Figure 3-24.

Step 2: $\angle ACB + \angle BCE = 180°$ by the definition of an exterior angle.

Step 3: $\angle BAC + \angle ACB + \angle CBA = 180°$ by Example 3-8.

Step 4: Thus

$$\angle BCE = 180° - \angle ACB$$
$$= 180° - (180° - \angle BAC - \angle CBA)$$

or

$$\angle BCE = \angle BAC + \angle CBA$$

Figure 3-24

PROBLEM 3-4 Consider any triangle $\triangle ABC$, where the side opposite $\angle A$ has length a and the side opposite $\angle B$ has length b. (a) Prove that if $a < b$, then $\angle A < \angle B$. (b) Suggest a way to prove that if $\angle A \geq \angle B$, then $a \geq b$ in $\triangle ABC$.

Solution

(a) Although the *truth* of this proposition may seem obvious to you, its proof may not be so obvious. What you know about the properties of congruent triangles and the properties of right angles, etc., doesn't seem to be applicable because the problem, as stated, doesn't have anything that's congruent to or perpendicular to or parallel to anything else. In short, you don't have much to work with or reason from. The "trick" to solving a problem like this one in geometry is to *make* something to reason from—a congruence or a right angle or a transversal or whatever seems necessary—being sure you follow the rules of correct geometric construction. Once you have some relationships to work with, the proof becomes a simple matter of following one fact after another until you have proved what you want to prove.

Step 1: Draw $\triangle ABC$, so that $a < b$.

Step 2: You're given that $a < b$, which would be useful if this relation were in a place you could do something with it. Let's construct two triangles $\triangle DCP$ and $\triangle ACP$ such that $\triangle ACP \cong \triangle DPC$ and side DC is collinear with side BC.

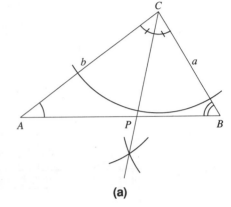

(a)

    (1) Using a compass, divide $\angle C$ into two congruent angles by constructing the angle bisector of $\angle C$ (see Example 1-6). Let P be the point where this bisector intersects AB, as in Figure 3-25a.

    (2) Extend the ray $\overrightarrow{CB}$ of angle $\angle C$. Then, with the sharp point of your compass on vertex C and the pencil point on vertex A, swing an arc from A through ray $\overrightarrow{CB}$. Let D be the point at which this arc intersects $\overrightarrow{CB}$, so that DC has length b.

    (3) Draw line segment DP. Now you have $\triangle DPC$. $\triangle ACP \cong \triangle DCP$ by SAS, because $CP \cong CP$, $\angle ACP \cong \angle DCP$, and $AC \cong DC$.

Step 3: Notice that in constructing $\triangle DPC$ you have also constructed a third triangle, $\triangle BDP$, such that $\angle PBC$ is an exterior angle of $\triangle BDP$. Thus you know from the results of Problem 3-3 that

$$\angle PBC = \angle BPD + \angle PDB$$

But since $\angle BPD$ exists, that is,

$$\angle BPD > 0$$

you can see that

$$\angle PBC > \angle PDB$$

Step 4: Angle $\angle PBC$ is the original $\angle B$, and $\angle PDB \cong \angle A$, so

$$\angle B > \angle A$$

or

$$\angle A < \angle B \qquad\qquad \text{QED}$$

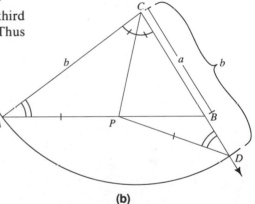

(b)

Figure 3-25

(b) Notice that "if $a < b$, then $\angle A < \angle B$" is an implication, p $\Rightarrow$ q where p $= a < b$ and q $= \angle A < \angle B$. You have just proved the truth of this implication. You can therefore be sure that the *contrapositive* of the implication is also true, that is $\sim$q $\Rightarrow$ $\sim$p or "if $\angle A \not< \angle B$, then $a \not< b$," which is the same as "if $\angle A \geq \angle B$, then $a \geq b$." Thus, you should be able to prove $\angle A \geq \angle B \Rightarrow a \geq b$ by indirect proof.

    note: This proves that if $\angle A < \angle B$, then $a < b$ also proves that the base angles of an isosceles triangle are equal. (Think about it!)

PROBLEM 3-5 Triangles $\triangle ABC$ and $\triangle EFG$ are two obtuse triangles that have congruent sides. Show that SSS established congruence between these two triangles.

Solution In Example 3-9 it is proved that SSS establishes congruence between two acute triangles that have congruent sides. If you wanted to, you could use the results of this example to prove that the triangles shown in Figure 3-26a are congruent. In this case, all you'd have to do would be to construct the altitude from the biggest angles $\angle B$ and $\angle F$ to sides AC and EG, respectively, and then perform the calculations used in Example 3-9. This proof, however, will take a different route.

Step 1: Extend sides BC and FG and drop the altitudes AD and EH, respectively, to these sides. Label the lengths of the altitudes h_1 and h_2 as in Figure 3-26b.

Step 2: Label the lengths of the corresponding congruent sides: BC and FG have length a, AC and EG have length b, and AB and EF have length c. Then you can say that BD has length a_1 and FH has length a_2, so that CD has length $a + a_1$ and GH has length $a + a_2$.

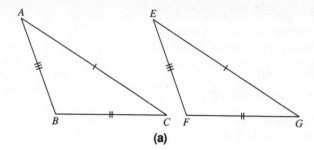

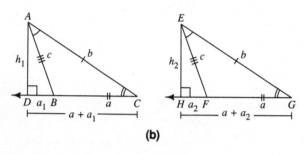

(a)

(b)

Figure 3-26

Step 3: Notice that the altitude expresses each original triangle as a difference of two right triangles: $\triangle ADB$ and $\triangle ADC$ in the first triangle and $\triangle EHF$ and $\triangle EHG$ in the second. Thus by the Pythagorean Theorem

$$a_1 = \sqrt{c^2 - h_1^2} \quad \text{and} \quad a_2 = \sqrt{c^2 - h_2^2}$$
$$b^2 = h_1^2 + (a + a_1)^2 \quad \text{and} \quad b^2 = h_2^2 + (a + a_2)^2$$

Thus

$$b^2 = h_1^2 + (a + \sqrt{c^2 - h_1^2})^2 \quad \text{and}$$
$$b^2 = h_2^2 + (a + \sqrt{c^2 - h_2^2})^2$$

Then

$$h_1^2 + (a + \sqrt{c^2 - h_1^2})^2 = h_2^2 + (a + \sqrt{c^2 - h_2^2})^2$$
$$h_1^2 + a^2 + 2a\sqrt{c^2 - h_1^2} + c^2 - h_1^2 = h_2^2 + a^2 + 2a\sqrt{c^2 - h_2^2} + c^2 - h_2$$
$$2a\sqrt{c^2 - h_1^2} = 2a\sqrt{c^2 - h_2^2}$$
$$\sqrt{c^2 - h_1^2} = \sqrt{c^2 - h_2^2}$$
$$c^2 - h_1^2 = c^2 - h_2^2$$
$$h_1^2 = h_2^2$$
$$h_1 = h_2$$

Step 4: Since $h_1 = h_2$, $\triangle ABD \cong \triangle EFH$ by hypotenuse-leg and $\triangle ADC \cong \triangle EHG$ also by hypotenuse-leg. Therefore, $\triangle ABC \cong \triangle EFG$ because their pairs of right triangles are congruent.   QED

PROBLEM 3-6   Prove that ASA establishes congruence when the triangles are obtuse, as in Figure 3-27a.

Solution

Step 1: Draw the triangles with altitudes dropped from one pair of congruent angles to the extended bases. Label the altitudes and angles as in Figure 3-27b.

Step 2: $\triangle ADC$ and $\triangle EHG$ are right triangles with congruent hypotenuses ($AC \cong EG$ as given) and a congruent pair of acute angles ($\angle 6 \cong \angle 5$, as given). Thus $\triangle ADC \cong \triangle EHG$ by hypotenuse-leg.

Step 3: $AD \cong EH, DC \cong HG$, and $\angle DAC \cong \angle HEG$ since these are corresponding parts of congruent triangles.

Step 4: Since $\angle DAC = \angle 1 + \angle 3$ and $\angle HEG = \angle 2 + \angle 4$ by construction and since $\angle 1 \cong \angle 2$ is given, you have $\angle 3 \cong \angle 4$.

Step 5: Thus right triangles $\triangle ADB$ and $\triangle EHF$ have an acute angle and adjacent leg congruent and therefore $\triangle ADB \cong \triangle EHF$.

Step 6: $DB \cong HF$ and $AB \cong EF$ since these are corresponding sides of congruent triangles.

Step 7: Since $DC = DB + BC$ and $HG = HF + FG$ and since $DC \cong HG$, and $DB \cong HF$, then $BC \cong FG$.

Step 8: $\triangle ABC \cong \triangle EFG$ by SSS.

note: The combined results of this problem and Example 3-9 provide proof of the ASA rule for *any* triangle.

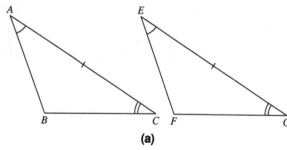

(a)

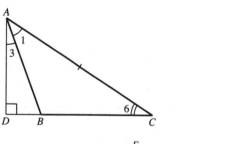

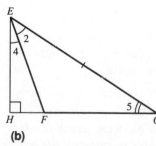

(b)

Figure 3-27

PROBLEM 3-7 Prove that the altitude to the base of an isosceles triangle bisects the base and the apex angle.

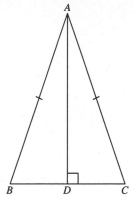

Figure 3-28

Solution

Step 1: Draw an isosceles triangle △*ABC* as in Figure 3-28 with altitude *AD*.

Step 2: *AB* ≅ *AC* by definition, since these are the equal sides of the isosceles triangle.

Step 3: *AD* ≅ *AD* by the reflexive property.

Step 4: △*ABD* and △*ACD* are right triangles with congruent hypotenuses (*AB* ≅ *AC*) and legs (*AD* ≅ *AD*). Thus △*ABD* ≅ △*ACD*.

Step 5: Since ∠*BAD* and ∠*CAD* are corresponding parts of congruent triangles, ∠*BAD* ≅ ∠*CAD*; and since ∠*A* = ∠*BAD* + ∠*CAD*, ∠*A* is bisected.

Step 6: Since *BD* and *CD* are also corresponding parts, *BD* ≅ *CD*; and since *BC* = *BD* + *CD*, the base *BC* is bisected.

PROBLEM 3-8 Point *C* is the midpoint of line segments *AD* and *BE* in Figure 3-29. Show that △*ABC* ≅ △*DEC*.

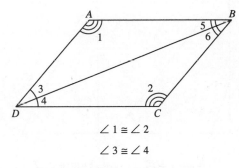

Figure 3-29

Solution

Step 1: Since *AD* and *BE* are (straight) line segments, ∠*ACB* + ∠*BCD* = 180° and ∠*DCE* + ∠*BCD* = 180° by the definition of a supplementary angle. Thus ∠*ACB* ≅ ∠*DCE*.

Step 2: Given that *C* is a midpoint of both segments *AD* and *BE*, then *AC* ≅ *CD* and *BC* ≅ *CE*.

Step 3: △*ABC* ≅ △*DEC* by SAS.

PROBLEM 3-9 Given that ∠1 ≅ ∠2 and ∠3 ≅ ∠4 in Figure 3-30, show that △*ABD* ≅ △*CBD*.

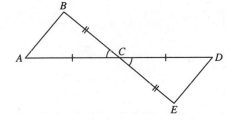

∠1 ≅ ∠2

∠3 ≅ ∠4

Figure 3-30

Solution

Step 1: Since the angles of all triangles sum to 180°, ∠5 = 180° − ∠1 − ∠3 and ∠6 = 180° − ∠2 − ∠4. Then, since you are given that ∠1 ≅ ∠2 and ∠3 ≅ ∠4, you can see that ∠5 ≅ ∠6.

Step 2: *BD* is a member of both △*ABD* and △*CDB* and is, of course, congruent to itself by the reflexive property.

Step 3: △*ABD* ≅ △*CBD* by ASA, since ∠3 ≅ ∠4, *BD* ≅ *BD*, and ∠5 ≅ ∠6.

PROBLEM 3-10 Find the length *y* in the triangle shown in Figure 3-31.

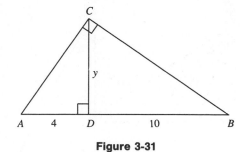

Figure 3-31

Solution Notice that △*ABC* is a right triangle and that line segment *CD* is an altitude having length *y*. Now, from the results of Example 3-12, you know that an altitude to the hypotenuse of a right triangle produces two similar right triangles, whose sides are proportional. Thus ∠*A* ≅ ∠*BCD* and ∠*B* ≅ ∠*ACD* and

$$\frac{AD}{CD} = \frac{CD}{BD}$$

or

$$\frac{4}{y} = \frac{y}{10}$$

Thus

$$y^2 = 40$$

$$y = \sqrt{40} = 2\sqrt{10} = 6.32$$

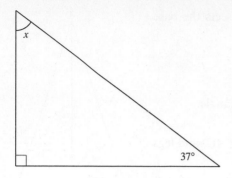

Figure 3-32

Review Exercises

EXERCISE 3-1   Find the acute angle x of the right triangle shown in Figure 3-32.

EXERCISE 3-2   Find the two acute angles x and $x + 8°$ of the right triangle shown in Figure 3-33.

EXERCISE 3-3   Find the hypotenuse x of the triangle shown in Figure 3-34.

EXERCISE 3-4   Find the hypotenuse x of the triangle shown in Figure 3-35.

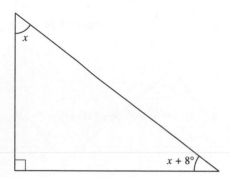

Figure 3-33

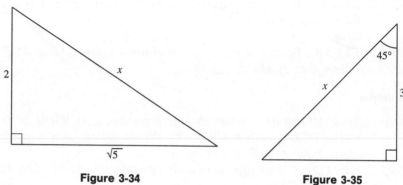

Figure 3-34   **Figure 3-35**

EXERCISE 3-5   Find the sine, cosine, and tangent of $\angle \alpha$ shown in Figure 3-36.

EXERCISE 3-6   Find the lengths x and y and the measure z of the angle shown in Figure 3-37.

EXERCISE 3-7   Find the length x and the measure y of the angle shown in Figure 3-38.

EXERCISE 3-8   Find x and y in Figure 3-39.

EXERCISE 3-9   Find x and y in Figure 3-40.

EXERCISE 3-10   Find x in Figure 3-41.

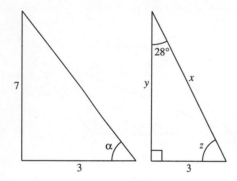

Figure 3-36   **Figure 3-37**

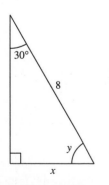   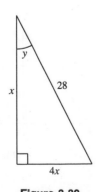   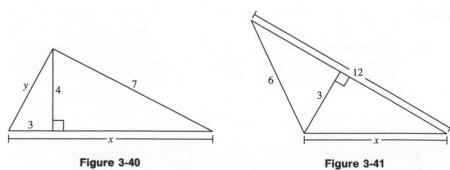

Figure 3-38   **Figure 3-39**   **Figure 3-40**   **Figure 3-41**

EXERCISE 3-11   Find x and y in Figure 3-42.

EXERCISE 3-12   Find x in Figure 3-43.

EXERCISE 3-13   Find x, y, and z in Figure 3-44.

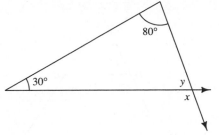

Figure 3-42

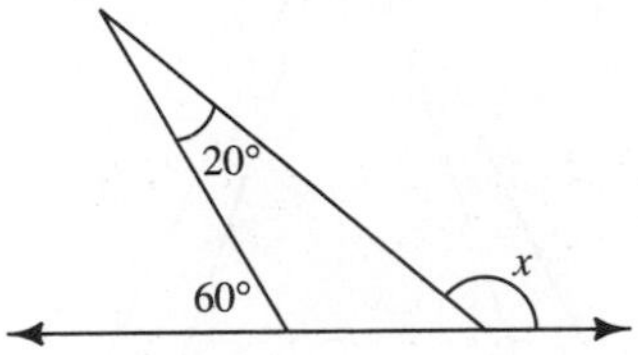

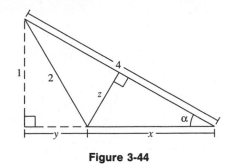

Figure 3-43

Figure 3-44

EXERCISE 3-14 In Figure 3-45, $DA \cong EC$, $\angle A \cong \angle C$, and B is the midpoint of AC. Show that $\triangle ADB \cong \triangle CEB$.

EXERCISE 3-15 In Figure 3-46, AC bisects $\angle A$, and $\angle B$ and $\angle D$ are right angles. Show that $BC \cong DC$.

EXERCISE 3-16 In Figure 3-47, $AD \cong AE$ and $EC \cong DB$. Prove that $\angle B \cong \angle C$.

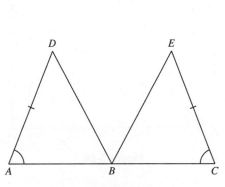

Figure 3-45

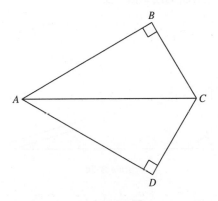

Figure 3-46

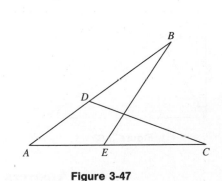

Figure 3-47

EXERCISE 3-17 In Figure 3-48, AD and BC bisect each other. Show that $\triangle ABE \cong \triangle DCE$.

EXERCISE 3-18 In Figure 3-49, $BF \cong DE$, $\angle AED \cong \angle CFB$, and $\angle CDE \cong \angle ABF$. Show that $\triangle ABE \cong \triangle CDF$.

EXERCISE 3-19 In Figure 3-50, $\triangle ABC$ is isosceles with $AB \cong AC$. Prove that $BF \cong CE$.

EXERCISE 3-20 In Figure 3-50, prove that $\triangle DEB \cong \triangle DFC$.

EXERCISE 3-21 In Figure 3-51, $\triangle BDE$ and $\triangle BAC$ are both isosceles. Show that $\triangle ADB \cong \triangle CEB$.

EXERCISE 3-22 In Figure 3-51, show that $\triangle BDC \cong \triangle BEA$.

EXERCISE 3-23 In Figure 3-52, AB is parallel to DE. Show that $\triangle CDE$ is similar to $\triangle CAB$.

EXERCISE 3-24 In Figure 3-52, $DE = 3$, $AB = 7$, $CE = 6$, and $AC = 18$. Find CB and CD.

EXERCISE 3-25 Show that $\triangle ACD$ and $\triangle ABC$ of Figure 3-22 are similar.

EXERCISE 3-26 Find x in Figure 3-53.

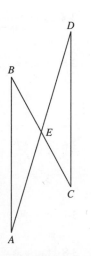

Figure 3-48

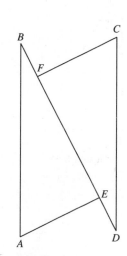

Figure 3-49

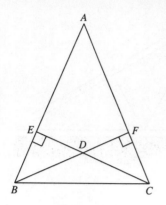

Figure 3-50

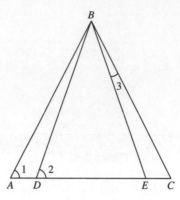

Figure 3-51

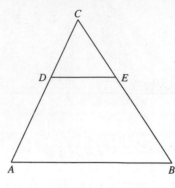

Figure 3-52

EXERCISE 3-27   Find x in Figure 3-54.

EXERCISE 3-28   Show that $\triangle AEC$ in Figure 3-55 is similar to $\triangle CDB$.

EXERCISE 3-29   Given that $CD = 6$, $AE = 3$, and $BD = 8$ in Figure 3-55, find AC, CE, and CB.

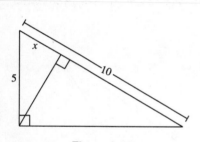

Figure 3-53

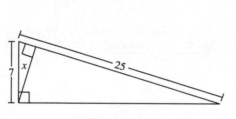

Figure 3-54

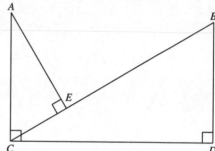

Figure 3-55

Answers to Review Exercises

3-1 $x = 53°$

3-2 $x = 41°, \qquad x + 8° = 49°$

3-3 $x = 3$

3-4 $x = \dfrac{3}{\cos 45°} = 3\sqrt{2} \approx 4.24$

3-5 $\tan \alpha = \dfrac{7}{3}, \qquad \sin \alpha = \dfrac{7}{\sqrt{49 + 9}} \approx 0.919,$

$\cos \alpha = \dfrac{3}{\sqrt{49 + 9}} \approx 0.39$

3-6 $z = 62°, \qquad x = \dfrac{3}{\sin 28°} \approx 6.39,$

$y = \dfrac{3}{\tan 28°} \approx 5.64$

3-7 $y = 60°, \qquad x = 4$

3-8 $\tan y = 4, \qquad y = 76°, \qquad x = \dfrac{28}{\sqrt{17}} \approx 6.79$

3-9 $y = 5, \qquad x = 3 + \sqrt{49 - 16} \approx 8.74$

3-10 $x \approx 7.44$

3-11 $x = 30° + 80° = 110°, \qquad y = 70°$

3-12 $x = 140°$ [or $x = (180° - 60°) + 20° = 140°$]

3-13 $y = \sqrt{3} \approx 1.73, \qquad x = \sqrt{15} - \sqrt{3} \approx 2.14$

$z = 2.14(\sin \alpha) = {\sim}2.14 \times \dfrac{1}{4} \approx 0.54$

3-24 $CB = 14, \qquad CD = 7.71$

3-26 $x = 2.5$

3-27 $x = 6.72$

3-29 $AC = 5, \qquad CE = 4, \qquad CB = 10$

TRIANGLE PROPERTIES

THIS CHAPTER IS ABOUT

☑ **Area**
☑ **Area of a Triangle**
☑ **Law of Sines**
☑ **Proof of the Pythagorean Theorem**
☑ **Law of Cosines**
☑ **Lines in Triangles**
☑ **Properties of Special Triangles**

4-1. Area

Triangles have one property that is common to many geometric figures: They enclose space. The space enclosed by a two-dimensional geometric figure or region is known as the *interior* of the region and consists of a set of points in the plane. Just as distance is a measure of the separation between two points on a line, *area* is a measure of the space occupied by a set of points in a plane. As with distance, we define area by its properties.

A. Properties of area

If R stands for a set of points on the plane, then an **area** A is an assignment to R of a number $A(R)$ that has the following properties:

(a) $A(R) \geq 0$	Area is never negative.
(b) If $R_1 \cap R_2 = \varnothing$, then $A(R_1 + R_2) = A(R_1) + A(R_2)$	If R_1 and R_2 are two sets of points on the plane that have no points in common (i.e., $R_1 \cap R_2 = \varnothing$), then the combined area of $R_1 + R_2$ (i.e. $R_1 \cup R_2$) is equal to the area of R_1 plus the area of R_2.
(c) If $R_1 \cong R_2$, then $A(R_1) = A(R_2)$	If R_1 is congruent to R_2, then the area of R_1 is equal to the area of R_2.

Any way of assigning numbers to sets in a plane that satisfies these properties is an "area."

> *note:* If two methods of assigning numbers to sets of points satisfy these properties, but differ on the same sets, then each of these is a valid "area" (sometimes called a "measure" in more advanced courses). Their differences reflect different models of space.

B. Determining area

The Euclidean method of determining area is based on distance. Thus, in order to understand area, we begin with the Cartesian plane with coordinate axes x and y (see Chapter 1). We mark off points one unit distance

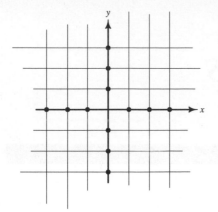

Figure 4-1

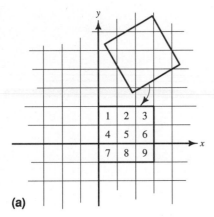

(a)

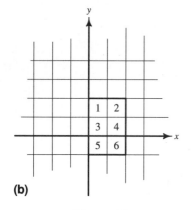

(b)

Figure 4-2

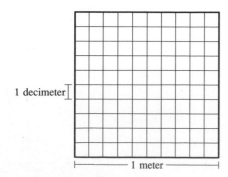

Figure 4-3

from each other on both the *x* axis and the *y* axis. Next, we construct a line perpendicular to the axis at each of these points. The result of this procedure is a grid like that shown in Figure 4-1, which looks—roughly—like standard, dimestore graph paper. The plane is now divided into piles of two-dimensional "boxes," called *squares*.

- A **square** is a four-sided figure in which all four sides are of equal length and each side is perpendicular to the two sides it joins.
- The space contained in each square is one **square unit** of area.

If, for example, inches are used as units of distance, then each square would contain one square inch (1 in^2); and if meters are used, then each square would contain one square meter (1 m^2).

We can find the area of any given square figure by drawing the figure on the grid so that its sides are parallel to the *x* and *y* axes, as in Figure 4-2a. The square shown in Figure 4-2a contains 9 unit squares, so its area is 9 square units. We can use the same method to find the area of a *rectangle*.

- A **rectangle** is a four-sided figure whose sides meet at right angles and whose opposite sides are equal and parallel.

If we draw a given rectangle on the grid, its area will be equal to the number of unit squares it contains. For example, the rectangle shown in Figure 4-2b has an area of 6 square units.

note: All squares are rectangles, but not all rectangles are squares. Notice that the definition of a square meets all the requirements for the definition of a rectangle; thus, a square is a special type of rectangle.

Now let's imagine an arbitrary rectangle with one side lined up on the *x* axis and the perpendicular side lined up on the *y* axis. If the two sides are not both an integer (whole) number of units long, we can choose a smaller unit of distance as a scale, just as we can when measuring a line's length (see Chapter 1). Once we have a fine enough resolution of the unit distance, the sides become integer multiples of the new, finer unit and we can count up the new unit squares inside the rectangle. If, for instance, we start with a square whose sides are each 1 meter (m) long, we can see that the area of this square is 1 square meter (1 m^2). But if we use the decimeter (dm: 1 dm = 0.1 m; 1 m = 10 dm) as the unit distance, the area of this square is 10 dm × 10 dm = 10^2 dm^2 = 100 dm^2 (see Figure 4-3). Then if we have a rectangle that measures, say, 1.2 m by 1.8 m, we can use a distance scale of decimeters and find that the area of this rectangle is 12 dm × 18 dm = 216 dm^2. And since 100 dm^2 = 1 m^2, we could also say that this rectangle has an area of 2.16 m^2. (The units of metric measure are given in Table 4-1.)

TABLE 4-1: Metric Measure*

1 m	= 10 dm	= 100 cm	= 1000 mm
1 m^2	= 10^2 dm^2	= 100^2 cm^2	= 1000^2 mm^2
	= 100 dm^2	= 10,000 cm^2	= 1,000,000 mm^2
1 mm	= 0.1 cm	= 0.01 dm	= 0.001 m
1 mm^2	= 0.1^2 cm^2	= 0.01^2 dm^2	= 0.001^2 m^2
	= 0.01 cm^2	= 0.0001 dm^2	= 0.000001 m^2

* Meter m, decimeter dm, centimeter cm, millimeter mm.

EXAMPLE 4-1: Find the area of the rectangle shown in Figure 4-4.

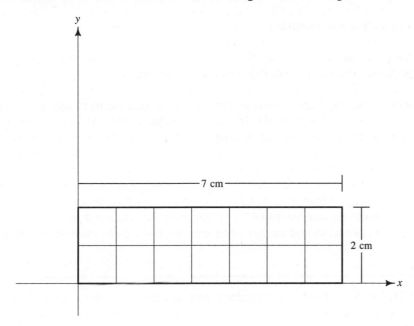

Figure 4-4

Solution: The rectangle measures 7 cm by 2 cm. In Figure 4-4 the rectangle has been placed on a Cartesian grid with a 1-cm distance scale. The rectangle contains 14 1-cm unit squares, so the area is 14 cm².

EXAMPLE 4-2: Express the area in Example 4-1 in units of (**a**) square meters and (**b**) square millimeters.

Solution

(**a**) Since 100 cm = 1 m, then 1 cm = 0.01 m. Thus 1 cm² = 0.01² m² = 0.0001 m² (that is, 0.0001 m² per 1 cm²). So

$$14 \text{ cm}^2 \times \frac{0.0001 \text{ m}^2}{1 \text{ cm}^2} = 0.0014 \text{ m}^2$$

(**b**) Since 100 cm = 1000 mm, then 1 cm = 10 mm. Thus 1 cm² = 10² mm² = 100 mm², and

$$14 \text{ cm}^2 \times \frac{100 \text{ m}^2}{1 \text{ cm}^2} = 1400 \text{ mm}^2$$

C. Formula for the area of a rectangle

The results of Examples 4-1 and 4-2 give us the clues that we need to develop a procedure for finding the area of a rectangle without having to go through the bother of counting squares. In brief, for any rectangle, such as the one shown in Figure 4-5, we choose a convenient unit of distance, using whatever ruler is lying around. Then we measure the perpendicular sides of the rectangle. If one side measures *a* and the perpendicular side measures *b*,

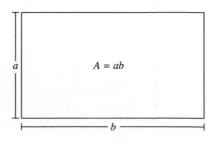

Figure 4-5

the area of the rectangle is always the product of the perpendicular sides $a \times b$, or

AREA OF A RECTANGLE $A = ab$

And once we've established ab as the Euclidean area of a rectangle, property **(b)** forces the area for all other figures in the plane.

> *note:* You may view a line segment as a degenerate rectangle where the length of one pair of sides has been reduced to 0, while the other two sides collapse on top of each other to form a line segment of length a. Since this rectangle measures a by 0, its area is $a \times 0 = 0$. Similarly, you can treat a point as a degenerate rectangle where both sets of sides shrink to length 0. The area of a point, then, is $0 \times 0 = 0$. Thus the Euclidean area of a line segment or point is always 0. This seems quite reasonable, since "area" measures the enclosed space of a figure (its interior) and neither a line segment nor a point encloses any space!

EXAMPLE 4-3:   Find the Euclidean area contained in the L-shaped figure shown in Figure 4-6.

Solution: The dotted line segment CD gives you the clue you need to find the area of this L-shaped figure. By property **(b)**, $A(R_1) + A(R_2) = A(R_1 + R_2)$; so all you need to do is to find the areas of the two rectangles $ABEC$ and $CDGF$ and add them.

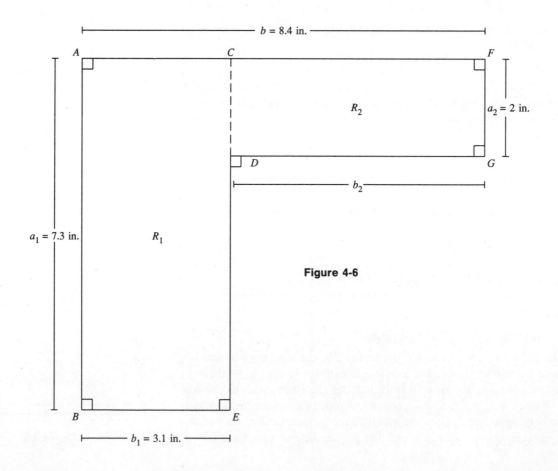

Figure 4-6

One side of rectangle $ABEC$ has length $a_1 = 7.3$ inches, and the other side has length $b_1 = 3.1$ inches. So

$$A(ABEC) = A(R_1) = a_1 b_1$$
$$= (7.3 \text{ in.})(3.1 \text{ in.})$$
$$= 22.63 \text{ in}^2$$

Then one side of rectangle $CDGF$ has length $a_2 = 2$ inches, and the other side has length $b_2 = b - b_1 = 8.4$ inches $- 3.1$ inches. So

$$A(CDGF) = A(R_2) = a_2 b_2 = a_2(b - b_1)$$
$$= (2 \text{ in.})(8.4 - 3.1 \text{ in.})$$
$$= (2 \text{ in.})(5.3 \text{ in.})$$
$$= 10.6 \text{ in}^2$$

Thus the whole figure has

$$A(R_1) + A(R_2) = A(R_1 + R_2)$$
$$22.63 \text{ in}^2 + 10.6 \text{ in}^2 = 33.23 \text{ in}^2$$

Example 4-3 demonstrates that property (**b**)—the additive property—is a very important property to know. You will frequently need to use this property to find the area of a geometric figure that is composed of other, smaller figures. But property (**b**) also has an important condition: $R_1 \cap R_2 = \phi$; that is, in order for their areas to be additive, two figures may *not* have any points in common. And in Figure 4-6 the two rectangles whose areas we found do have points in common—the line segment CD. How, then, can we say that the area of an L-shaped figure is equal to the area of its component rectangles?

Actually, the answer to this question is fairly easy. Look at Figure 4-6 again; then let the first rectangle $ABEC$ be R_1 where R_1 includes CD and let the second rectangle $CDGF$ be $R_2 + CD$ where R_2 includes every point in $CDGF$ *except* CD. Then, since CD has an area of 0, we can say the $A(R_2) + 0 = A(CDGF)$, or $A(R_2) = A(CDGF)$. Now we can see that $A(ABEC) + A(CDGF) = A(R_1) + A(R_2) = A(R_1 + R_2)$, as if the two rectangles $ABEC$ and $CDGF$ have no points in common.

This reasoning can also be applied to other geometric figures:

- When two figures combine to form a third figure, the area of the third figure is equal to the sum of the areas of the two other figures as long as the first two figures have no *interior* points in common.

4-2. Area of a Triangle

We can use what we know about the area of a rectangle to find the area of a triangle. Let's start with the special case of the right triangle.

A. Right triangles

Any rectangle may be cut into two right triangles by drawing the **diagonal**—a line segment that joins any two opposite vertices. Figure 4-7, for example, shows the rectangle $ABCD$, which has been cut by the diagonal BD into two right triangles $\triangle ABD$ and $\triangle CDB$. Since their corresponding legs are congruent ($AB \cong CD$ and $AD \cong BC$) and they share a common hypotenuse ($BD \cong BD$), the triangles $\triangle ABD$ and $\triangle CBD$ are congruent by SSS.

Now, according to property (**c**), $A(\triangle ABD) = A(\triangle CDB)$, since $\triangle ABD$ and $\triangle CDB$ are congruent. And since we can combine $\triangle ABD$ and $\triangle CDB$

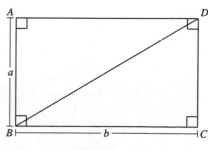

Figure 4-7

to form rectangle $ABCD$, we can use property **(b)** to find that

$$A(\text{rectangle } ABCD) = ab = A(\triangle ABD) + A(\triangle CDB)$$
$$= A(\triangle ABD) + A(\triangle ABD)$$
$$= 2A(\triangle ABD)$$

Thus, $ab = 2A(\triangle ABD)$, or

AREA OF A RIGHT TRIANGLE
$$A(\triangle ABD) = \frac{ab}{2}$$

where a and b are the lengths of the legs of the right triangle.

note: By the reasoning used in Section 4-1C, $\triangle ABD$ and $\triangle BCD$—which share a common hypotenuse—have no interior points in common. Although you do not need to include this reasoning when finding an area, it's wise to remember it.

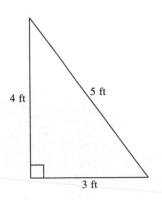

Figure 4-8

EXAMPLE 4-4: Find the area of the triangle shown in Figure 4-8.

Solution: The legs are 3 feet and 4 feet, so the area of the triangle is

$$\frac{(3 \text{ ft})(4 \text{ ft})}{2} = \frac{12 \text{ ft}^2}{2} = 6 \text{ ft}^2$$

Once we know the formula for finding the area of a right triangle, we can find the area of any triangle.

B. Any triangle

Any triangle may be broken down into two right triangles by drawing the altitude (the perpendicular dropped from any vertex to the opposite side; see Chapter 3).

Acute Case: If we cut up acute $\triangle ABC$ by drawing the altitude AD of length h as in Figure 4-9, we get two right triangles, $\triangle ABD$ and $\triangle ACD$, where h is the length of the leg common to both right triangles, b_1 is the length of the second leg of $\triangle ABD$, and b_2 is the length of the second leg of $\triangle ACD$. Then reasoning from property **(b)** and neglecting the area of the common line segment AD, we find the area

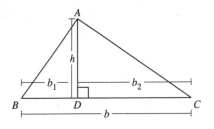

Figure 4-9

$$A(\triangle ABC) = A(\triangle ABD) + A(\triangle ACD) = \frac{hb_1}{2} + \frac{hb_2}{2}$$

$$= \frac{h(b_1 + b_2)}{2}$$

$$= \frac{hb}{2}$$

Obtuse Case: If we drop an altitude AD from an acute angle $\angle A$ in an obtuse triangle $\triangle ABC$ and extend the base b as in Figure 4-10, we get two right triangles, $\triangle ADB$ and $\triangle ADC$, where h is the length of the leg common to both right triangles, b_1 is the length of the second leg of $\triangle ADB$, and b_2 is the length of the second leg of $\triangle ADC$. Then reasoning from property **(b)** and neglecting the area of the common side AD, we can find the area of $\triangle ABC$, which is the difference between the areas of $\triangle ADC$ and $\triangle ADB$:

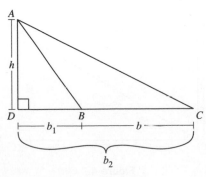

Figure 4-10

$$A(\triangle ABC) = \frac{hb_2}{2} - \frac{hb_1}{2}$$

Then, since $b = b_2 - b_1$,

$$A(\triangle ABC) = \frac{h(b_2 - b_1)}{2} = \frac{hb}{2}$$

General Case: Since the formulas for finding the area of an acute triangle and the area of an obtuse triangle are the same, we can see that the formula for finding the area of any triangle is always

**AREA OF
ANY TRIANGLE**
$$A(\triangle) = \frac{hb}{2}$$

that is, the area of any triangle is always one-half (1/2) the length of any altitude (the height h) times the length of the side to which the altitude is drawn (the base b).

note: The additive property (**b**) now allows us to find the area of any two-dimensional geometric figure that can be cut up into triangles.

EXAMPLE 4-5: Using a metric ruler, find the area of $\triangle ABC$ shown in Figure 4-11 to two decimal places.

Solution: The triangle $\triangle ABC$ has sides of 3, 6, and 7 cm. Draw any one of the altitudes and measures its length. In Figure 4-11a, AD is probably the easiest altitude to draw, but you may also draw BE or CF, as shown in Figure 4-11b. In any case, the area is

$$A(\triangle ABC) = \frac{hb}{2} = \frac{AD \times BC}{2} = \frac{BE \times AC}{2} = \frac{CF \times BA}{2}$$

where $AD \approx 2.6$ cm, $BE \approx 3.0$ cm, and $CF \approx 6.0$ cm. Thus

$$A(\triangle ABC) \approx \frac{(2.6\text{ cm})(7\text{ cm})}{2} \approx \frac{(3.0\text{ cm})(6\text{ cm})}{2} \approx \frac{(6.0\text{ cm})(3\text{ cm})}{2}$$

$$\approx 9\text{ cm}^2$$

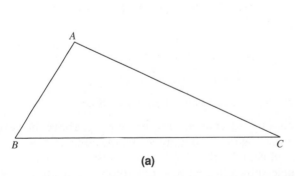

(a)   (b)

Figure 4-11

note: As Example 4-5 shows, your accuracy in finding the area of a triangle with a ruler is limited by your measuring device. As you've probably already guessed, there are ways to calculate the area of a triangle if you're given certain information. In Section 4-3 we'll discuss the Law of Sines, which enables you to find the area from the angles and the sides, and in Section 4-5 we'll discuss the Law of Cosines, which allows you to find an angle of a triangle from the three sides.

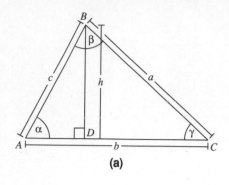

(a)

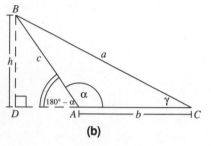

(b)

Figure 4-12

4-3. Law of Sines

note: The **Law of Sines** is just a fancy name for restating the area formula for triangles. Although it probably shouldn't be dignified with a special name, it's been around for a while and you should be familiar with it.

If we have a triangle $\triangle ABC$ with angles α, β, and γ opposite sides a, b, and c, respectively, we can pick a vertex, say B, and draw an altitude h to the opposite side b. Thus we know that in Figure 4-12a, $A(\triangle ABC) = hb/2$. But we also know that the altitude h cuts $\triangle ABC$ into two right triangles, $\triangle ADB$ and $\triangle CDB$, and that the sine of an angle in a right triangle is equal to the length of the opposite leg divided by the length of the hypotenuse. Thus $\sin\alpha = h/c$ and $\sin\gamma = h/a$, or $h = c\sin\alpha$ and $h = a\sin\gamma$. Now we have another expression for the area of $\triangle ABC$:

LAW OF SINES

$$A(\triangle ABC) = \frac{b(c\sin\alpha)}{2} = \frac{b(a\sin\gamma)}{2}$$

$$= \frac{cb\sin\alpha}{2} = \frac{ab\sin\gamma}{2}$$

In short,

- The area of a triangle is one-half the product of the lengths of any two sides and the sine of the angle between them.

note: This is one form of the Law of Sines, which is the one we'll use here. But we can express this law in a different form, which you may encounter in your textbooks. Since $\frac{1}{2}cb\sin\alpha = \frac{1}{2}ab\sin\gamma$, we can divide through by $b/2$ (i.e., multiply through by $2/b$) to get $c\sin\alpha = a\sin\gamma$, or

$$\frac{\sin\alpha}{a} = \frac{\sin\gamma}{c} \qquad \text{or} \qquad \frac{a}{\sin\alpha} = \frac{c}{\sin\gamma}$$

Then since we can also show that $A(\triangle ABC) = \frac{1}{2}ac\sin\beta$, we find that $\frac{1}{2}ab\sin\gamma = \frac{1}{2}ac\sin\beta$; so $(\sin\gamma)/c = (\sin\beta)/b$. Thus we get a three-way equality

LAW OF SINES
(alternative form)

$$\frac{\sin\alpha}{a} = \frac{\sin\beta}{b} = \frac{\sin\gamma}{c}$$

or

$$\frac{a}{\sin\alpha} = \frac{b}{\sin\beta} = \frac{c}{\sin\gamma}$$

which is another way of expressing the Law of Sines.

The Law of Sines applies to any triangle. The proof given above, however, depends on the fact that the angles α and γ are acute. (As long as β is also acute, we could repeat the argument for that angle as well.) But what happens if α is obtuse, as in the triangle shown in Figure 4-12b? Indeed, our model for space doesn't know what to do with the sine of an obtuse angle!

Fortunately, there is a way around the obtuse-angle problem. The area of $\triangle ABC$ in Figure 4-12b is still $hb/2$, and from the right triangle $\triangle BDC$ h is still $a\sin\gamma$. Thus $A(\triangle ABC) = hb/2 = (ab\sin\gamma)/2$. Then from right triangle $\triangle BDA$, $h = c\sin(180° - \alpha)$, since $\angle BAD$ and $\angle BAC$ are supplementary. Since $0° \leq 180° - \alpha \leq 90°$, this makes sense, and

$$A(\triangle ABC) = \frac{bc\sin(180° - \alpha)}{2}$$

This leaves us with two possibilities: We can either make a special case in the Law of Sines for obtuse angles, or we can extend the model of space by defining the sine of an obtuse angle as $\sin \alpha = \sin(180° - \alpha)$. Let's choose the latter. Thus

- When $90° \leq \alpha \leq 180°$ (i.e., when α is an obtuse angle),

  SINE OF AN OBTUSE ANGLE $\sin \alpha = \sin(180° - \alpha)$

That way, we still have $A(\triangle ABC) = (bc \sin \alpha)/2$.

note: Whenever you choose to extend a mathematical model by adding new properties as we have done here, you must make sure that the new properties do not conflict with any properties you already know to be true. You should really check out *all* previous work for its compatibility with a new fact. In this case, no difficulties arise.

The Law of Sines is most convenient for finding the area of a triangle when the lengths of some of the sides and the measure of some of its angles are known. If we use the Law of Sines, we don't have to go to the bother of finding the height of a particular altitude.

EXAMPLE 4-6: Find the area of the triangle shown in Figure 4-13.

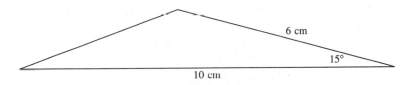

Figure 4-13

Solution: By the Law of Sines, the area is

$$\frac{(6 \text{ cm})(10 \text{ cm})(\sin 15°)}{2}$$

And from Table 3-1 $\sin 15° = 0.2588$, so

$$A = \frac{(6 \text{ cm})(10 \text{ cm})(0.2588)}{2} = 7.764 \text{ cm}^2$$

4-4. Proof of the Pythagorean Theorem

Now we can prove the Pythagorean Theorem ($a^2 + b^2 = c^2$, where a and b are the legs of a right triangle and c is the hypotenuse), a task we postponed from Section 3-3. The ancient proofs of this theorem are expressed in terms of area, so let's begin by restating the theorem as follows:

- The Pythagorean Theorem says that if we have a right triangle $\triangle ABC$, as shown in Figure 4-14a, the area $A(R_3)$ formed by drawing a square with hypotenuse c as one side is equal to the combined areas $A(R_1) + A(R_2)$ of the two squares built on the legs of the right triangle, as shown in Figure 4-14b. Thus

$$A(R_1) + A(R_2) = A(R_3)$$

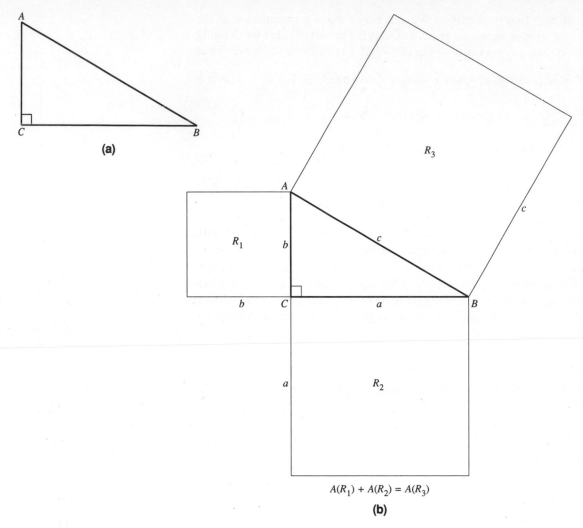

$$A(R_1) + A(R_2) = A(R_3)$$

(b)

Figure 4-14

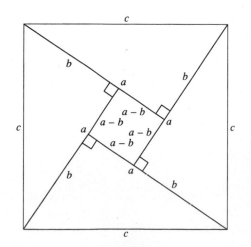

Figure 4-15

Proof: Draw four triangles congruent to $\triangle ABC$ positioned in such a way that the hypotenuse c of each triangle forms the side of a four-sided figure, as shown in Figure 4-15. In this four-sided figure the four right triangles, which are congruent to $\triangle ABC$ and hence to each other, are arranged so that the shorter leg of one triangle abuts the longer leg of another and the vertices of the two acute angles coincide. The four-sided figure with sides of length c thus contains the four right triangles with sides of a, b, and c plus a smaller four-sided figure with sides of $a - b$. In each corner of the larger four-sided figure two sides of length c meet at an angle that is the sum of two acute angles. And these two angles that make up the corner are congruent to the two different acute angles of the right triangle $\triangle ABC$. Then, since the two acute angles of a right triangle are always complementary, each corner angle of the larger four-sided figure must be 90°. Moreover, each side of the larger four-sided figure has sides of length c. Thus we can say that the larger four-sided figure is a square. Similarly, each side of the internal four-sided figure has length $a - b$ and each corner angle, which is supplementary to a right angle, is a right (90°) angle. Therefore, the internal four-sided figure is a square.

The larger square has area $c \times c$, or c^2, which is equal to the sum of the area of the four congruent right triangles $(4ab/2)$ plus the area of the internal square $[(a - b)^2]$. Thus

$$c^2 = \frac{4ab}{2} + (a - b)^2$$

$$= 2ab + a^2 - 2ab + b^2$$

$$= a^2 + b^2 \qquad\qquad\qquad \text{QED}$$

4-5. Law of Cosines

The Pythagorean Theorem is really a special case of a more general rule for relating the sides of any triangle. This general rule is called the Law of Cosines, which works as follows.

Acute Case: Given a triangle $\triangle ABC$ with sides of length a, b, and c and acute angle γ, we first draw the altitude AD of length h as in Figure 4-16a. The triangle is thus cut into two right triangles, $\triangle ADC$ and $\triangle ADB$. Then, since the cosine of an angle in a right triangle is equal to the length of the adjacent side divided by the length of the hypotenuse, we have $\cos \gamma = CD/b$, or $CD = b \cos \gamma$. Thus we can see that $BD = a - CD = a - b \cos \gamma$. Now applying the Pythagorean Theorem to each of these right triangles, we get

$$b^2 = (b \cos \gamma)^2 + h^2$$

and

$$c^2 = (a - b \cos \gamma)^2 + h^2$$
$$= a^2 - 2ab \cos \gamma + (b \cos \gamma)^2 + h^2$$

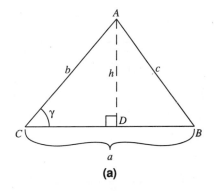

(a)

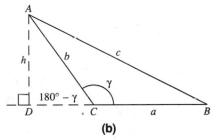

(b)

Figure 4-16

Then using the first equation, we can replace $(b \cos \gamma)^2 + h^2$ in the second equation with b^2 to get the Law of Cosines:

LAW OF COSINES

$$c^2 = a^2 + b^2 - 2ab \cos \gamma$$

note: If $\gamma = 90°$, then $\cos \gamma = 0$ (see Table 3-1) and the Law of Cosines reduces to the Pythagorean Theorem.

Obtuse Case: As we did in Section 4-3, we need to see what happens if γ is an obtuse angle: Does the Law of Cosines still hold? To find this out, we need to figure out what the cosine of an obtuse angle is. Let's begin by drawing an obtuse triangle $\triangle ABC$ with an altitude AD exterior to the triangle as in Figure 4-16b. Thus we have two right triangles $\triangle ADC$ and $\triangle ADB$. Now we can see that γ is supplementary to $\angle ACD$, so $\angle ACD$ must be $180° - \gamma$. Thus, by the reasoning above, side $CD = b \cos(180° - \gamma)$ and side $DB = a + b \cos(180° - \gamma)$. We can see that $\cos(180° - \gamma)$ makes sense since $0 \le 180° - \gamma \le 90°$. Now if we use the Pythagorean Theorem on each of these triangles, we get

$$b^2 = [b \cos(180° - \gamma)]^2 + h^2$$

and

$$c^2 = [a + b \cos(180° - \gamma)]^2 + h^2$$
$$= a^2 + 2ab \cos(180° - \gamma) + [b \cos(180° - \gamma)]^2 + h^2$$

Then, if we substitute the first equation into the second, we wind up with

$$c^2 = a^2 + b^2 + 2ab \cos(180° - \gamma)$$

The two equations for c^2 differ according to whether γ is acute or obtuse. So we have the same dilemma that we had with obtuse angles and the Law of

Sines (see Section 4-3). But instead of making a special formula for obtuse angles, let's solve the dilemma by defining the cosine of an obtuse angle as

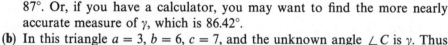

COSINE OF AN OBTUSE ANGLE $\cos \gamma = -\cos(180° - \gamma)$ where $90 \leq \gamma \leq 180°$

If we use this definition, the formula $c^2 = a^2 + b^2 - 2ab \cos \gamma$ does hold.

EXAMPLE 4-7: Use the Law of Cosines to find the measure of $\angle C$ in the triangles shown in Figure 4-17a and Figure 4-17b.

Solution

(a) In this triangle $a = 4$, $b = 6$, $c = 7$, and the unknown angle $\angle C = \gamma$. Thus

$$c^2 = a^2 + b^2 - 2ab \cos \gamma$$

$$49 = 4^2 + 6^2 - 2(4 \times 6) \cos \gamma$$

$$= 16 + 36 - 48 \cos \gamma = 52 - 48 \cos \gamma$$

$$48 \cos \gamma = 52 - 49 = 3$$

$$\cos \gamma = \frac{3}{48} = \frac{1}{16}$$

$$= 0.0625$$

Using Table 3-1, you can see that the measure of γ is between 86° and 87°. Or, if you have a calculator, you may want to find the more nearly accurate measure of γ, which is 86.42°.

(b) In this triangle $a = 3$, $b = 6$, $c = 7$, and the unknown angle $\angle C$ is γ. Thus

$$c^2 = a^2 + b^2 - 2ab \cos \gamma$$

$$7^2 = 3^2 + 6^2 - 2(3 \times 6) \cos \gamma$$

$$49 = 45 - 36 \cos \gamma$$

$$\cos \gamma = -\frac{4}{36} = -\frac{1}{9}$$

$$= -0.111 \ldots$$

But this leaves you stuck with a negative value for the cosine of γ! Obviously, γ must be obtuse, so you have to apply the definition of the cosine for an obtuse angle. To do this, use your calculator to find the angle whose cosine is $+0.111 \ldots$ (that is, find $\cos^{-1}$ for $0.111 \ldots$). This angle is 83.62°—but by the definition of an obtuse angle, γ is $180° - 83.62 = 96.38°$. Notice that the definition $-\cos(180° - \gamma)$ is a negative quantity, which takes care of the negative sign in $-0.111 \ldots$. Now you can say that $\angle C = 96.38°$.

(a)

(b)

Figure 4-17

4-6. Lines in Triangles

When we develop proofs that involves triangles, we often need to draw extra lines in the figures. There are four special kinds of lines that are frequently used: *medians*, *angle bisectors*, *perpendicular bisectors*, and *altitudes*.

A. Medians

A **median** of a triangle is a line segment drawn from a vertex to the midpoint of the opposite side. (The *midpoint of a line segment* is the point halfway between the two endpoints of the line segment.) For example, AD is a median of $\triangle ABC$ in Figure 4-18.

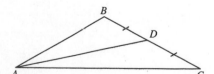

Figure 4-18

caution: It's important not to read more into a median than is really there!
A median just bisects the side to which it is drawn—nothing else.
It's all too easy to make serious mistakes by wanting a median
to do something it doesn't do and assuming that the desired
property is really there!

- A median does *not* in general bisect the angle from which it is drawn.
- A median is *not* in general perpendicular to the opposite side.

There are three possible medians for any triangle:

- The three medians of a triangle all intersect at a common point, called
  the **centroid**, which is two-thirds (2/3) of the way along each median
  from the vertex angle to the opposite side.

note: If you made a triangle out of wood or metal, you could balance the
triangle by placing a support under the centroid.

EXAMPLE 4-8: Show that the medians of a triangle meet at a common point
which is two-thirds of the distance along each median from the vertex.

Solution: Let's use the coordinate geometry discussed in Section 1-7.

Step 1: Draw a triangle $\triangle ABC$ in the coordinate plane as in Figure 4-19a.
Place one vertex, say B, at the origin, and rotate the triangle so that
one side falls on the x axis. The coordinates of vertices A, B, and C
are thus respectively (a, b), $(0, 0)$, and $(c, 0)$.

Step 2a: Draw median BD to side AC. Since point D falls halfway between A
and C, D's x coordinate is $a + \dfrac{c-a}{2} = \dfrac{a+c}{2}$ and D's y coordinate is
$\dfrac{b}{2}$.

Step 2b: Let P be the point two-thirds of the distance from B to D on BD.
Thus P's x coordinate is $\dfrac{2}{3}\left(\dfrac{a+c}{2}\right) = \dfrac{a+c}{3}$ and P's y coordinate is
$\dfrac{2}{3}\left(\dfrac{b}{2}\right) = \dfrac{b}{3}$.

Step 3a: Draw median CE to side AB in $\triangle ABC$ as in Figure 4-19b, so that
E's coordinates are $\left(\dfrac{a}{2}, \dfrac{b}{2}\right)$.

Step 3b: Let Q be the point two-thirds of the distance from C to E. Then Q's
x coordinate is $c - \left(\dfrac{2}{3}\right)\left(c - \dfrac{a}{2}\right) = \dfrac{a+c}{3}$ and Q's y coordinate is
$\dfrac{2}{3}\left(\dfrac{b}{2}\right) = \dfrac{b}{3}$.

Step 4a: Draw median AF to side BC as in Figure 4-19c, so that F's coordinates
are $\left(\dfrac{c}{2}, 0\right)$.

Step 4b: Let R be the point two-thirds of the distance from A to F. Then R's
x coordinate is $a - \left(\dfrac{2}{3}\right)\left(a - \dfrac{c}{2}\right) = \dfrac{a+c}{3}$ and R's y coordinate is $\dfrac{b}{3}$.

Step 5: Thus points P, Q, and R have the same coordinates $\left(\dfrac{a+c}{3}, \dfrac{b}{3}\right)$ and
are the same point! QED

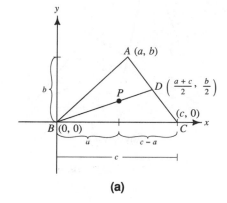

(a)

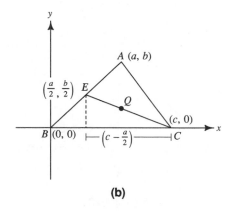

(b)

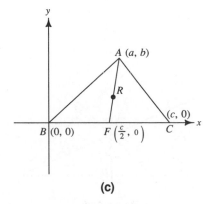

(c)

Figure 4-19

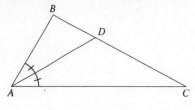

Figure 4-20

B. Angle bisectors of a triangle

An **angle bisector** of a triangle is a line segment that is drawn from a vertex to the opposite side and bisects the vertex angle. It is not necessarily perpendicular to the opposite side, and it does not generally bisect the opposite side. For example, AD is an angle bisector of $\triangle ABC$ in Figure 4-20.

- There are three possible angle bisectors for any triangle, and these line segments intersect at a common point.

EXAMPLE 4-9:　Show that the three angle bisectors of a triangle intersect at a common point.

Solution

Step 1:　Draw a triangle $\triangle ABC$ as in Figure 4-21 with angle bisectors AD and BE. These two bisectors intersect at point P.

Step 2:　Draw perpendiculars PF, PG, and PH to sides BC, AC, and AB respectively.

Step 3:　$\angle PBF \cong \angle PBH$ and $\angle PAG \cong \angle PAH$ by the definition of an angle bisector.

Step 4:　$\triangle BFP \cong \triangle BHP$, since they are two right triangles with a common hypotenuse and congruent acute angles.

Step 5:　$PF \cong PH$, since they are corresponding parts of congruent triangles.

Step 6:　Similarly, $\triangle APG \cong APH$ and $PG \cong PH$.

Step 7:　$PF \cong PG$, since they are both congruent to PH.

Step 8:　Draw PC.

Step 9:　$\triangle CPF \cong \triangle CPG$, since they are right triangles with a common hypotenuse CP and congruent legs PF and PG.

Step 10:　$\angle PCF \cong \angle PCG$, as corresponding parts of congruent triangles; that is, PC bisects $\angle ACB$.

Step 11:　Thus the bisector of $\angle ACB$ must also pass through point P.　QED

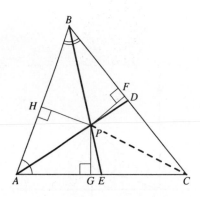

Figure 4-21

C. Perpendicular bisectors

A **perpendicular bisector** of the side of a triangle is a line that is perpendicular to the side and passes through its midpoint. In Figure 4-22, for example, $\overleftrightarrow{DE}$ is a perpendicular bisector of side AC. The perpendicular bisector of a side need *not* pass through the opposite vertex. Thus it will not in general be a median or an angle bisector of the triangle.

- There are three possible perpendicular bisectors for the sides of a triangle, and these bisectors intersect at one common point.

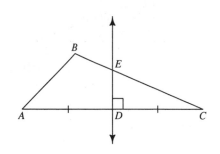

Figure 4-22

EXAMPLE 4-10:　Show that the three perpendicular bisectors of a triangle intersect at a common point.

Solution

Step 1:　Draw a triangle $\triangle ABC$ as in Figure 4-23 where $\overleftrightarrow{PF}$ and $\overleftrightarrow{PD}$ are the respective perpendicular bisectors of AC and BC, which meet at a point P.

　　note: If the perpendicular bisectors of AC and BC did *not* meet at some point P, then they would be parallel. In that case the lines perpendicular to them, $\overleftrightarrow{AC}$ and $\overleftrightarrow{BC}$, would also be parallel and could not meet at vertex C.

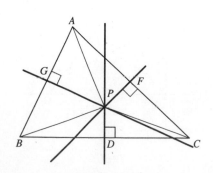

Figure 4-23

Step 2: Draw *PB*, *PC*, and *PA*.

Step 3: $BD \cong DC$ and $AF \cong FC$ by the definition of a perpendicular bisector.

Step 4: $\triangle PBD \cong \triangle PCD$, since they are right triangles with a common leg *PD* and congruent legs $BD \cong CD$.

Step 5: $BP \cong CP$, as corresponding parts of congruent triangles.

Step 6: $\triangle PAF \cong \triangle PCF$, since they are right triangles with a common leg *PF* and congruent legs $AF \cong FC$.

Step 7: $PC \cong AP$ as corresponding parts of congruent triangles.

Step 8: $BP \cong AP$, since they are both congruent to *PC*.

Step 9: Draw a perpendicular *PG* from *P* to *AB*.

Step 10: $\triangle APG \cong \triangle BPG$, since they are right triangles with common leg *PG* and congruent hypotenuses $BP \cong AP$.

Step 11: $AG \cong GB$ as corresponding parts of congruent triangles.

Step 12: Thus *GP* lies on the perpendicular bisector of *AB*, and hence this bisector must pass through *P* also!

> ***note:*** In this example the common point at which the perpendicular bisectors meet is in the interior of $\triangle ABC$. But perpendicular bisectors may also meet outside the triangle. Can you draw a triangle for which this is the case? [*Hint:* Try an obtuse triangle.]

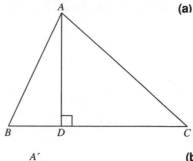

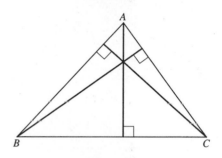

Figure 4-24

D. Altitudes of a triangle

An altitude of a triangle (discussed in Section 3-4) is a line segment that is drawn from a vertex to the opposite side and is perpendicular to the opposite side (possibly extended). For example, *AD* is an altitude of $\triangle ABC$ in Figure 4-24a and *A'D'* is an altitude of $\triangle A'B'C'$ in Figure 4-24b. An altitude is merely perpendicular to the opposite side. It need *not* bisect the opposite side and it need *not* bisect the angle from whose vertex it was drawn!

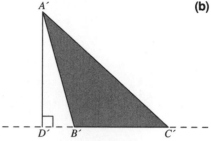

Figure 4-25

- There are three altitudes for every triangle and these three altitudes intersect at a common point. If the triangle is acute as in Figure 4-25, the altitudes intersect at a common point inside the triangle. If the triangle is obtuse, the lines through the altitudes still intersect at a common point, as shown in Figure 4-26, but this point is outside the triangle.

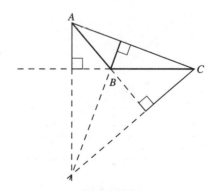

Figure 4-26

EXAMPLE 4-11: Show that the altitudes of a triangle meet at a common point.

Solution: This is a little tricky and needs a lot of extra constructions!

Step 1: Draw $\triangle ABC$ with altitudes *AD*, *BE*, and *CF*, as shown in Figure 4-27.

Step 2: As in Example 1-8, construct a line L_1 through *A* parallel to *BC*, a line L_2 through *B* parallel to *AC*, and a line L_3 through *C* parallel to *AB*. Lines L_1 and L_2 intersect at a point *G*, L_1 and L_3 intersect at a point *H*, and L_2 and L_3 intersect at a point *K*. These lines generate $\triangle GHK$.

Step 3: Use the fact that parallel lines such as L_1 and $\overleftrightarrow{BC}$ cut by a transversal such as $\overleftrightarrow{AB}$ have equal (congruent) alternate interior angles. Thus $\angle GAB \cong \angle ABC$. Similarly, $\overleftrightarrow{AB}$ is also a transversal cutting the parallel lines L_2 and $\overleftrightarrow{AC}$. Thus $\angle GBA \cong \angle BAC$.

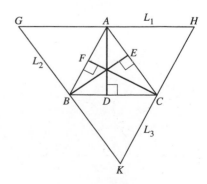

Figure 4-27

Step 4:   Since $BA \cong AB$, $\triangle GBA \cong \triangle CAB$ by ASA.

Step 5: $GA \cong BC$, as corresponding parts of congruent triangles.

Step 6:   As in Step 3, $\angle BCA \cong \angle HAC$ and $\angle BAC \cong \angle HCA$.

Step 7:   Since $AC \cong AC$, $\triangle ABC \cong \triangle CHA$ by ASA.

Step 8: $HA \cong BC$, as corresponding parts of congruent triangles.

Step 9: $GA \cong AH$, since both are congruent to BC.

Step 10: Since GH is parallel to BC and AD is perpendicular to BC, AD is also perpendicular to GH.

Step 11: $\overleftrightarrow{AD}$ is a perpendicular bisector of GH.

Step 12: Similarly $\overleftrightarrow{BE}$ is a perpendicular bisector of GK, and $\overleftrightarrow{CF}$ is a perpendicular bisector of HK.

Step 13: The lines through the altitudes of $\triangle ABD$ are the perpendicular bisectors of $\triangle KHG$ and thus meet in a common point.   QED!

4-7.   Properties of Special Triangles
A. Isosceles triangles

An isosceles triangle has two sides of equal length; the side that is not equal to the other two is considered the base, which is opposite the apex angle (see Section 3-2A). In Solved Problem 3-7, we established an important property of isosceles triangles: The altitude to the base bisects the base. Therefore, the altitude to the base is also a median and a perpendicular bisector of an isosceles triangle. In addition, the altitude to the base also bisects the apex angle, so this altitude is an angle bisector of the triangle as well.

- In an isosceles triangle, the median to the base, the perpendicular bisector of the base, the altitude to the base, and the bisector of the apex angle are collinear; that is, they are all the same line segment.

B. Special right triangles
1. 45° Right triangles

A **45° right triangle** as shown in Figure 4-28 is an isosceles right triangle that has two acute angles of 45°. Its legs have length s; therefore, from the Pythagorean Theorem its hypotenuse has length $\sqrt{s^2 + s^2} = \sqrt{2}\,s$. Its area is $s^2/2$. The altitude to the hypotenuse is a median and thus bisects the hypotenuse. This altitude divides the triangle into two congruent right triangles whose hypotenuses have length s and whose legs have length $s/\sqrt{2}$.

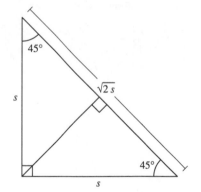

Figure 4-28

EXAMPLE 4-12:   Right triangle $\triangle ABC$ has $\angle A = \angle B = 45°$. Without drawing the triangle, find **(a)** AC when $BC = 4$, **(b)** AB when $AC = 6$, **(c)** BC when $AB = x$, and **(d)** AB when $AC = 3$.

Solution: Since $C = 90°$, AC and BC are the legs and AB is the hypotenuse of an isosceles right triangle. Thus $AC = BC$. Then

(a) if $BC = 4$,

$$AC = BC = 4$$

(b) if $AC = 6$,

$$AB = \sqrt{2}AC = \sqrt{2} \times 6 = 6\sqrt{2}$$

(c) if $AB = x$,

$$BC = \frac{AB}{\sqrt{2}} = \frac{x}{\sqrt{2}}$$

(d) if $AC = 3$,

$$AB = \sqrt{2}AC = \sqrt{2} \times 3 = 3\sqrt{2}$$

2. 30°-60°-90° Right triangles

A **30°-60°-90° right triangle** is a right triangle with acute angles of 30° and 60°, as shown in Figure 4-29.

note: This triangle occurs in a lot of applications, so it's a useful one to know.

The hypotenuse of the 30°-60°-90° right triangle has legnth s. And since $\sin 30° = 0.5000 = \frac{1}{2}$ (see Table 3-1), the length of the leg opposite the 30° angle is $s/2$, which is one-half the length of the hypotenuse. Then the Pythagorean Theorem shows that the length of leg opposite the 60° angle is $\sqrt{s^2 - (s/2)^2} = \sqrt{3}s/2$.

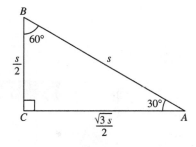

Figure 4-29

EXAMPLE 4-13: Right triangle $\triangle ABC$ has $\angle A = 30°$ and $\angle B = 60°$. Without drawing the triangle, find **(a)** AB when $BC = 4$, **(b)** AC when $AB = 7$, and **(c)** AC when $BC = 5$.

Solution: Since $\angle C = 90°$, AC is the leg opposite the 60° angle, BC is the leg opposite the 30° angle, and AB is the hypotenuse.

(a) Since $BC = AB/2$, then $AB = 2BC$. Thus if $BC = 4$,

$$AB = 2BC = 2 \times 4 = 8$$

(b) If $AB = 7$, then

$$AC = \sqrt{3}\left(\frac{AB}{2}\right) = \sqrt{3}\left(\frac{7}{2}\right) = 3.5\sqrt{3}$$

(c) When $BC = 5$, $AB = 2 \times 5 = 10$; thus

$$AC = \sqrt{3}\left(\frac{AB}{2}\right) = \sqrt{3}\left(\frac{10}{2}\right) = 5\sqrt{3}$$

C. Equilateral triangles

An **equilateral triangle** has three sides of equal length. And from Solved Problem 3-4, we know that all of its interior angles must also be equal. Then, since the interior angles of any triangle must sum to 180°, each of the interior angles of an equilateral triangle must be $180°/3 = 60°$.

The equilateral triangle is a special case of an isosceles triangle in which any side can be the base. Thus,

- The medians, altitudes, perpendicular bisectors, and angle bisectors of the equilateral triangle all coincide, forming one set of three lines that meet at a common point.

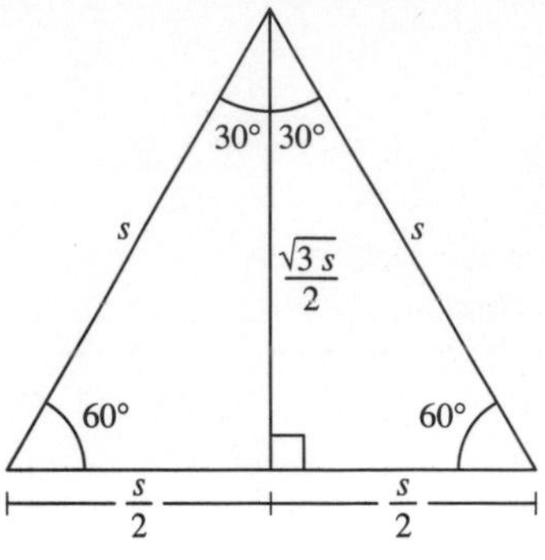

Figure 4-30

Any altitude of an equilateral triangle divides the triangle into two congruent 30°-60°-90° right triangles, as shown in Figure 4-30. Thus, if the equilateral triangle has sides of length s, the height of the altitude is $\sqrt{3}\,s/2$, and the area of the equilateral triangle is

AREA OF AN EQUILATERAL TRIANGLE $A = \dfrac{s(\sqrt{3}\,s/2)}{2} = \dfrac{\sqrt{3}\,s^2}{4}$

And, since the altitude of an equilateral triangle is also a median, the centroid of an equilateral triangle is located at a distance of $\dfrac{2}{3}\dfrac{\sqrt{3}\,s}{2} = \dfrac{\sqrt{3}\,s}{3}$ units from each vertex.

SUMMARY

1. Area is a measure of the space interior to a two-dimensional figure or region; i.e., area A is a measure of the space occupied by a set of points R in the plane.

    (a) Area is never negative:

    $$A(R) \geq 0$$

    (b) The area of a region consisting of two (or more) figures with no interior points in common is the sum of the areas of the two (or more) figures:

    $$A(R_1 + R_2) = A(R_1) + A(R_2) \qquad \text{if} \qquad R_1 \cap R_2 = \varnothing$$

    (c) If one two-dimensional figure is congruent to another, then the area of the first figure is equal to the area of the second:

    $$A(R_1) = A(R_2) \qquad \text{if} \qquad R_1 \cong R_2$$

2. A rectangle is a four-sided figure whose sides meet at right angles and whose opposite sides are equal and parallel.

    (a) A square is a rectangle whose sides are equal.
    (b) The area of a rectangle is the product ab of the lengths of the two perpendicular sides, a and b.
    (c) The area of a square is the square a^2 of the length a, one of its sides.

3. The area of a line segment or point is 0.
4. The area of any triangle $A(\triangle ABC)$ is one-half the length (height) of any altitude h times the length of the corresponding base b:

    $$A(\triangle ABC) = \frac{hb}{2}$$

    If $\triangle ABC$ is a right triangle, where ab is the product of the lengths of the two legs, $A(ABC) = ab/2$.

5. The Law of Sines is a restatement of the area formula for a triangle: The area of a triangle is equal to one-half the product of the lengths of any two sides a and b times the sine of the angle γ between them:

    $$A(\triangle ABC) = \frac{ab \sin \gamma}{2}$$

or

$$\frac{\sin \alpha}{a} = \frac{\sin \beta}{b} = \frac{\sin \gamma}{c}$$

6. The Pythagorean Theorem, $a^2 + b^2 = c^2$ where a and b are the lengths of the two legs of a right triangle and c is the length of the hypotenuse, is a special case of the Law of Cosines.

7. The Law of Cosines relates the lengths, a, b, and c, of the sides of a triangle. The square of any side c of a triangle equals the sum of the squares of the other two sides, a and b, minus twice the product of these two sides times the cosine of the angle γ between them:

$$c^2 = a^2 + b^2 - 2ab \cos \gamma$$

8. When an angle γ is obtuse; that is, when $90° \leq \gamma \leq 180°$,

$$\sin \gamma = \sin(180° - \gamma) \quad \text{and} \quad \cos \gamma = -\cos(180° - \gamma)$$

9. The three medians of a triangle, each of which bisects a side, meet at a common point called the centroid.
10. The three altitudes of a triangle meet at a common point.
11. The three angle bisectors of a triangle meet at a common point.
12. The three perpendicular bisectors of the sides of a triangle meet at a common point.
13. In an isosceles triangle, the altitude, median, and perpendicular bisector to the base as well as the bisector of the apex angle are all the same line.
14. In an equilateral triangle, the set of the altitudes, medians, perpendicular bisectors, and angle bisectors is a single set of three lines that meet at a common point.
15. The medians of any triangle meet at a point that is two-thirds the distance from any vertex.
16. The length of the hypotenuse of an isosceles right triangle is $\sqrt{2}$ times the length of the legs.
17. The length of the side opposite the 30° angle in a 30°-60°-90° right triangle is one-half the length of the hypotenuse.

RAISE YOUR GRADES

Can you ... ?

☑ find the area of a rectangle
☑ find the area of a right triangle
☑ find the height of any altitude of a triangle
☑ find the area of any triangle given the height and base
☑ find the area of any triangle given the length of two sides and the measure of the included angle
☑ find the sine and cosine of an obtuse angle
☑ find the length of the third side of a triangle given the two order sides and the included angle
☑ find the interior angles of a triangle given the length of all three sides
☑ draw the median of a triangle
☑ find the centroid of a triangle
☑ find where the medians of a triangle meet
☑ find the side opposite the 30° angle in a 30°-60°-90° right triangle given the hypotenuse
☑ find the hypotenuse of a 45° right triangle given the length of one leg
☑ find the area of an equilateral triangle given the length of one side

SOLVED PROBLEMS

PROBLEM 4-1 Find the area of the shaded region inside the rectangle shown in Figure 4-31.

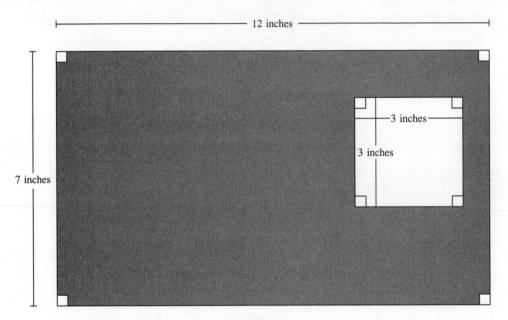

Figure 4-31

Solution You know that the area of a rectangle is the product of the lengths of its two perpendicular sides. You also know, from property (**b**), that the area of a region composed of two figures is the sum of the areas of the two figures. So you begin by finding the area of the entire rectangle and the area of the square inside the rectangle:

$$A(\text{entire rectangle}) = 7 \text{ in.} \times 12 \text{ in.} = 84 \text{ in}^2$$

$$A(\text{square}) = 3 \text{ in.} \times 3 \text{ in.} = 9 \text{ in}^2$$

Now, by the additive property (**b**),

$$A(\text{entire rectangle}) = A(\text{shaded region}) + A(\text{square})$$

or

$$A(\text{entire rectangle}) - A(\text{square}) = A(\text{shaded region})$$
$$84 \text{ in}^2 - 9 \text{ in}^2 = 75 \text{ in}^2$$

PROBLEM 4-2 Find the area of a right triangle whose hypotenuse has length $c = 7$ cm and whose leg has length $b = 3$ cm.

Solution First, use the Pythagorean Theorem to find the length of the other leg a:

$$a^2 + b^2 = c^2$$
$$a^2 + (3 \text{ cm})^2 = (7 \text{ cm})^2$$
$$a^2 = 49 \text{ cm}^2 - 9 \text{ cm}^2 = 40 \text{ cm}^2$$
$$a = \sqrt{40 \text{ cm}^2} = 6.32 \text{ cm}$$

Now use the formula for finding the area of a right triangle:

$$A(\text{right triangle}) = \frac{ab}{2}$$

$$= \frac{(6.32 \text{ cm})(3 \text{ cm})}{2}$$

$$= 9.48 \text{ cm}^2$$

PROBLEM 4-3 **(a)** Find the area of each small region enclosed by the large triangle shown in Figure 4-32. **(b)** Find the area of the large triangle.

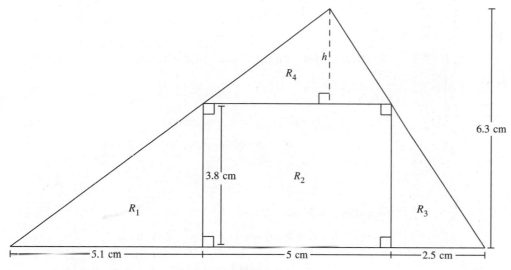

Figure 4-32

Solution

(a) The large triangle is composed of four regions: R_1 and R_3, which are right triangles; R_2, which is a rectangle; and R_4, which is a scalene triangle. Use the appropriate formulas to find the area of each region.

$$A(R_1) = A(\text{right triangle}) = \frac{ab}{2}$$

$$= \frac{(5.1 \text{ cm})(3.8 \text{ cm})}{2}$$

$$= 9.69 \text{ cm}^2$$

$$A(R_2) = A(\text{rectangle}) = ab$$

$$= (5 \text{ cm})(3.8 \text{ cm})$$

$$= 19 \text{ cm}^2$$

$$A(R_3) = A(\text{right triangle}) = \frac{(2.5 \text{ cm})(3.8 \text{ cm})}{2}$$

$$= 4.75 \text{ cm}^2$$

$$A(R_4) = A(\text{scalene triangle}) = \frac{hb}{2}$$

$$= \frac{(6.3 \text{ cm} - 3.8 \text{ cm})(5 \text{ cm})}{2}$$

$$= 6.25 \text{ cm}^2$$

(b) By the additive property **(b)**,

$$A(\text{large triangle}) = A(R_1) + A(R_2) + A(R_3) + A(R_4)$$
$$= 9.69 \text{ cm}^2 + 19 \text{ cm}^2 + 4.75 \text{ cm}^2 + 6.25 \text{ cm}^2$$
$$= 39.69 \text{ cm}^2$$

Check:

$$A(\text{large triangle}) = \frac{hb}{2}$$

$$= \frac{(6.3 \text{ cm})(5.1 \text{ cm} + 5 \text{ cm} + 2.5 \text{ cm})}{2}$$

$$= 39.69 \text{ cm}^2$$

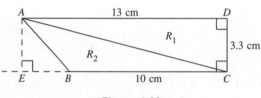

Figure 4-33

PROBLEM 4-4   Find the area of the region *ABCD* shown in Figure 4-33.

Solution

$$A(ABCD) = A(R_1) + A(R_2)$$

$$= \frac{(3.3 \text{ cm})(13 \text{ cm})}{2} + \frac{(3.3 \text{ cm})(10 \text{ cm})}{2}$$

$$= 37.95 \text{ cm}^2$$

Alternative Solution

$$A(ABCD) = A(AECD) - A(\triangle AEB)$$

$$= (13 \text{ cm})(3.3 \text{ cm}) - \frac{(3.3 \text{ cm})(3 \text{ cm})}{2}$$

$$= 37.95 \text{ cm}^2$$

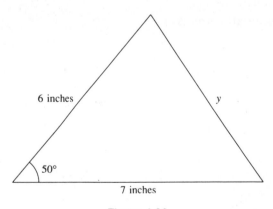

Figure 4-34

PROBLEM 4-5   Find the area of the triangle shown in Figure 4-34.

Solution   Since you're given the lengths of two sides and the measure of the angle between these two sides, you can use the Law of Sines to find this area. Let $a = 6$ in., $b = 7$ in., and $\gamma = 50°$. Then

$$A(\text{triangle}) = \frac{ab \sin \gamma}{2}$$

$$= \frac{(6 \text{ in.} \times 7 \text{ in.}) \sin 50°}{2}$$

$$= 16.09 \text{ in}^2$$

PROBLEM 4-6   Find the length *y* in Figure 4-34.

Solution   Use the Law of Cosines, which allows you to find the length of an unknown side given the other two sides and the measure of the angle between them. Let $y = c$, $a = 6$ in., $b = 7$ in., and $\gamma = 50°$. Then

$$c^2 = a^2 + b^2 - 2ab \cos \gamma$$

$$y^2 = (6 \text{ in.})^2 + (7 \text{ in.})^2 - 2(6 \text{ in.} \times 7 \text{ in.}) \cos 50°$$

$$= 31.01 \text{ in}^2$$

$$y = \sqrt{31.01 \text{ in}^2} = 5.57 \text{ in.}$$

PROBLEM 4-7 (a) Find the centroid of $\triangle ABC$ shown in Figure 4-35a. (b) Given the Cartesian coordinates of the vertices $A = (1, 2)$, $B = (3, 7)$, and $C = (4, 3)$, find the coordinates of the centroid.

Solution

(a) You can find the centroid by drawing the three medians as in Figure 4-35b; each of these medians bisects a side of the triangle. These three medians, AE, BD, and CF, meet at a common point P, which is the centroid of $\triangle ABC$.

(b) Pick any of the medians, say BD, and find the coordinates of the point at which the median bisects the side. Thus, point D, which is exactly halfway between points $A(1, 2)$ and $C(4, 3)$, must have coordinates

$$\overset{\text{add the } x \text{ coordinates} \quad \text{add the } y \text{ coordinates}}{\left(\frac{4 + 1}{2}, \frac{3 + 2}{2}\right) = (2.5, 2.5)}$$

The centroid P is 2/3 of the distance along BD from B, or 1/3 of the distance along BD from D. Thus P's x coordinate is

$$2.5 + \tfrac{1}{3}(3 \overset{\text{subtract the } x \text{ coordinates}}{-} 2.5) = 2.67$$

and P's y coordinate is

$$2.5 + \tfrac{1}{3}(7 \overset{\text{subtract the } y \text{ coordinates}}{-} 2.5) = 4$$

So the coordinates of the centroid P are $(2.67, 4)$.

PROBLEM 4-8 Show that if $\triangle ABC$ is similar to $\triangle EFG$ such that

$$\frac{AB}{EF} = \frac{BC}{FG} = \frac{CA}{GE} = k$$

then the altitudes of these triangles stand in the same ratio.

Solution

Step 1: Draw $\triangle ABC$ and $\triangle EFG$ with altitudes AD and EH, as in Figure 4-36.

Step 2: By definition, the corresponding angles of similar triangles are congruent. Thus $\angle A \cong \angle E$, $\angle B \cong \angle F$, and $\angle C \cong \angle G$.

Step 3: Since $\angle ADC \cong \angle EHG = 90°$ and $\angle C \cong \angle G$, then $\angle 1 \cong \angle 2$.

Step 4: $\triangle ADC$ is similar to $\triangle EHG$ because their corresponding angles are congruent (AAA $\cong$ AAA).

Step 5: $\dfrac{AD}{EH} = \dfrac{CA}{GE} = k$.

PROBLEM 4-9 Show that the areas of the triangles in Problem 4-8 stand in ratio k^2.

Solution

$$\begin{aligned}
A(\triangle ABC) &= \frac{AD \times BC}{2} \\
&= \frac{(kEH) \times (kFG)}{2} \\
&= \frac{k^2(EH \times FG)}{2} \\
&= k^2 A(\triangle EFG)
\end{aligned}$$

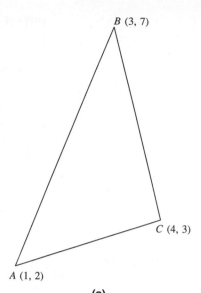

(a)

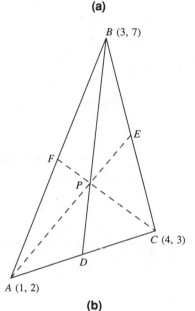

(b)

Figure 4-35

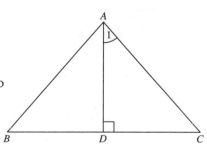

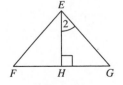

Figure 4-36

PROBLEM 4-10   Find the area of an isosceles triangle whose base angles are 45° and whose base is 5 units in length.

Solution

Step 1: The base angles of an isosceles triangle are equal. Thus you are dealing with a 45°-45°-90° right triangle with hypotenuse of length 5.

Step 2: The lengths of the legs of this triangle are each $5/\sqrt{2}$.

Step 3: The area is

$$\frac{(5/\sqrt{2})^2}{2} = \frac{25}{4} = 6.25 \text{ square units}$$

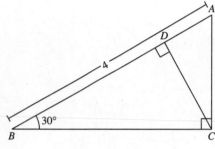

Figure 4-37

PROBLEM 4-11 $\triangle ABC$ of Figure 4-37 is a 30°-60°-90° right triangle with hypotenuse $AB = 4$. Find the length of altitude CD.

Solution

Step 1: AC is opposite the 30° angle in $\triangle ACB$, so $AC = 4/2 = 2$.

Step 2: AC is the hypotenuse of the 30°-60°-90° right triangle $\triangle ADC$.

Step 3: CD is opposite the 60° angle in $\triangle ADC$. Therefore

$$CD = \frac{2\sqrt{3}}{2} = \sqrt{3}$$

Review Exercises

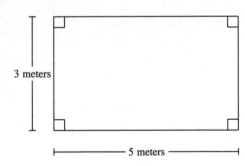

Figure 4-38

EXERCISE 4-1   Find the area of the region shown in Figure 4-38.

EXERCISE 4-2   Find the area of the region shown in Figure 4-39.

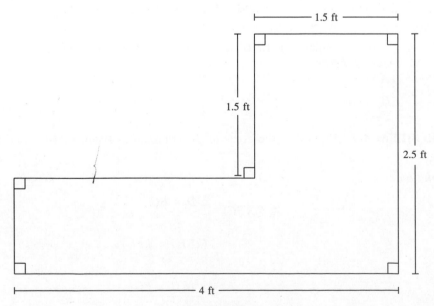

Figure 4-39

EXERCISE 4-3 Find the area of the region shown in Figure 4-40.

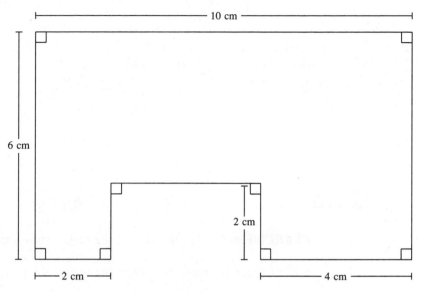

Figure 4-40

EXERCISE 4-4 Find the area of the region shown in Figure 4-41.

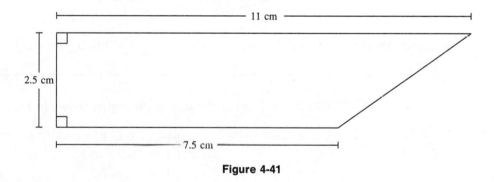

Figure 4-41

EXERCISE 4-5 Find the area of the region shown in Figure 4-42.

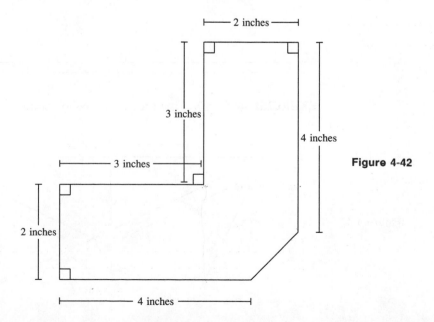

Figure 4-42

EXERCISE 4-6   Find the area of the region shown in Figure 4-43.

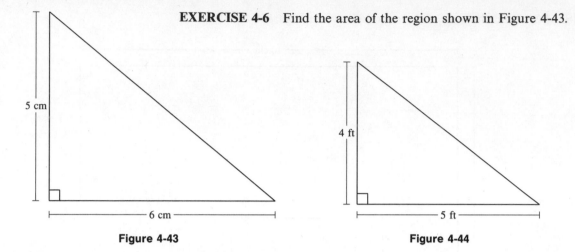

Figure 4-43   Figure 4-44

EXERCISE 4-7   Find the area of the region shown in Figure 4-44.

EXERCISE 4-8   Find the area of the region shown in Figure 4-45.

Figure 4-45

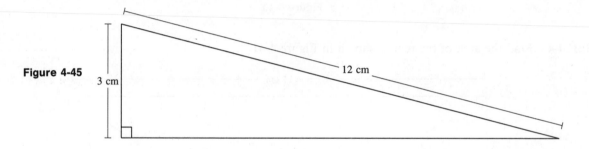

EXERCISE 4-9   Find the area of the region shown in Figure 4-46.

Figure 4-46

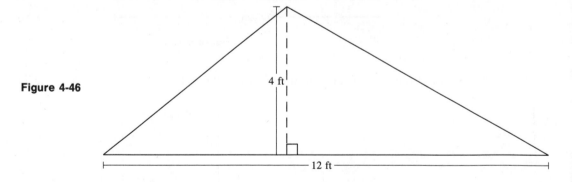

EXERCISE 4-10   Find the area of the region shown in Figure 4-47.

Figure 4-47

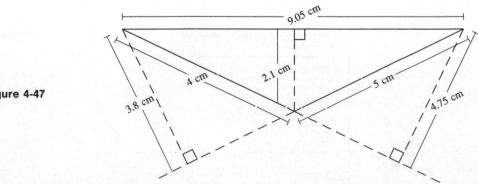

EXERCISE 4-11 Find the area of the region shown in Figure 4-48.

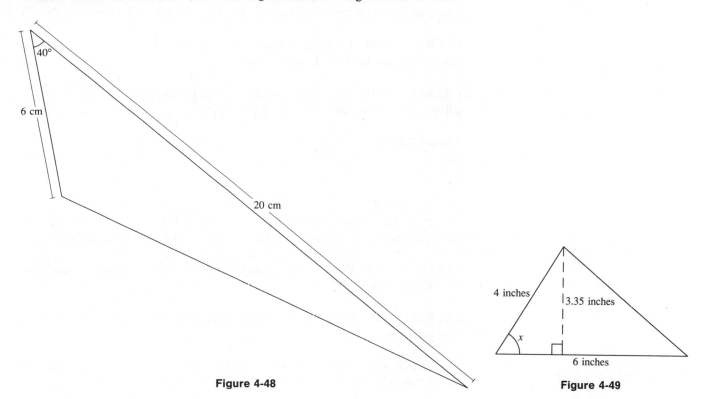

Figure 4-48

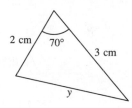

Figure 4-49

EXERCISE 4-12 Find the measure x in the triangle shown in Figure 4-49.

EXERCISE 4-13 Find the length y in the triangle shown in Figure 4-50.

EXERCISE 4-14 Find sine and cosine of **(a)** 120°, **(b)** 135°, and **(c)** 170°.

EXERCISE 4-15 Find the length x in the triangle shown in Figure 4-51.

EXERCISE 4-16 Find the measure x in the triangle shown in Figure 4-52.

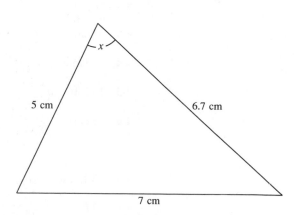

Figure 4-50

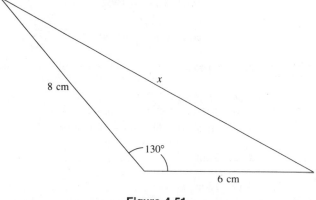

Figure 4-51

Figure 4-52

EXERCISE 4-17 Consider three triangles with sides **(a)** 18, 24, 30; **(b)** 5, 11, 13; and **(c)** 2, 3, 4. Determine whether each triangle is acute, right, or obtuse.

EXERCISE 4-18 Find the length of the altitude drawn to the base of an isosceles triangle whose base is 8 cm in length and whose sides are 5 cm.

EXERCISE 4-19   The median to the base of an isosceles triangle is 8 inches long and the equal legs are 10 inches. Find the length of the base.

EXERCISE 4-20   Find the altitude of the equilateral triangles the lengths of whose sides are (a) 4, (b) 7, (c) $2\sqrt{3}$, (d) $5/\sqrt{3}$.

EXERCISE 4-21   Find the sides of the equilateral triangles the heights of whose altitudes are (a) 4, (b) 7, (c) $2\sqrt{3}$, (d) $5/\sqrt{3}$.

EXERCISE 4-22   The measure of one acute angle of a right triangle is double that of the other. The shorter leg has a length of 2.5. Find the lengths of the hypotenuse and longer leg.

EXERCISE 4-23   The length of one leg of a right triangle is double that of the other. The altitude to the hypotenuse has a length of $2\sqrt{3}$. Find (a) the lengths of the legs, (b) the length of the hypotenuse, and (c) the area of the triangle.

EXERCISE 4-24   For an acute triangle, draw all three medians and find their intersection point.

EXERCISE 4-25   For an acute triangle, draw all three altitudes and find where they intersect.

EXERCISE 4-26   For an obtuse triangle, draw all three altitudes and find where they intersect. [*Hint:* Extend the altitudes.]

EXERCISE 4-27   Draw the three angle bisectors of an acute triangle and find where they intersect. Repeat this for an obtuse triangle.

Answers to Review Exercises

4-1   15 m^2

4-2   6.25 ft^2

4-3   52 cm^2

4-4   23.125 cm^2

4-5   15.5 in^2

4-6   15 cm^2

4-7   10 ft^2

4-8   17.43 cm^2

4-9   24 ft^2

4-10   9.5 cm^2

4-11   38.57 cm^2

4-12   56.9°

4-13   2.98 cm

4-14   (a) $\sin(120°) = 0.8660$, $\cos(120°) = -0.5000$
(b) $\sin(135°) = 0.7071$, $\cos(135°) = -0.7071$
(c) $\sin(170°) = 0.1736$, $\cos(170°) = -0.9848$

4-15   12.72 cm

4-16   71.8°

4-17   (a) right   (b) obtuse   (c) obtuse

4-18   3 cm

4-19   12 in.

4-20   (a) 3.46   (b) 6.06   (c) 3   (d) 2.5

4-21   (a) 4.62   (b) 8.08   (c) 4   (d) 10/3

4-22   hypotenuse = 5, longer leg = 4.33

4-23   (a) 3.87, 7.75   (b) 8.66   (c) 15

4-24

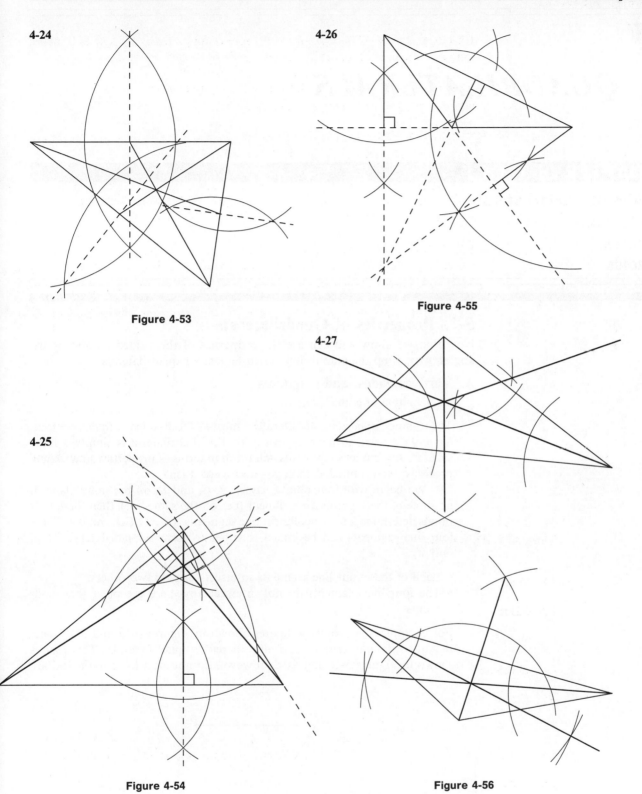

Figure 4-53

4-25

Figure 4-54

4-26

Figure 4-55

4-27

Figure 4-56

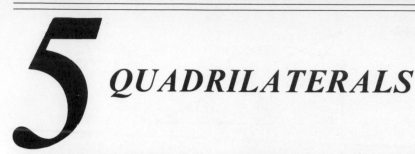

5 QUADRILATERALS

THIS CHAPTER IS ABOUT

☑ **Properties of Quadrilaterals**
☑ **Parallelograms**
☑ **Rhombuses**
☑ **Trapezoids**

5-1. Properties of Quadrilaterals

Now that we know and can use the properties of three-sided figures, or triangles, we can up the ante to four-sided figures, or quadrilaterals.

A. Vertices, sides, and diagonals

1. Definition of a quadrilateral

The name *quadrilateral* literally means "four-sided" (*quadri* = four; *lat*[*eral*] = side), so it's easy to say that a quadrilateral is simply a four-sided figure. But let's refine this definition in terms of properties so we know precisely what a quadrilateral is—and what it isn't.

We begin with four points, say A, B, C, and D, on the same plane. If no three of these points are collinear (i.e., on the same line), then these four points determine four noncollinear line segments AB, BC, CD, and DA. These four line segments can be joined together to form a quadrilateral if and only if:

• Each of these four line segments intersects exactly two others.
• The four line segments do not intersect anywhere except at their endpoints.

For example, the geometric figures shown in Figures 5-1a and 5-1b meet the quadrilateral criteria and are therefore quadrilaterals. The figures shown in Figures 5-1c and 5-1d, however, do not meet the criteria and are

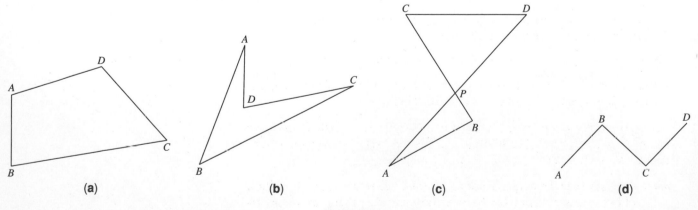

| | | | |
| (a) | (b) | (c) | (d) |

Figure 5-1

not quadrilaterals: In Figure 5-1c the line segments *DA* and *CB* intersect at a point *P* that is not an endpoint of either *DA* or *CB*; and in Figure 5-1d the line segments *AB* and *CD* do not intersect two other line segments at both of their endpoints. Knowing the criteria for the formation of a quadrilateral allows us to come up with a formal definition of a quadrilateral:

- A **quadrilateral** is the *union* of four line segments *AB*, *BC*, *CD*, and *DA* if the points *A*, *B*, *C*, and *D* are coplanar points no three of which are collinear and if each of the line segments *AB*, *BC*, *CD*, and *DA* intersects exactly two others, one at each endpoint.

2. Parts of quadrilaterals

The points *A*, *B*, *C*, and *D* where the line segments intersect are called the **vertices** of the quadrilateral, and the line segments *AB*, *BC*, *CD*, and *DA* are its **sides**.

We refer to quadrilaterals by their vertices, but we have to be careful when we do this. In order to correctly name a quadrilateral, we have to list the vertices in the order in which they would appear if we were to walk around the figure. For example, the quadrilaterals shown in Figure 5-1a and b could be named □*ABCD*, □*BCDA*, □*CDAB*, □*DABC* or □*ADCB*, □*DCBA*, □*CBAD*, □*BADC*; but these quadrilaterals could *not* be named □*ADBC* or any other combination of vertices that differs from the order we'd encounter if we walked around the figure.

- Two vertices that are encountered directly after one another as we "walk around the figure" are called **consecutive vertices**.

Consecutive vertices, then, are really the two endpoints of a side. For example, points *A* and *B* in Figure 5-1a and b are consecutive vertices. Similarly,

- Two sides that are encountered directly after one another as we "walk around the figure" are called **consecutive sides**.

Consecutive sides always have a common endpoint. For example, *AD* and *CD* are consecutive sides in Figure 5-1a and b. Then, in contrast,

- Two sides or vertices that are *not* consecutive are called **opposite sides** or **opposite vertices**, respectively.

In Figure 5-1a and b, for example, sides *AD* and *BC* are opposite sides, as are *AB* and *CD*; and vertices *A* and *C* are opposite vertices, as are *B* and *D*.

We can see that the four vertices of a quadrilateral determine the four sides, but we can also see that these same four points determine two additional lines—the *diagonals*:

- The line segment determined by two opposite vertices of a quadrilateral is called a **diagonal of the quadrilateral**.

There are two diagonals in every quadrilateral, and

- Each diagonal of a quadrilateral results in the formation of two triangles, two of whose sides are sides of the quadrilateral.

For example, in Figure 5-2a the diagonal *AC* divides □*ABCD* into two triangles △*ADC* and △*ABC*; and in Figure 5-2b the diagonal *A'C'* results in two triangles △*A'D'C'* and △*A'B'C'*.

note: The fact that the diagonals of a quadrilateral result in the formation of two triangles is our key to understanding the properties of quadrilaterals.

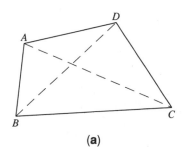

(a)

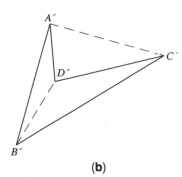

(b)

Figure 5-2

EXAMPLE 5-1:   Which of the figures shown in Figure 5-3 are quadrilaterals and which are not? Explain why the figures that are not quadrilaterals are not quadrilaterals.

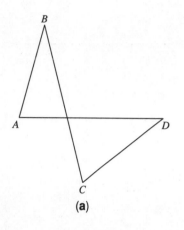
(a)

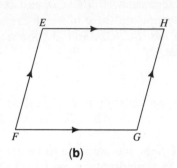
(b)

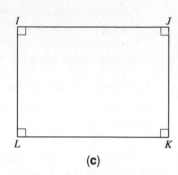
(c)

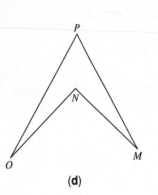
(d)

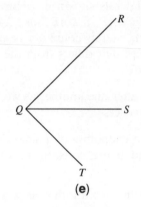
(e)

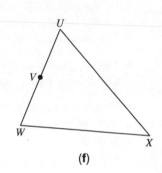
(f)

Figure 5-3

Solution

(a) *ABDC* is not a quadrilateral because the line segments *AD* and *CB* intersect at a point that is not an endpoint for either of the line segments.

(b) □*EFGH* is a quadrilateral.

(c) □*IJKL* is a quadrilateral; in fact, this figure is a *rectangle*, which is a quadrilateral whose opposite sides are parallel and whose consecutive sides are perpendicular (see Section 4-2).

(d) □*MNOP* is a quadrilateral.

(e) *QRST* is not a quadrilateral because the line segments *RQ*, *SQ*, and *TQ* do not intersect two other line segments at both endpoints.

(f) *UVWX* is not a quadrilateral because three of the given points, *U*, *V*, and *W*, are collinear. In fact, this figure is a triangle.

B. Angles

We know that triangles have interior and exterior angles and that the interior angles of a triangle sum to 180°. Like triangles, quadrilaterals also have interior and exterior angles, but the interior and exterior angles of a quadrilateral are trickier than those of triangles. In order to understand the angles of a quadrilateral, we first need to understand something about convexity and concavity.

1. Convexity and concavity

- If every side of a quadrilateral is extended as a line and the two vertices not on each line fall on the same side of this line, the quadrilateral is **convex**.
- If any side of a quadrilateral may be extended to a line so that the vertices not on the line fall on opposite sides of the line, the quadrilateral is **concave**. Any quadrilateral that is not convex is concave.

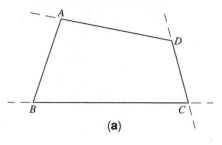

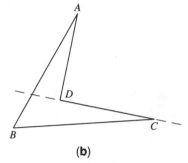

(a)

For example, the quadrilateral shown in Figure 5-4a is convex—when any one of its sides is extended, the two vertices not on the line fall on the same side of the line. Thus, when AB is extended, vertices D and C are on the same side of $\overleftrightarrow{AB}$; when BC is extended, A and D are on the same side; and so on. The quadrilateral shown in Figure 5-4b is concave. When side CD is extended, the vertices A and B fall on opposite sides of the line $\overleftrightarrow{CD}$.

(b)

Figure 5-4

2. Interior of a quadrilateral and interior angles

Like triangles, quadrilaterals enclose space (see Section 3-1). Thus we can say that

- The **interior of a quadrilateral** is the region (set of points) bounded by the four sides.

This definition, which is true for both convex and concave quadrilaterals, seems obvious enough; but what does it mean with respect to the interior angles? Let's examine the two types of quadrilaterals and see.

Convex Case: For a convex quadrilateral, $\square ABCD$ as shown in Figure 5-5, the interior angles are $\angle ABC$, $\angle BCD$, $\angle CDA$, and $\angle DAB$. These angles all have a measure between $0°$ and $180°$. Any point P that lies in the interior of the quadrilateral must also lie in the interior of all the vertex angles. Thus:

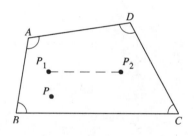

Figure 5-5

- The interior of the convex quadrilateral is the set of points that lie in the interior of all its vertex angles.

So we can see that if P_1 and P_2 are any points interior to a convex quadrilateral, then all the points of the line segment P_1P_2 are also interior to the quadrilateral and to all of the vertex angles.

Concave Case: For a concave quadrilateral, $\square ABCD$ as shown in Figure 5-6, the interior angles are also $\angle ABC$, $\angle BCD$, $\angle CDA$, and $\angle DAB$. In this case, $\angle ABC$, $\angle BCD$, and $\angle DAB$ each have a measure between $0°$ and $180°$, but $\angle CDA$ must have a measure greater than $180°$ if it is to be an interior angle.

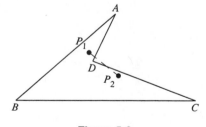

Figure 5-6

digression: Every two rays that meet at their endpoints really define two angles. One of these angles ($\angle 1$) is the space between the rays as shown in Figure 5-7a, where the measure of the angle is less than $180°$. The other angle ($\angle 2$) is the space between the rays on the other side, as shown in Figure 5-7b, where the measure of the angle is greater than $180°$. Both of these angles can be measured with a protractor. Angle 1 is no problem—we just put the protractor on it and read off the number of degrees less than $180°$ in the space. Angle 2 is a little trickier. In order to measure this angle we must first extend $\overline{CD}$ past D to form ray $\overrightarrow{DE}$, as in Figure 5-7c. This gives the straight angle, $\angle 3$, which has a $180°$ measure, plus

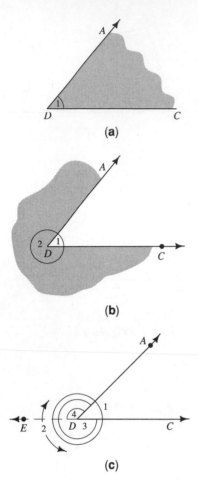

(a)

(b)

(c)

Figure 5-7

angle 4. Then we can see that $\angle 4$ is supplementary to $\angle 1$, so $\angle 4 = 180° - \angle 1$. Thus,

$$\angle 2 = 180° + (180° - \angle 1)$$
$$= 360° - \angle 1$$

The space in the plane that is not on the rays or interior to $\angle 1$ is the interior of $\angle 2$.

Now, returning to Figure 5-6, we can see that since one of the interior angles ($\angle CDA$) is greater than 180°, the interior of this concave quadrilateral $\square ABCD$ must include only the set of points that lie both in $\angle CDA$ *and* in the opposite angle $\angle ABC$. Also, because the quadrilateral shown in Figure 5-6 is concave, the other two interior angles $\angle BCD$ and $\angle DAB$ each have one ray that separates two of the vertices of the quadrilateral, cutting through its interior. As a result, we can find points such as P_1 and P_2 in Figure 5-6 that lie in the interior of the quadrilateral but do not lie in the interior of $\angle BCD$ or $\angle DAB$, respectively. Furthermore, the line segment P_1P_2 that joins these points does not lie entirely in the quadrilateral. Thus we can say that

• When two sides of a quadrilateral form an angle such that the defining rays of this angle can be intersected by a line segment defined by points interior to the other interior angles less than 180°, then the interior of this quadrilateral is the set of points that lie interior to the larger angle and at least one other interior angle.

General Case: Now that we can recognize the interior angles of a convex and a concave quadrilateral and can describe the interiors of the two kinds of quadrilaterals, we can say that

• The interior of any quadrilateral is the set of points lying in at least three of the four interior angles.

Moreover, we are ready to prove the most important property of quadrilaterals:

• **The measures of the interior angles of any quadrilateral sum to 360°.**

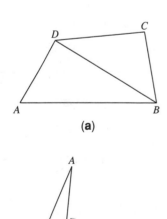

(a)

(b)

Figure 5-8

EXAMPLE 5-2: Show that the measures of the interior angles of a quadrilateral sum to 360°.

Solution: Whether a quadrilateral is convex, as in Figure 5-8a, or concave, as in Figure 5-8b, you can always draw one diagonal that lies entirely interior to the quadrilateral.

Step 1: Draw a convex and a concave quadrilateral $\square ABCD$; then draw an interior diagonal BD for each, as in Figure 5-8.

Step 2: $\angle DAB + \angle ABD + \angle BDA = 180°$ and $\angle DBC + \angle BCD + \angle CDB = 180°$, since the interior angles of any triangle always sum to 180°.

Step 3: $\angle ABC = \angle ABD + \angle DBC$ and $\angle CDA = \angle CDB + \angle BDA$.

Step 4: Add the two equations in Step 2.

$$\angle DAB + \angle ABD + \angle BDA + \angle DBC + \angle BCD + \angle CDB = 180° + 180°$$

or

$$\angle DAB + (\angle ABD + \angle DBC) + \angle BCD + (\angle BDA + \angle CDB) = 360°$$

Step 5: Substitute from Step 3.

$$\angle DAB + \angle ABC + \angle BCD + \angle CDA = 360° \qquad \text{QED}$$

3. Exterior angles

If we extend one side of □*ABCD* in Figure 5-9, we see that another angle supplementary to the interior angle is formed.

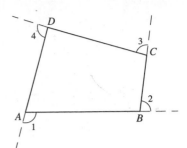

Figure 5-9

- In a quadrilateral, the angle whose rays are the extended side and the consecutive nonextended side (which it shares with the supplementary interior angle) is called an **exterior angle**.

For example, $\angle 1$, $\angle 2$, $\angle 3$, and $\angle 4$ are the exterior angles of the convex quadrilateral □*ABCD* shown in Figure 5-9.

- The measures of the exterior angles of any quadrilateral sum to 360°.

The idea of exterior angles for convex quadrilaterals is completely straightforward, as these angles are literally *exterior*. And for three of the four angles in a concave quadrilateral this literal description holds nicely. For example, $\angle 1$, $\angle 2$, and $\angle 4$ in Figure 5-10 meet the definition of an exterior angle and are, in fact, exterior. But what about $\angle 3$? Although $\angle 3$ does fit the definition of an exterior angle, its own interior is inside the quadrilateral! And, to make matters worse, we know that the interior angle $\angle BCD$ is greater than 180°, but the exterior angle $\angle 3$ must be supplementary to $\angle BCD$. Fortunately, there is a way out of this problem. We simply tag the exterior angle with a minus sign; that is, we consider the exterior angle negative. This way, the measures of the interior angle and the exterior angle can still add up to 180°, and the sum of the measures of all the exterior can still be 360°.

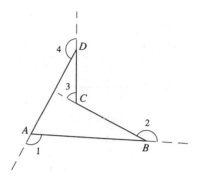

Figure 5-10

note: In this book (and in most geometry books) you will usually be dealing with convex quadrilaterals. But you must remember that when you are trying to develop general theorems about quadrilaterals, the proofs have to hold for both types.

C. Areas

There is no general formula for finding the area of an arbitrary quadrilateral. Although some of the special quadrilaterals, which are discussed in the rest of this chapter, do have area formulas, the best we can do for an arbitrary quadrilateral is to break it down into two triangles and add those areas.

EXAMPLE 5-3: Find the area of the quadrilateral shown in Figure 5-11.

Solution: Draw one of the diagonals, say *AC*. Then use the Law of Sines to find the area of each resulting triangle and add these areas:

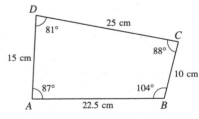

Figure 5-11

$$A(\square ABCD) = A(\triangle ABC) + A(\triangle ADC)$$

$$= \frac{(BA)(BC)(\sin \angle B)}{2} + \frac{(DA)(DC)(\sin \angle D)}{2}$$

$$= \frac{(22.5 \text{ cm})(10 \text{ cm})(\sin 104°)}{2} + \frac{(15 \text{ cm})(25 \text{ cm})(\sin 81°)}{2}$$

$$= 294.35 \text{ cm}^2$$

5-2. Parallelograms

- A quadrilateral is a **parallelogram** if both sets of opposite sides are parallel. For example, the quadrilateral shown in Figure 5-12 is a parallelogram.

note: Squares and rectangles are special cases of parallelograms.

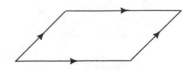

Figure 5-12

A. Properties of parallelograms

Parallelograms have several important properties that can be stated as theorems. We can readily show that these properties are true from what we know about triangles.

- **A diagonal of a parallelogram divides the parallelogram into two congruent triangles.**

EXAMPLE 5-4:   Show that the diagonal of a parallelogram divides the parallelogram into two congruent triangles.

Solution

Step 1: Draw a parallelogram $\square ABCD$ as in Figure 5-13.

Step 2: By the definition of a parallelogram, AB is parallel to DC and AD is parallel to BC.

Step 3: Draw one of the diagonals, say AC, as in Figure 5-13. AC is a segment of the transversal $\overleftrightarrow{AC}$, which intersects both sets of parallel line segments. Since the alternate interior angles of two parallel lines cut by a transversal are congruent, $\angle DCA \cong \angle CAB$ and $\angle DAC \cong \angle ACB$.

Step 4: $AC \cong AC$.

Step 5: $\triangle ADC \cong \triangle CBA$ by ASA.

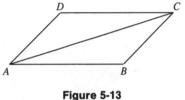

Figure 5-13

- **Opposite sides of a parallelogram are congruent** (i.e., opposite sides of a parallelogram have equal length).

EXAMPLE 5-5:   Show that opposite sides of a parallelogram are congruent.

Solution: Use Figure 5-13 and the results of Example 5-4. Since $\triangle ADC \cong \triangle CBA$, $CD \cong AB$ and $AD \cong CB$ because these sides are corresponding sides of congruent triangles.

- **Opposite interior angles of a parallelogram are congruent** (i.e., have equal measure or, more loosely speaking, are equal).

EXAMPLE 5-6:   Show that the opposite interior angles of a parallelogram are congruent.

Solution: Use Figure 5-13 and the results of Example 5-4 again. Since $\triangle ADC \cong \triangle CBA$, $\angle D \cong \angle B$ as corresponding angles of congruent triangles.

note: If you'd drawn diagonal BD in Figure 5-13, then you'd have, similarly, $\triangle BAD \cong \triangle DCB$ and $\angle A \cong \angle C$.

- **The two diagonals of a parallelogram bisect each other.**

EXAMPLE 5-7:   Show that the two diagonals of a parallelogram bisect each other.

Solution

Step 1: Draw a parallelogram $\square ABCD$ as in Figure 5-14, with diagonals AC and DB meeting at P.

Step 2: $\angle 1 \cong \angle 4$ and $\angle 2 \cong \angle 3$, since they are the alternate interior angles cut from two parallel lines $\overleftrightarrow{AD}$ and $\overleftrightarrow{BC}$ by transversals $\overleftrightarrow{DB}$ and $\overleftrightarrow{AC}$, respectively.

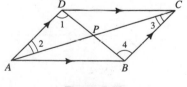

Figure 5-14

Step 3: $AD \cong BC$, as opposite sides of a parallelogram.

Step 4: $\triangle ADP \cong \triangle CBP$ by ASA.

Step 5: $DP \cong BP$ and $AP \cong CP$, as corresponding parts of congruent triangles.

- **If the two diagonals of a parallelogram are congruent, then the parallelogram is a rectangle.**

EXAMPLE 5-8: Show that if the two diagonals of a parallelogram are congruent, then the parallelogram is a rectangle.

Solution

Step 1: Assume the hypothesis is true and draw the parallelogram $\square ABCD$ as in Figure 5-15 with diagonals AC and BD.

Step 2: $AC \cong BD$ by the given information.

Step 3: $AB \cong DC$ since they are opposite sides of a parallelogram.

Step 4: $AD \cong AD$.

Step 5: $\triangle ABD \cong \triangle DCA$ by SSS.

Step 6: $\angle BAD \cong \angle CDA$ as corresponding angles of congruent triangles.

Step 7: Since opposite angles of a parallelogram are congruent, all four interior angles $\angle BAD$, $\angle CDA$, $\angle DCB$, $\angle CBA$ have a common measure, x.

Step 8: The sum of the interior angles of a quadrilateral is 360°; so $4x = 360°$, or $x = 90°$.

Step 9: The parallelogram is a rectangle because all four interior angles measure 90°.

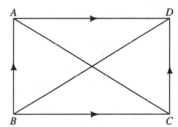

Figure 5-15

B. Area of a parallelogram

A perpendicular can be dropped from one vertex, say A, of a parallelogram $\square ABCD$ to the opposite side BC, as in Figure 5-16. The length of the perpendicular AE is called the **height h of the parallelogram**; and the side to which the perpendicular is drawn, BC, is called the **base of the parallelogram**, whose length is b. Then the diagonal drawn from this vertex, AC, divides the parallelogram into two congruent triangles, $\triangle ABC$ and $\triangle CDA$—whose areas are, of course, equal. Thus the area of a parallelogram is twice the area of $\triangle ABC$ or $2(\frac{1}{2}) \times$ base $\times$ height = base $\times$ height; that is,

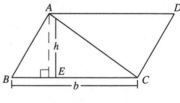

Figure 5-16

AREA OF A PARALLELOGRAM $A(\text{parallelogram}) = bh$

EXAMPLE 5-9: Find the area of the parallelogram shown in Figure 5-17.

Solution: Use either height ($h_1 = 3.30$ cm; $h_2 = 5.47$ cm) and the corresponding base ($b_1 = 6.30$ cm; $b_2 = 3.80$ cm):

$$A(\text{parallelogram}) = bh = b_1 h_1 = b_2 h_2$$
$$= (3.30 \text{ cm})(6.30 \text{ cm}) = (5.47 \text{ cm})(3.80 \text{ cm})$$
$$= 20.79 \text{ cm}^2$$

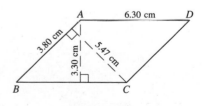

Figure 5-17

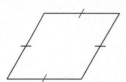

Figure 5-18

5-3. Rhombuses

Figure 5-18 shows a *rhombus*.

● A **rhombus** is a parallelogram with four congruent sides.

That is, the four sides of a rhombus have equal measure.

note: Just as you might describe a parallelogram as a rectangle that's been sat on—either it bends at the joints like a cheap beach chair or it holds up like a right-angled church pew—you might describe a rhombus as a square that's been sat on. Remember, though, that the rectangle is a special case of the parallelogram and the square is a special case of the rhombus: They don't have to be "bent."

Since a rhombus is a parallelogram, it has all the properties of a parallelogram. But a rhombus also has two special properties:

● **The diagonals of a rhombus are orthogonal; that is, they are mutually perpendicular.**
● **The diagonals of a rhombus bisect its interior angles.**

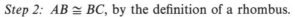

EXAMPLE 5-10: Show that the diagonals of a rhombus are orthogonal.

Solution

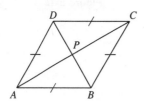

Figure 5-19

Step 1: Draw rhombus □*ABCD*, as in Figure 5-19, with diagonals *AC* and *DB*.
Step 2: $AB \cong BC$, by the definition of a rhombus.
Step 3: $AP \cong CP$, since the diagonals of a parallelogram bisect each other.
Step 4: $PB \cong PB$.
Step 5: $\triangle APB \cong \triangle CPB$ by SSS.
Step 6: $\angle APB \cong \angle CPB$ as corresponding angles of congruent triangles.
Step 7: Since $\angle APB$ and $\angle CPB$ are cut from diagonal *AC* by *DB*, they are supplementary angles.
Step 8: $\angle APB = \angle CPB = 90°$, since $\angle APB$ and $\angle CPB$ are congruent and sum to 180°.
Step 9: *DB* is perpendicular to *AC* because they meet at right angles.

EXAMPLE 5-11: Show that the diagonals of a rhombus bisect the interior angles.

Solution

Step 1: Refer to Figure 5-19 again.
Step 2: $\triangle APB \cong \triangle CPB$ from Example 5-10.
Step 3: $\angle PBA \cong \angle PBC$ as corresponding angles of congruent triangles, so diagonal *DB* bisects the interior angle $\angle B$.
Step 4: Repeat the arguments of Example 5-10 to show that $\triangle APB \cong \triangle APD$, $\triangle APD \cong \triangle CPD$, and $\triangle CPD \cong \triangle CPB$. Thus, as in Step 3, the other interior angles are also bisected by their respective diagonals.

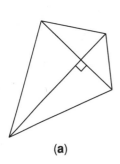

(a)

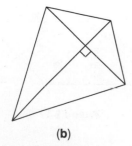

(b)

Figure 5-20

note: If the diagonals of a quadrilateral intersect at right angles, the figure is *not necessarily* a rhombus. For example, the two quadrilaterals shown in Figure 5-20 are not rhombuses—their four sides are not congruent and their diagonals do not bisect all of the interior angles. In Figure 5-20b, however, two of the sides are congruent and two opposite interior angles are bisected by one of the diagonals. This figure is known as a **kite**, defined as a quadrilateral in which one—and only one—of the diagonals is the perpendicular bisector of the other.

EXAMPLE 5-12: Show that any quadrilateral must be a rhombus if its diagonals bisect the interior angles. This is the converse of the theorem proved in Example 5-11.

Solution

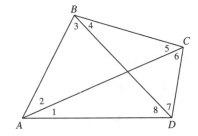

Figure 5-21

Step 1: Draw quadrilateral *ABCD* as in Figure 5-21 with diagonals *AC* and *BD*. Label the angles cut from the interior angles as in the figure.

Step 2: $\angle 1 \cong \angle 2$ and $\angle 5 \cong \angle 6$ since interior angles $\angle BAD$ and $\angle BCD$ are given as bisected by diagonal *AC*.

Step 3: $AC \cong AC$.

Step 4: $\triangle ABC \cong \triangle ADC$ by ASA.

Step 5: $AB \cong AD$ and $CB \cong CD$ as corresponding sides of congruent triangles.

Step 6: $\angle 3 \cong \angle 4$ and $\angle 8 \cong \angle 7$ since diagonal *BD* is given as bisecting interior angles $\angle ABC$ and $\angle ADC$.

Step 7: $BD \cong BD$.

Step 8: $\triangle ABD \cong \triangle CBD$ by ASA.

Step 9: $AB \cong CB$ and $AD \cong CD$ as corresponding sides of congruent triangles.

Step 10: Thus $AB \cong CB \cong CD \cong AD$ and the quadrilateral is a rhombus.

note: We see that if a quadrilateral is a rhombus its diagonals bisect the interior angles, and if its diagonals bisect the interior angles it is a rhombus. We say that the property of diagonals bisecting interior angles is *necessary and sufficient* for the quadrilateral to be a rhombus. That is, one property (having congruent sides) cannot be present without the other (diagonals bisecting angles). These properties are equivalent in quadrilaterals and a rhombus could have been defined by the diagonals bisecting interior angles property. This is in marked contrast to the situation where the diagonals of quadrilateral are perpendicular. Example 5-11 shows that a rhombus must have perpendicular diagonals. This condition is *necessary* for a quadrilateral to be a rhombus. Orthogonal diagonals however are *not sufficient* to assure that a quadrilateral is a rhombus. A kite has perpendicular diagonals but is not a rhombus.

It is important to distinguish whether properties are *necessary*, are *sufficient*, or are *necessary and sufficient*.

5-4. Trapezoids

Figure 5-22 shows a trapezoid.

• A trapezoid is a quadrilateral that has only one pair of parallel sides.

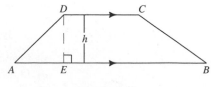

Figure 5-22

Each of the parallel sides is called a **base of the trapezoid**. A perpendicular dropped from one of the bases to its intersection with the other base is called the **altitude of the trapezoid**. The length of the altitude, that is, the distance between the two bases, is the **height h of the trapezoid**. In Figure 5-22, for example, the bases of the trapezoid are *DC* and *AB* and its altitude, of height h, is *DE*.

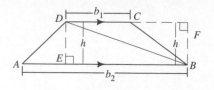

Figure 5-23

note: If the two nonparallel sides of a trapezoid are equal in length, the figure is called an **isosceles trapezoid**.

Finding the area of a trapezoid is almost as easy as finding the area of a parallelogram. First we draw a diagonal, say DB, of the trapezoid $\square ABCD$, as in Figure 5-23. Then we call DC's length b_1 and AB's length b_2. Now we can see that the altitude DE, which can be dropped from any point on either base, is an altitude for $\triangle ADB$ with base b_2 and is also an altitude for $\triangle DBC$ with base b_1. So the area of the trapezoid must be the sum of the areas of these two triangles, or

$$A(\text{trapezoid}) = A(\triangle DCB) + (\triangle ADB)$$

**AREA OF A
TRAPEZOID**

$$= \frac{hb_1}{2} + \frac{hb_2}{2}$$

$$= \frac{h(b_1 + b_2)}{2}$$

That is:

- The area of a trapezoid is one-half its height times the sum of its two bases.

EXAMPLE 5-13:   Find the area of a trapezoid whose parallel sides measure 4 and 6 inches and whose altitude is 5 inches.

Solution

$$A(\text{trapezoid}) = \frac{(5 \text{ in.})(4 \text{ in.} + 6 \text{ in.})}{2} = 25 \text{ in}^2$$

SUMMARY

1. A quadrilateral is a two-dimensional four-sided figure; its vertices are coplanar, no three of its vertices are collinear, and each of its sides intersects exactly two others, one at each endpoint.
2. A diagonal of a quadrilateral connects two opposite vertices and divides the quadrilateral into two triangles.
3. The area of a quadrilateral is the sum of the areas of the two triangles determined by one of the quadrilateral's diagonals.
4. The measures of the interior angles of a quadrilateral sum to 360°, and the measures of the exterior angles of a quadrilateral sum to 360°.
5. In a convex quadrilateral, each of the interior angles is less than 180°.
6. In a concave quadrilateral, one interior angle is greater than 180°.
7. Line segments joining two interior points of a convex quadrilateral lie completely interior to the quadrilateral, but line segments joining two interior points of a concave quadrilateral do not necessarily lie completely interior to the quadrilateral.
8. A parallelogram is a quadrilateral whose opposite sides are parallel:

   (a) Opposite sides of a parallelogram are congruent.
   (b) Opposite angles of a parallelogram arc congruent.
   (c) The diagonals of a parallelogram bisect each other.
   (d) The diagonals of a parallelogram divide the parallelogram into two congruent triangles.
   (e) A rectangle is a special case of a parallelogram, in which consecutive sides are perpendicular.

(f) The area of a parallelogram is the base times the height, i.e.,

$$A(\text{parallelogram}) = bh$$

9. A rhombus is a parallelogram with all sides congruent:

   (a) The diagonals of a rhombus are orthogonal (mutually perpendicular).
   (b) The diagonals of a rhombus bisect the interior angles.

10. A kite is a quadrilateral in which one and only one diagonal is the perpendicular bisector of the other.

11. A trapezoid is a quadrilateral that has only one set of opposite sides parallel:

    (a) An isosceles trapezoid has nonparallel sides of equal length.
    (b) The area of a trapezoid is one-half the height (h) times the sum of the bases (b_1 and b_2), or

$$A(\text{trapezoid}) = \frac{h(b_1 + b_2)}{2}$$

RAISE YOUR GRADES

Can you . . . ?

☑ identify the two types of quadrilaterals
☑ find the area of a quadrilateral, given the measure of its sides and interior angles
☑ recognize when a quadrilateral is a parallelogram
☑ find the measures of all the interior angles of a parallelogram, given the measure of one interior angle
☑ find an exterior angle for a quadrilateral
☑ find the length of one side of a parallelogram from the opposite side
☑ recognize when a parallelogram is a rhombus
☑ find the length of the side of a rhombus, given the length of its two diagonals
☑ find the area of a rhombus from the length of its diagonals
☑ recognize a trapezoid
☑ find the area of a trapezoid

SOLVED PROBLEMS

PROBLEM 5-1 Show that the sum of the exterior angles of any quadrilateral is 360°.

Solution Each exterior angle at a vertex of the quadrilateral is supplementary to an interior angle at the same vertex; therefore, each interior angle plus its exterior angle has a combined measure of 180°. (Remember that in concave figures the exterior angle may be considered negative.) Thus, if you add up the four exterior angles along with their associated interior angles, you really add 180° four times. This yields

$$4(180°) = \text{sum of exterior angles} + \text{sum of interior angles}$$

or

$$720° = \text{sum of exterior angles} + 360°$$

So

$$\text{sum of exterior angles} = 720° - 360° = 360°$$

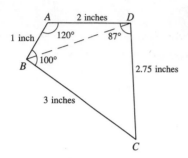

Figure 5-24

PROBLEM 5-2 Find **(a)** the missing angle and **(b)** the area of the quadrilateral shown in Figure 5-24.

Solution

(a) You know that $\angle C + \angle D + \angle A + \angle B = 360°$. Therefore

$$\angle C = 360° - 87° - 120° - 100° = 53°$$

(b) The diagonal BD divides the quadrilateral $\square ABCD$ into two triangles, $\triangle BAD$ and $\triangle DCB$. The sum of the areas of these two triangles is the area of the quadrilateral. Use the Law of Sines to find the areas of the two triangles and add the areas:

$$A(\square ABCD) = A(\triangle BAD) + A(\triangle DCB)$$

$$= \frac{(BA)(AD)(\sin \angle A)}{2} + \frac{(BC)(CD)(\sin \angle C)}{2}$$

$$= \frac{(1 \text{ in.})(2 \text{ in.})(\sin 120°)}{2} + \frac{(3 \text{ in.})(2.75 \text{ in.})(\sin 53°)}{2}$$

$$= 4.16 \text{ in}^2$$

PROBLEM 5-3 Angle $\angle A$ of parallelogram $\square ABCD$ has a measure of $47°$. Find the measures of the other angles.

Solution Since opposite interior angles of a parallelogram are congruent, $\angle C \cong \angle A$. So

$$\angle C \cong \angle A = 47°$$

Now, if you let $\angle B \cong x$, by the same reasoning

$$\angle B \cong \angle D = x$$

Thus, since all the angles of a parallelogram sum to $360°$,

$$\angle C + \angle A + \angle B + \angle D = 360°$$

or

$$2(47°) + 2x = 360°$$

$$2x = 266°$$

$$x = 133°$$

So $\angle B \cong \angle D = 133°$.

PROBLEM 5-4 Show that the consecutive angles of parallelogram $\square ABCD$ are supplementary angles.

Solution

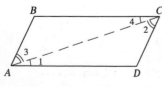

Figure 5-25

Step 1: Draw parallelogram $\square ABCD$ as in Figure 5-25 with a diagonal, say AC, that divides the parallelogram into two congruent triangles, $\triangle ADC$ and $\triangle CBA$.

Step 2: For convenience, label angles $\angle CAB$, $\angle BCA$, $\angle CAD$, and $\angle DCA$ as angles $\angle 1$, $\angle 2$, $\angle 3$, and $\angle 4$, respectively. Thus $\angle 1 \cong \angle 4$ as congruent angles of congruent triangles.

Step 3: Since the angles of a triangle always sum to $180°$,

$$\angle B + \angle 3 + \angle 4 = 180°$$

Step 4: $\angle B = 180° - (\angle 3 + \angle 4) = 180° - (\angle 3 + \angle 1)$, since $\angle 4 \cong \angle 1$.

Step 5: $\angle 1 + \angle 3 = \angle A$, so $\angle B = 180° - \angle A$. Therefore, $\angle B$ is supplementary to $\angle A$.

Step 6: Then, since $\angle B \cong \angle D$ and $\angle A \cong \angle C$, any two consecutive angles in a parallelogram must be supplementary angles.

Alternative Solution

Step 1: Let $\angle A = x$ and $\angle B = y$. Then, since the opposite interior angles of a parallelogram are congruent, $\angle A \cong \angle C = x$ and $\angle B \cong \angle D = y$.

Step 2: The sum of any two consecutive angles must always be $x + y$, so

$$\angle A + \angle B + \angle C + \angle D = 360°$$

or

$$2x + 2y = 360°$$

Step 3: Factoring the equation in Step 2 gives

$$2x + 2y = 2(x + y) = 360°$$

$$x + y = 180°$$

And since the sum of any two consecutive angles is $180°$, any two consecutive angles must be supplementary angles.

PROBLEM 5-5 Show that if a quadrilateral has one set of opposite sides both congruent and parallel, then the quadrilateral is a parallelogram.

Solution

Step 1: Draw a quadrilateral $\square ABCD$ as in Figure 5-26, and let $DC \cong AB$ and DC be parallel to AB.

Step 2: Draw diagonal AC.

Step 3: $\angle DCA$ and $\angle CAB$ are alternate interior angles cut from the parallel lines $\overleftrightarrow{DC}$ and $\overleftrightarrow{AB}$ by transversal $\overleftrightarrow{AC}$. Thus $\angle DCA \cong \angle CAB$.

Step 4: $AC \cong AC$.

Step 5: $\triangle ADC \cong \triangle CBA$ by SAS.

Step 6: $\angle DAC \cong \angle BCA$ as corresponding parts of congruent triangles.

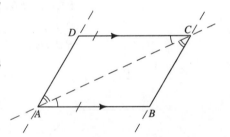

Figure 5-26

Step 7: Since the congruent angles $\angle DAC$ and $\angle BCA$ are also the alternate interior angles cut from lines $\overleftrightarrow{AD}$ and $\overleftrightarrow{BC}$ by transversal $\overleftrightarrow{AC}$, then AD must be parallel to BC.

Step 8: The quadrilateral is a parallelogram by definition, since both sets of opposite sides are parallel.

PROBLEM 5-6 Show that if the diagonals of a quadrilateral bisect each other, then the quadrilateral is a parallelogram. This is the converse of the theorem proved in Example 5-7 that says the two diagonals of a parallelogram bisect each other.

Solution

Step 1: Draw a quadrilateral $\square ABCD$ as in Figure 5-27 such that diagonals AC and DB bisect each other at P.

Step 2: $DP \cong PB$ and $AP \cong PC$ as given.

Step 3: $\angle APB \cong \angle CPD$ as vertical angles formed by two intersecting lines.

Step 4: $\triangle APB \cong \triangle CPD$ by SAS.

Step 5: $\angle PAB \cong \angle PCD$ as corresponding parts of congruent triangles.

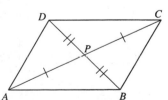

Figure 5-27

Step 6: DC is parallel to AB since the congruent angles $\angle PAB$ and $\angle PCD$ are the alternate interior angles cut from $\overleftrightarrow{DC}$ and $\overleftrightarrow{AB}$ by transversal $\overleftrightarrow{AC}$.

Step 7: $DC \cong BA$ as corresponding parts of congruent triangles.

Step 8: Since DC and AB are both parallel and congruent, the quadrilateral is a parallelogram by the results of Problem 5-5.

PROBLEM 5-7   A certain rhombus has diagonals 6 and 12 cm long. (**a**) Find the area. (**b**) Write a formula for finding the area of a rhombus from the lengths of its diagonals.

Solution

(**a**) Since the diagonals of a rhombus are perpendicular bisectors of each other, they cut the rhombus into four congruent right triangles whose legs have lengths equal to one-half the lengths of the diagonals. Thus, in this case, the legs of the right triangles have lengths of 3 and 6 cm. The area of this rhombus is therefore four times the areas of these triangles:

$$\frac{4(3 \text{ cm})(6 \text{ cm})}{2} = 36 \text{ cm}^2$$

(**b**) Using the reasoning in part (**a**) and letting d_1 and d_2 be the lengths of the diagonals,

$$A(\text{rhombus}) = 4\left(\frac{1}{2}\right)\left(\frac{d_1}{2}\right)\left(\frac{d_2}{2}\right)$$

$$= \frac{d_1 d_2}{2}$$

Thus, for the rhombus in this problem, the area is $(6 \text{ cm})(12 \text{ cm})/2 = 36 \text{ cm}^2$.

PROBLEM 5-8   In an isosceles trapezoid the lengths of the bases are 3 and 9 while the height is 4 units. Find the length of the two nonparallel sides.

Solution   You can probably do this problem quickly by using what you know about isosceles trapezoids and the Pythagorean Theorem, but let's lay out the reasoning anyway, just to be sure.

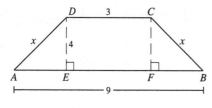

Figure 5-28

Step 1: Draw the given trapezoid $\square ABCD$ as in Figure 5-28, where $CD = 3$, $AB = 9$, and $DA \cong BC$ (given). Let x be the length of the two nonparallel sides $DA \cong BC$.

Step 2: Since the height of a trapezoid is defined as the length of a perpendicular dropped from any point on one base to another, you can drop heights from the vertices D and C to points E and F, respectively, on AB such that $\angle AED$ and $\angle BFC$ are right angles (by definition) and $DE \cong CF = 4$ (by definition and given).

Step 3: $\triangle AED$ and $\triangle BFC$ are right triangles and $\triangle AED \cong \triangle BFC$ by hypotenuse-leg.

Step 4: $AE \cong BF$ as corresponding parts of congruent triangles. Let their length be y.

Step 5: Since $\square EFCD$ is a rectangle, $EF \cong CD$ and has length 3. Thus

$$y + 3 + y = 9$$

$$y = 3$$

Step 6: Since x is the length of the hypotenuse of the triangles $\triangle AED \cong \triangle BFC$, from the Pythagorean Theorem

$$x^2 = 4^2 + 3^2 = \sqrt{25} = 5$$

So the two nonparallel sides $DA \cong BD$ are 5 units in length.

Review Exercises

EXERCISE 5-1 Which of the figures in Figure 5-29 are quadrilaterals?

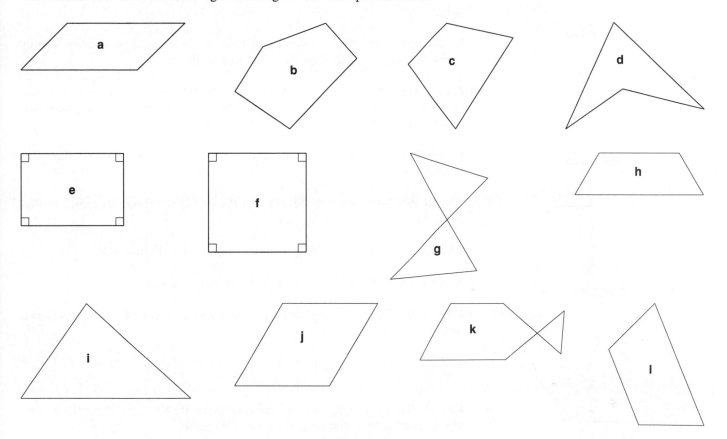

Figure 5-29

EXERCISE 5-2 Which of the figures in Figure 5-29 are rectangles?

EXERCISE 5-3 Which of the figures in Figure 5-29 are parallelograms?

EXERCISE 5-4 Which of the figures in Figure 5-29 are trapezoids?

EXERCISE 5-5 Which of the figures in Figure 5-29 are rhombuses?

EXERCISE 5-6 Which of the figures in Figure 5-29 are squares?

EXERCISE 5-7 Find the missing interior angle in the quadrilateral of Figure 5-30.

EXERCISE 5-8 Find the area of the quadrilateral in Figure 5-30.

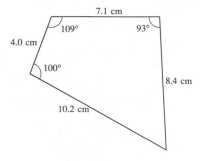

Figure 5-30

EXERCISE 5-9 Prove that if a quadrilateral $\square ABCD$ has $AB \cong CD$ and $AD \cong BC$, then it is a parallelogram. This fact is the converse of the theorem proved in Example 5-5 that says the opposite sides of a parallelogram are congruent.

EXERCISE 5-10 For isosceles trapezoid $\square ABCD$ in Figure 5-31, show that the base angles $\angle DAB \cong \angle CBA$.

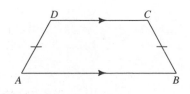

Figure 5-31

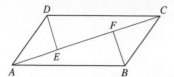

Figure 5-32

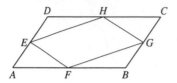

Figure 5-33

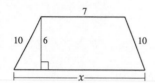

Figure 5-34

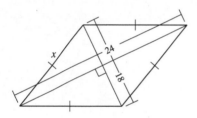

Figure 5-35

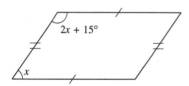

Figure 5-36

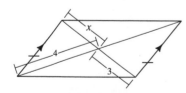

Figure 5-37

EXERCISE 5-11   Prove that the diagonals of an isosceles trapezoid are congruent.

EXERCISE 5-12   Show that a kite has two pairs of congruent consecutive sides.

EXERCISE 5-13   Quadrilateral $\square ABCD$ of Figure 5-32 is a parallelogram. Show that if $AE \cong FC$, then DE is parallel to BF.

EXERCISE 5-14   In Figure 5-33, quadrilateral $\square ABCD$ is a parallelogram. Points E, F, G, and H are the midpoints of the respective sides. Show that quadrilateral $\square EFGH$ is also a parallelogram.

EXERCISE 5-15   Find the length x of the base of the trapezoid shown in Figure 5-34.

EXERCISE 5-16   Find the length x of the side of the rhombus shown in Figure 5-35.

EXERCISE 5-17   Find the area of the trapezoid in Figure 5-34.

EXERCISE 5-18   Find the area of the rhombus in Figure 5-35.

EXERCISE 5-19   Find the measure of the angle x in the parallelogram shown in Figure 5-36.

EXERCISE 5-20   Find the length x in the parallelogram shown in Figure 5-37.

EXERCISE 5-21   The sides of a parallelogram are 8 cm and 6 cm. They meet at a 60° angle. Find the area of the parallelogram.

EXERCISE 5-22   Find the area of a parallelogram with height 12 in. and base for this altitude of 18 in.

EXERCISE 5-23   The area of a parallelogram is 48 in.² Find the altitude's length if the base for this altitude is 6 inches.

EXERCISE 5-24   A parallelogram has sides of 14 inches and 20 inches and an area of 168.5 in.² Find the interior angles for this parallelogram.

EXERCISE 5-25   A rhombus has sides of 3 and an interior angle of 42°. Find its area.

EXERCISE 5-26   The diagonals of a rhombus are 5 inches and 8 inches. (**a**) Find its area. (**b**) Find its side.

EXERCISE 5-27   A rhombus has sides of 4 cm. The longer diagonal is twice the length of the shorter diagonal. (**a**) Find the length of the diagonals. (**b**) Find the area of the rhombus.

EXERCISE 5-28   The bases of a trapezoid measure 3.2 inches and 2.8 inches. Its altitude measures 7.0 inches. Find its area.

EXERCISE 5-29   An isosceles trapezoid has base angle 45°. The bases measure 12 meters and 20 meters. Find (**a**) the length of the other two sides and (**b**) the area.

EXERCISE 5-30 The perimeter (combined sides) of a rhombus measures 48 cm and one diagonal measures 8 cm. (**a**) Find the other diagonal. (**b**) Find the area of the rhombus.

EXERCISE 5-31 The bases of an isosceles trapezoid measure 28 and 8 inches. One base angle measures 47°. (**a**) Find the height. (**b**) Find the area. (**c**) Find the length of the diagonals.

Answers to Review Exercises

5-1	a, c, d, e, f, h, j, l		**5-20**	3	
5-2	e, f		**5-21**	41.57 cm²	
5-3	a, e, f, j		**5-22**	216 in²	
5-4	h, l		**5-23**	8 in.	
5-5	f, j		**5-24**	37°, 143°	
5-6	f		**5-25**	6.02	
5-7	58°		**5-26**	(**a**) 20 in² (**b**) 4.72	
5-8	49.76 cm²		**5-27**	(**a**) 3.58 cm (**b**) 12.8 cm²	
5-15	23		**5-28**	21 in²	
5-16	15		**5-29**	(**a**) $4\sqrt{2}$ or 5.66 m (**b**) 64 m²	
5-17	90 square units		**5-30**	(**a**) 22.63 cm (**b**) 90.5 cm²	
5-18	216 square units		**5-31**	(**a**) 10.72 in. (**b**) 193.03 in² (**c**) 13.38 in.	
5-19	55°				

6 POLYGONS

THIS CHAPTER IS ABOUT

☑ *n*-Sided Figures: Polygons
☑ Regular Polygons
☑ Inscribed and Circumscribed Circles

6-1. *n*-Sided Figures: Polygons

Having studied first three-sided and then four-sided figures, it would seem logical to go on to five-sided figures, then six-sided ones, and so on. But we won't. Instead, we're going to drop back and consider the general case of many-sided figures—polygons—which includes the special cases of triangles and quadrilaterals.

A. Defining and describing a polygon

The name *polygon* means, literally, "many-angled" (*poly* = many; *gon* = angle), but we think of a polygon as a many- or *n*-sided figure, where $n \geq 3$. More formally,

- A **polygon** $E_1E_2, E_2E_3, \ldots, E_{n-1}E_n, E_nE_1$ is a figure formed by joining n line segments at their coplanar endpoints (E) such that the line segments do not intersect anywhere except at their endpoints, and no three of the endpoints are collinear.

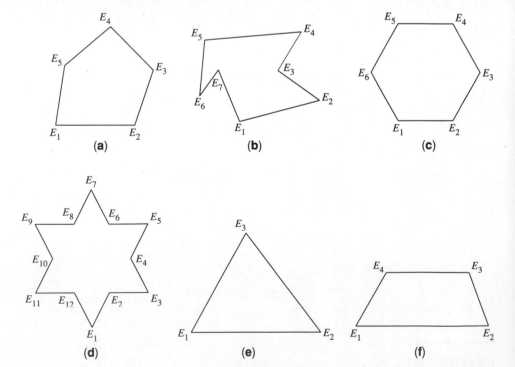

Figure 6-1

The figures shown in Figure 6-1 are all polygons—including the triangle (**e**) and the quadrilateral (**f**).

The figures shown in Figure 6-2, however, are not polygons because they do not meet the criteria: (**a**) and (**c**) have line segments intersecting at points other than their endpoints, and (**b**) has three collinear endpoints, *C*, *E*, and *D*.

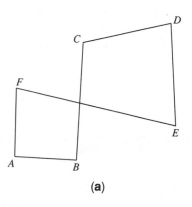
(a)

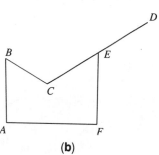
(b)

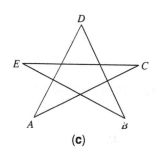
(c)

Figure 6-2

We can describe polygons with the same terms that describe quadrilaterals. The line segments making up the polygon are its sides, and the endpoints of these line segments are its *vertices*. *Consecutive vertices* are the endpoints of a side. (We meet consecutive vertices one after the other as we "walk around" the edge of the polygon.) *Consecutive sides* are sides that have a common endpoint. (We also meet these one after the other as we "walk around" the edge of the polygon.) A *diagonal* of a polygon is a line segment that joins two nonconsecutive vertices. For example, *AC*, *AD*, and *AE* are the diagonals of the polygon shown in Figure 6-3. Finally, the actual sides together make up the *perimeter* of the polygon—although sometimes the word "perimeter" is also used for the combined lengths of all the sides.

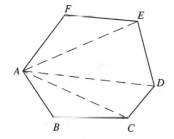

Figure 6-3

B. Naming polygons

The name for a three-sided polygon is a triangle, and the name for a four-sided polygon is a quadrilateral. We're used to those names, so we don't usually refer to these figures any other way. But if we have a polygon where $n \geq 5$, we can name it as an ***n*-gon**, where *n*- is replaced by the Greek prefix for that number. Thus, for sides five through twelve, we have

pentagon	five sides	**nonagon**	nine sides
hexagon	six sides	**decagon**	ten sides
heptagon	seven sides	**undecagon**	eleven sides
octagon	eight sides	**dodecagon**	twelve sides

note: After twelve, our Greek tends to fail us, so we resort to saying 13-gon, 14-gon, and so on.

In Figure 6-1, for example, (**a**) is a pentagon, (**b**) is a heptagon, (**c**) is a hexagon, and (**d**) is a dodecagon.

EXAMPLE 6-1: Which of the figures in Figure 6-4 are polygons? Give the proper name for each of the polygons.

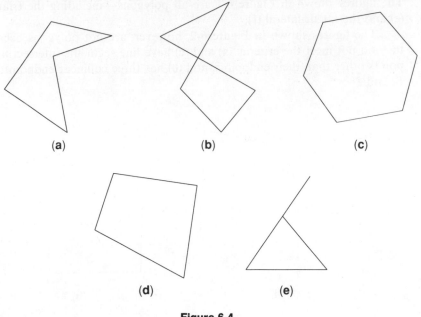

Figure 6-4

Solution: Figures (**a**), (**c**), and (**d**) are polygons. Figure (**a**) is a pentagon; figure (**c**) is a hexagon; and figure (**d**) is a 4-gon, which is usually called a quadrilateral.

C. Classifying polygons

1. Convexity and concavity

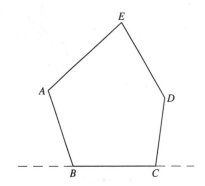

Figure 6-5

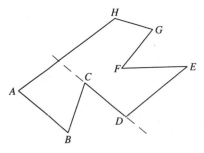

Figure 6-6

Consider a polygon line *ABCDE* shown in Figure 6-5. If we extend one side, say *BC*, to a line $\overleftrightarrow{BC}$, all the remaining vertices lie on one side of the line. If this happens for every side, then the polygon is a *convex* polygon; that is,

- If all the vertices of a polygon lie on or on one side of an extension of every side, then the polygon is **convex**.

Now consider a polygon like *ABCDEFGH* shown in Figure 6-6. This figure is a concave polygon.

- When at least one side of a polygon can be extended so that the remaining vertices fall on both sides of the line, then the polygon is **concave**.

If we were to walk around the edge of a convex polygon, we would turn left at every vertex, or else we would always turn right. But a walk around the edge of a concave polygon would require that we turn *both* left and right.

note: Convex polygons are the ones you'll deal with most frequently. Concave polygons are much less common—and they can be a real pain, involving such awkward concepts as negative exterior angles and interior angles greater than 180°.

    WE WILL TREAT ONLY *CONVEX* POLYGONS IN THE REST OF THIS CHAPTER!

If a concave polygon does come up, you can handle it by drawing its diagonals, thereby cutting it up into lots of (manageable) triangles.

EXAMPLE 6-2: Classify the polygons shown in Figure 6-4 as convex or concave.

Solution: Figure (a) is concave, and figures (c) and (d) are convex.

2. Angles of a convex polygon

At each vertex of a convex polygon two sides form an angle between 0° and 180°; this angle is an **interior angle of the polygon**. For example, $\angle ABC$, $\angle BCD$, $\angle CDE$, $\angle DEA$, and $\angle EAB$ are the interior angles for the pentagon shown in Figure 6-7. The **interior of the convex polygon** is the set of all the points common to the interiors of the interior angles. This tells us that if two points P_1 and P_2 lie in the interior of a convex polygon, then the line segment joining these two points $P_1 P_2$ also lies completely in the interior of the polygon.

Points that are not on the perimeter or in the interior of a polygon lie in the exterior of the polygon. Thus, if we extend one side of a polygon past a vertex, the angle at the vertex between the extended side and the nonextended side is called an **exterior angle of the polygon**. For example, $\angle 1$, $\angle 2$, $\angle 3$, $\angle 4$, and $\angle 5$ are the exterior angles for the pentagon shown in Figure 6-7.

note: As Figure 6-7 shows, there are really two possible ways of forming an exterior angle at each vertex of a polygon. For example, $\angle 1'$, $\angle 2'$, $\angle 3'$, $\angle 4'$, and $\angle 5'$ also fit our description of an exterior angle. But since such exterior angles as $\angle 1$ and $\angle 1'$ are both supplementary to the same interior angle, these exterior angles are always congruent. For this reason, we usually treat the two possible exterior angles at one vertex as one and the same angle, speaking only of "*the* exterior angle."

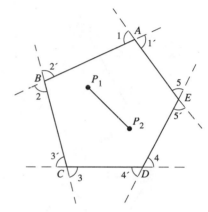

Figure 6-7

If a convex polygon has n sides—and, of course, n vertices—then for any vertex there are $n - 3$ vertices that are not consecutive with it. That means that we can draw $n - 3$ diagonals from any one vertex, as shown in Figure 6-8. These diagonals cut the polygon up into $n - 2$ triangles. Then, if we add up all the interior angles of all these triangles, we have the sum of the interior angles of the convex polygon. And since the interior angles of any triangle sum to 180°, and there are $n - 2$ such triangles, we see that

- The measures of the interior angles of an n-sided convex polygon always sum to $(n - 2)180°$.

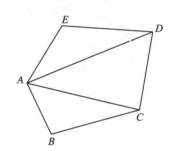

Figure 6-8

EXAMPLE 6-3: (a) Classify the angles of the hexagon shown in Figure 6-9. (b) What is the sum of the measures of the interior angles?

Solution

(a) $\angle 1$, $\angle 4$, and $\angle 3$ are exterior angles; $\angle 2$ and $\angle 5$ are interior angles.
(b) There are six vertices, so the sum of the measures of the interior angles must be

$$(6 - 2)180° = 720°$$

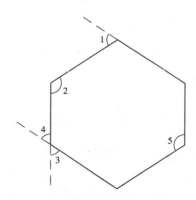

Figure 6-9

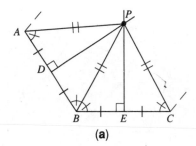

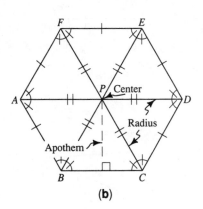

Figure 6-10

6-2. Regular Polygons

When we think of polygons, we most often think of regular polygons:

- A polygon is a **regular polygon** if all of its sides are congruent and all of its angles are congruent.

That is, a regular polygon is one in which all of the sides are of equal length and all of the angles have equal measure. Thus, for example, a square is a regular polygon, while the rhombus—in general—is not. We see examples of regular polygons everywhere, from the architecture of buildings to the structures of crystals.

A. Properties of a regular polygon

If we draw the two perpendicular bisectors of two consecutive sides, say AB and BC, of a regular polygon as in Figure 6-10a, we see that these bisectors must meet at some point P. Then, if we draw the line segments PA, PB, and PC, we see that we have four right triangles, $\triangle PAD$, $\triangle PBD$, $\triangle PBE$, and $\triangle PCE$, which make up two larger triangles, $\triangle PAB$ and $\triangle PCB$. We know that (**1**) $BE = EC$ and $\angle PEB \cong \angle PEC$ by construction and (**2**) $PE \cong PE$, so $\triangle PBE \cong \triangle PCE$ by SAS. Therefore, $PB \cong PC$ as congruent sides of congruent triangles, so $\triangle PBC$ must be an isosceles triangle with $\angle PBC \cong \angle PCB$. By the same reasoning, $\triangle PAB$ must also be an isosceles triangle, where $PA \cong PB$. Hence $PA \cong PC$, since both of these line segments are congruent to PB. And since $AB \cong BC$ as sides of a regular polygon, $\triangle PAB \cong \triangle PCB$ by SSS. This means that we have two congruent isosceles triangles, where $\angle PAB \cong \angle PBA \cong \angle PBC \cong \angle PCB$.

Now we know that all of the angles of a regular polygon are congruent by definition. Thus, $\angle A \cong \angle B \cong \angle C \cong \angle D$, etc. So if we were to draw the next consecutive side (and the next, and the next, . . .) and repeat the constructions above, we would see that *all* the angles formed by a line segment drawn from P to a vertex are also congruent—and they will all be equal in measure to $\angle PBA$.

When the perpendicular bisectors of every side of an n-sided regular polygon are drawn, they meet at a point P, which forms the common apex of n congruent isosceles triangles. This apex point is called the **center of the regular polygon**. The sides of the n isosceles triangles are all of equal length; this length is called the **radius of the regular polygon**. Finally, any perpendicular bisector drawn from the center of the polygon to the side is called the **apothem** (or **altitude**) of the regular polygon. (See Figure 6-10b.)

EXAMPLE 6-4: Find the number of degrees in one of the interior angles of a regular n-sided polygon.

Solution

Step 1: Let any interior angle of an n-sided regular polygon have a measure of x degrees.

Step 2: In any n-sided polygon there are n interior angles, which sum to $(n-2)180°$. Thus $nx = (n-2)180° = 180n° - 360°$

Step 3: All of the angles in a regular polygon are of equal measure, so

$$x = 180° - \frac{360°}{n}$$

- Each interior angle of an *n*-sided regular polygon has a measure of $180° - \dfrac{360°}{n}$.

EXAMPLE 6-5: Find the measure of the interior angles for a regular polygon of (**a**) five sides, (**b**) six sides, (**c**) eight sides, and (**d**) ten sides.

Solution

(**a**) For a pentagon, the measure of the interior angle is $180° - \dfrac{360°}{5} = 108°$.

(**b**) For a hexagon, the measure of the interior angle is $180° - \dfrac{360°}{6} = 120°$.

(**c**) For an octagon, the measure of the interior angle is $180° - \dfrac{360°}{8} = 135°$.

(**d**) For a decagon, the measure of the interior angle is $180° - \dfrac{360°}{10} = 144°$.

B. Area of a regular polygon

We know that we can cut an *n*-sided regular polygon into *n* congruent isosceles triangles. Thus the area of an *n*-sided regular polygon must be *n* times the area of these triangles. We also know that the apothem of the polygon is the altitude of one of these triangles, while the side of the polygon is the base of the triangle. Thus, if the apothem has length *h* and the side of the polygon has length *s*, the area of an *n*-sided regular polygon must be

AREA OF A REGULAR POLYGON
$$A(\text{polygon}) = n\,\frac{hs}{2}$$

Or we can say that $A(\text{polygon}) = hp/2$, where $p = ns$ is the perimeter of the regular polygon.

This is a handy formula for finding the area of a regular polygon if we know the length of the apothem *h* as well as the length of the side *s*. But more often we know only the number of sides (angles) *n* and the length of of a side. To find a formula for the area of a regular polygon that depends on these two numbers, let's look at Figure 6-11, which shows one of the isosceles triangles that make up a regular polygon.

Let the polygon have *n* sides of length *s* and apothem *AE* of length *h*. Then

AE is a perpendicular bisector of *BC*, so $BE \cong EC = s/2$;
r is the length of the radius of the polygon, so $AB \cong AC = r$;
$\angle \alpha$ is one-half of the interior angle of the polygon, so

$$\angle \alpha = \left(\frac{1}{2}\right)(\text{interior angle})$$

$$= \left(\frac{1}{2}\right)\left(180° - \frac{360°}{n}\right)$$

$$= 90° - \frac{180°}{n}$$

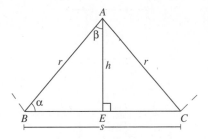

Figure 6-11

Then, since $\triangle ABE$ is a right triangle, the measure of $\angle \beta$ must be

$$\angle \beta = 180° - 90° - \left(90° - \frac{180°}{n}\right) = \frac{180°}{n}$$

And the tangent of $\angle \beta$ is a trigonometric ratio defined as the measure of the opposite leg divided by the measure of the adjacent leg, so

$$\frac{BE}{AE} = \tan \beta = \frac{s/2}{h}$$

Thus

$$h = \frac{s/2}{\tan \beta} = \frac{s/2}{\tan(180°/n)}$$

So the area of the triangle is

$$A(\triangle ABC) = \frac{hs}{2} = \left(\frac{s/2}{\tan(180°/n)}\right)\left(\frac{s}{2}\right)$$

$$= \frac{(s/2)^2}{\tan(180°/n)}$$

Then, since there are n triangles in the n-sided polygon, the area of the polygon is

**AREA OF A
REGULAR POLYGON** $A(\text{polygon}) = \dfrac{n(s/2)^2}{\tan(180°/n)}$

EXAMPLE 6-6:   Find the length of the perimeter and the area for (**a**) a regular hexagon of side 4 cm and (**b**) a regular octagon of side 4 cm.

Solution

(**a**) $p = ns = 6(4 \text{ cm}) = 24 \text{ cm}$

$$A(\text{hexagon}) = \frac{6(s/2)^2}{\tan(180°/6)} = \frac{6(4/2 \text{ cm})^2}{\tan 30°} = 41.57 \text{ cm}^2$$

(**b**) $p = ns = 8(4 \text{ cm}) = 32 \text{ cm}$

$$A(\text{octagon}) = \frac{8(s/2)^2}{\tan(180°/8)} = \frac{8(4/2 \text{ cm})^2}{\tan 22.5°} = 77.25 \text{ cm}^2$$

6-3. Inscribed and Circumscribed Circles

Regular polygons are closely related to circles. We see this close relationship when we draw one of these figures *in* (*in*scribe) or *around* (*circum*scribe) the other so that one touches the other at every possible point. We can say, in fact, that any regular polygon actually determines two circles—a circle that touches the polygon at the midpoint of all its sides and a circle that touches the polygon at all its vertices.

A. Inscribed circles

- The **inscribed circle** determined by a regular polygon just touches the polygon at the midpoint of each of its sides and is the largest circle that can be drawn completely *inside* that polygon.

EXAMPLE 6-7: (a) Draw the inscribed circle for a regular hexagon. Explain your drawing. (b) List the determining conditions of the inscribed circle for a regular polygon.

Solution

(a) Start by drawing a regular hexagon $E_1E_2E_3E_4E_5E_6$, as in Figure 6-12. Then, since the hexagon has six equal sides, you can cut it up into six congruent isosceles triangles with a common apex C. This apex C is the center of the hexagon—and is also one endpoint of the apothem CM, which is the perpendicular bisector of a side. Therefore, because the (shortest) distance between a point and a line is the length of the perpendicular line segment between them, you know that all the points on a side must lie as far away from C as M or farther. Thus the edge of the circle may not go beyond the midpoint M (of a side). Now place a compass point on C and spread the legs so that the pencil point is on any M. Swing the compass and draw the circle. This circle touches every side of the hexagon at its midpoint M.

(b) From this explanation you can easily list the two determining conditions of an inscribed circle for a regular polygon:

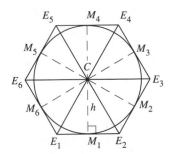

Figure 6-12

- The center of the polygon and the center of the inscribed circle must be the same point.
- The radius of the inscribed circle and the apothem of the polygon must be equal in length.

B. Circumscribed circles

- A **circumscribed circle** for a regular polygon touches the polygon at each of its vertices and is the smallest circle that can be drawn completely *around* the outside of the polygon.

EXAMPLE 6-8: (a) Draw the circumscribed circle for a regular hexagon. Explain your drawing. (b) List the determining conditions of a circumscribed circle for a regular polygon.

Solution

(a) Draw a regular hexagon as in Figure 6-13. Label the vertices E (E_1 through E_6) and the center C as in Figure 6-12. Then, if you cut this hexagon up into its six congruent isosceles triangles, you can see that each of the vertices lies at a distance r from C—while all the other points on any side must lie at a distance less than r from C. Now place a compass point on C, with the pencil point on one of the vertices. Swing the compass and draw the circle. This circle touches the hexagon at every vertex.

(b) The determining conditions of a circumscribed circle for a regular polygon are as follows:

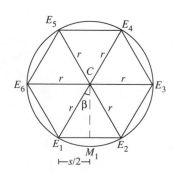

Figure 6-13

- The center of the polygon and the center of the circumscribed circle must be the same point.
- The radius of the polygon and the radius of the circumscribed circle must be equal.

note: If we let r be the radius of the polygon, then r is also the radius of the circle and

$$\sin \beta = \frac{s/2}{r}$$

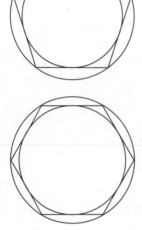

Figure 6-14

where β is $\angle E_1CM_1$ and s is the length of a side. Thus

$$r = \frac{s/2}{\sin \beta}$$

$$= \frac{s/2}{\sin(180°/n)}$$

(See Section 6-2B.)

---

caution: Don't let the words "inscribe" and "circumscribe" throw you. If you see a polygon with a circle inscibed in it, you may describe what you see as the inscribed circle for the polygon (as we have done here) or you may describe the whole thing as a polygon circumscribed around a circle. Similarly, a circumscribed polygon for a circle may be said to be a circle inscribed in a polygon. The thing to remember is this: While a given regular polygon can have only one circle inscribed in it and only one circle circumscribed around it, a given circle may have many different polygons inscribed in and circumscribed around it. Notice, for instance, that Figure 6-14 shows three outer circles of equal diameter, but each circle has a different polygon inscribed in it—a pentagon, a hexagon, and an octagon. Also notice that as the number of sides increases, the circumscribed circle of the polygon gets closer to the inscribed circle.

SUMMARY

1. A polygon is a figure formed from joining $n \geq 3$ nonintersecting noncollinear line segments at their endpoints; the line segments form the sides and their endpoints form the vertices of the polygon.
2. Diagonals connect nonconsecutive vertices of a polygon.
3. Line segments joining two interior points of a convex polygon lie completely inside the polygon.
4. A polygon with n sides and n vertices is an n-gon; where $5 \leq n < 13$, a Greek prefix is usually substituted for the integer n.
5. An n-sided convex polygon can be cut up into $n - 2$ triangles by diagonals.
6. Regular polygons have sides of equal length and congruent interior angles.
7. The perpendicular bisectors of all the sides of a regular polygon meet at one point—the center of the polygon.
8. The apothem of a regular polygon is the perpendicular line segment drawn from the center to the midpoint of any side.
9. The radius of a regular polygon is the distance from the center to any vertex.
10. Each interior angle of a regular polygon of n sides is equal to $180° - \dfrac{360°}{n}$.
11. The length of the perimeter of a regular polygon of n sides with sides of length s is ns.
12. The area of a regular polygon with n sides of length s is

$$\frac{n(s/2)^2}{\tan(180°/n)}$$

13. The inscribed circle for a regular polygon, which just touches the polygon at the midpoint of each side, is the largest circle contained inside the polygon.
14. The circumscribed circle for a regular polygon, which touches all of the vertices of the polygon, is the smallest circle that contains the polygon.

RAISE YOUR GRADES

Can you . . . ?

☑ recognize when a figure is or is not a polygon
☑ identify the proper name for a polygon
☑ draw a diagonal of a polygon
☑ recognize a regular polygon
☑ find the measure of the interior angle of an *n*-sided polygon
☑ find the measure of the exterior angle of an *n*-sided regular polygon
☑ find the apothem of a regular polygon
☑ find the length of the perimeter of a regular polygon, given the length of the side
☑ find the area of a regular polygon
☑ draw a circle circumscribing a regular polygon
☑ draw a circle inscribed in a regular polygon

SOLVED PROBLEMS

PROBLEM 6-1   Find the measure of $\angle EAB$ in Figure 6-15.

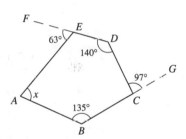

Figure 6-15

Solution

Step 1: $\angle AED$ is supplementary to $\angle AEF$ and $\angle BCD$ is supplementary to $\angle GCD$, so

$$\angle AED = 180° - 63° = 117° \quad \text{and} \quad \angle BCD = 180° - 97° = 83°$$

Step 2: The measures of all the angles in a polygon sum to $(n - 2)(180°)$. So, since this is a pentagon, $n = 5$ and all the interior angles sum to $(5 - 2)(180°) = 540°$.

Step 3: Let x be the measure of $\angle EAB$ in degrees. Then

$$x + 117° + 140° + 83° + 135° = 540°$$

$$x = 65°$$

PROBLEM 6-2   Show that the sum of the measures of the exterior angles (only one for each vertex) of a convex polygon is 360°.

Solution   Each exterior angle at a vertex is supplementary to an interior angle at that vertex. Then, since there are *n* vertices.

$$\text{sum of exterior angles} + \text{sum of interior angles} = n(180°)$$

$$\text{sum of exterior angles} + (n - 2)180° = n(180°)$$

or

$$\text{sum of exterior angles} = 2(180°) = 360°$$

PROBLEM 6-3   The **central angle** of a regular polygon is the angle formed at the center of the polygon by two radii (the plural of "radius") drawn to consecutive vertices. Find the measure of the central angle of an *n*-sided regular polygon.

Solution   The central angle is the apex angle of an isosceles triangle whose two base angles sum to $180° - \dfrac{360°}{n}$.

Thus the measure of the central angle $= \dfrac{360°}{n}$.

PROBLEM 6-4   Find **(a)** the length of the perimeter, **(b)** the length of the apothem, and **(c)** the area of a regular dodecagon whose sides measure 3 cm.

Solution

(a) The length of the perimeter p is

$$p = ns = 12(3 \text{ cm}) = 36 \text{ cm}$$

(b) The length of the apothem h is

$$h = \frac{s/2}{\tan(180°/n)} = \frac{3/2 \text{ cm}}{\tan(180°/12)} = \frac{3 \text{ cm}}{2 \tan 15°} = 5.6 \text{ cm}$$

(c) The area A of the dodecagon is

$$A(\text{dodecagon}) = \frac{hp}{2} = \frac{(5.6 \text{ cm})(36 \text{ cm})}{2} = 100.8 \text{ cm}^2$$

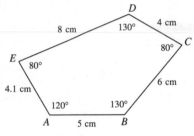

Figure 6-16

PROBLEM 6-5   Find the area of the pentagon shown in Figure 6-16.

Solution

Step 1:   Draw diagonals EB and EC to form three triangles, $\triangle AEB$, $\triangle BEC$, and $\triangle DEC$.

Step 2:   From the Law of Sines [$A(\text{triangle}) = (ab \sin \alpha)/2$, where a and b are sides of the triangle and α is the angle between a and b; see Section 4-3]

$$A(\triangle AEB) = \frac{(4.1 \text{ cm})(5 \text{ cm})(\sin 120°)}{2} = 8.78 \text{ cm}^2$$

and

$$A(\triangle DEC) = \frac{(8 \text{ cm})(4 \text{ cm})(\sin 130°)}{2} = 12.26 \text{ cm}^2$$

Step 3a: Let y be the length of EB and z be the length of EC.

Step 3b: From the Law of Cosines [$c^2 = a^2 + b^2 - 2ab \cos \gamma$, where a, b, and c are sides of the triangle and γ is the angle opposite c; see Section 4-5]

$$y^2 = (4.1 \text{ cm})^2 + (5 \text{ cm})^2 - 2(4.1 \text{ cm})(5 \text{ cm})(\cos 120°) = 62.31 \text{ cm}^2$$

$$z^2 = (4 \text{ cm})^2 + (8 \text{ cm})^2 - 2(4 \text{ cm})(8 \text{ cm})(\cos 130°) = 121.14 \text{ cm}^2$$

Thus, $y = \sqrt{62.31 \text{ cm}^2} = 7.89 \text{ cm}$ and $z = \sqrt{121.14 \text{ cm}^2} = 11.01 \text{ cm}$.

Step 3c: Again, from the Law of Cosines,

$$(6 \text{ cm})^2 = y^2 + z^2 - 2yz \cos \angle BEC$$

$$36 \text{ cm}^2 = (7.89 \text{ cm})^2 + (11.01 \text{ cm})^2 - 2(7.89 \text{ cm})(11.01 \text{ cm})(\cos \angle BEC)$$

Thus $\cos \angle BEC = 0.8487$ and $\angle BEC = 31.93°$.

Step 3d: From the Law of Sines,

$$A(\triangle BEC) = \frac{(7.89 \text{ cm})(11.01 \text{ cm})(\sin 31.93°)}{2} = 22.97 \text{ cm}^2$$

Step 4:

$$A(\text{pentagon}) = A(\triangle AEB) + A(\triangle DEC) + A(\triangle BEC)$$

$$= 8.88 \text{ cm}^2 + 12.26 \text{ cm}^2 + 22.97 \text{ cm}^2$$

$$= 44.11 \text{ cm}^2$$

Review Exercises

EXERCISE 6-1   Find the central angles of a regular polygon with (**a**) 5 sides, (**b**) 8 sides, and (**c**) 20 sides.

EXERCISE 6-2   Find the interior angles of a regular polygon with (**a**) 7 sides, (**b**) 12 sides, and (**c**) 24 sides.

EXERCISE 6-3   Find the exterior angles of a regular polygon with (**a**) 6 sides, (**b**) 8 sides, and (**c**) 12 sides.

EXERCISE 6-4   Find $\angle 6$ and $\angle 7$ of the pentagon shown in Figure 6-17.

EXERCISE 6-5   Find $\angle 4$ as shown in Figure 6-17.

EXERCISE 6-6   Find $\angle 8$ and $\angle 9$ as shown in Figure 6-17.

EXERCISE 6-7   Find $\angle 1$, $\angle 2$, and $\angle 3$ as shown in Figure 6-17.

EXERCISE 6-8   Find the area of the pentagon shown in Figure 6-17.

EXERCISE 6-9   Find the apothem of a regular hexagon whose side is 7 inches long.

EXERCISE 6-10   Find the area of a regular octagon whose side is 3 meters long.

EXERCISE 6-11   Find the length of the perimeter of a regular polygon whose side measures 6 inches if the polygon has (**a**) 3 sides, (**b**) 5 sides, (**c**) 6 sides, and (**d**) 8 sides.

EXERCISE 6-12   The radius of a regular polygon is 3 cm. Find the length of the perimeter of the polygon if it has (**a**) 6 sides, (**b**) 12 sides, and (**c**) 20 sides.

EXERCISE 6-13   The radius of a regular polygon is 3 cm. Find the area of the polygon if it has (**a**) 6 sides, (**b**) 12 sides, and (**c**) 20 sides.

EXERCISE 6-14   Find the length of the diagonals DA and DB of the pentagon shown in Figure 6-16.

EXERCISE 6-15   Find the area of the pentagon shown in Figure 6-15 if the lengths of the sides are $AB = 4.45$ cm, $BC = 3.80$ cm, $CD = 3.70$ cm, $DE = 1.90$ cm, and $EA = 5.40$ cm.

EXERCISE 6-16   Show that the bisector of the interior angles of a regular polygon also meet at one point and divide the regular polygon into n congruent isosceles triangles.

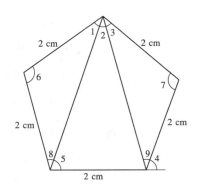

Figure 6-17

Answers to Review Exercises

6-1 (**a**) 72°   (**b**) 45°   (**c**) 18°

6-2 (**a**) 128.57°   (**b**) 150°   (**c**) 165°

6-3 (**a**) 60°   (**b**) 45°   (**c**) 30°

6-4 $\angle 6 = \angle 7 = 180°$

6-5 72°

6-6 $\angle 8 = \angle 9 = 36°$

6-7 $\angle 1 = \angle 2 = \angle 3 = 36°$

6-8 6.88 cm²

6-9 6.06 in.

6-10 43.46 m²

6-11 (a) 18 in. (c) 36 in.
          (b) 30 in. (d) 48 in.

6-12 (a) 18 cm (b) 18.63 cm (c) 18.77 cm

6-13 (a) 23.38 cm² (c) 27.81 cm²
          (b) 27 cm²

6-14 $DA = 8.33$ cm, $DB = 6.61$ cm

6-15 approximately 22.5 or 22.6 cm²

6-16 Draw angle bisectors from two adjacent vertices of the regular polygon. These bisectors meet at some point P. In Figure 6-18 AP and BP are two such bisectors. Next, draw a line segment from this point P to the next adjacent vertex C. Since BP bisects $\angle ABC$, $\angle ABP \cong \angle PBC$. $AB \cong BC$ by the definition of a regular polygon. But

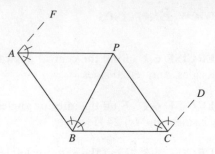

Figure 6-18

$PB \cong PB$ and we have $\triangle ABP \cong \triangle CBP$ by SAS.

Now $\angle PAB \cong \angle PBA$ since AP also bisects $\angle FAB$. As corresponding parts of congruent triangles $\angle PCB \cong \angle PAB$ since $\angle PAB \cong \angle PBA \cong \angle PBC$, we must also have $\angle PCB \cong \angle PBC$. Thus $\triangle APB$ and $\triangle BPC$ are congruent isosceles triangles. Further since $\angle ABC \cong \angle BCD$ by the definition of a regular polygon, we see that PC must also be the angle bisector of $\angle BCD$

Next draw a line segment from P to the next adjacent vertex (either D of F) and repeat this process. Thus all the angle bisectors pass through the common point P.

7 CIRCLES

THIS CHAPTER IS ABOUT

- ☑ **Terminology**
- ☑ **Circumference**
- ☑ **Area of a Circle**
- ☑ **Arc Measurement**
- ☑ **Theorems: Angles of a Circle**
- ☑ **Theorems: Lines of a Circle**
- ☑ **Analytic Expression of a Circle**

7-1. Terminology

A. Definition of a circle

The trouble with circles is that they contain no line segments—not even very short ones. Thus a circle can't be defined the same way a polygon is, even though an *n*-sided polygon looks increasingly like a circle as *n* increases. Instead, we have to think of a circle as a boundary of a round region in the plane, which we can define as a set of points related by distance:

- A **circle** is the set of points in the plane that lie a fixed distance from a specific point.

This fixed distance is the **radius of the circle**, and the specific point is the **center of the circle**. Then we say that the **interior of the circle** is the set of points that lie closer to the center than the radial distance, and the **exterior of the circle** is the set of points that lie farther away from the center than the radial distance.

note: When you refer to the radius or center or interior or exterior *of* a circle, you must be careful not to consider these characteristics as parts of the circle itself. Only those points that meet the conditions of the definition are actually part of the circle. You can, though, think of a circle as dividing the plane into three distinct subsets: the interior the circle itself, and the exterior.

Although not parts of the circle, there are various straight lines, line segments, and angles associated with circles. These are described below and illustrated in Figures 7-1 through 7-5.

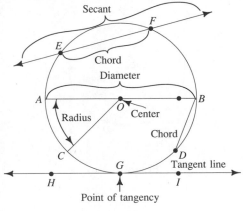

Figure 7-1

B. Lines and line segments

Figure 7-1 shows a circle *O* with the following lines and line segments marked: *radius, diameter, chord, secant,* and *tangent.*

- A line segment is a **radius** if one of its endpoints is the center of a circle and the other endpoint is a point on the circle.

Line segments *OA, OB,* and *OC* are radii of the circle *O* shown in Figure 7-1.

note: The term "radius" can be used two ways. It can mean the *line segment* drawn from the center of a circle to a point on the circle, or it can mean the *distance* (a number) between the center and a point on the circle (i.e., the length of the line segment).

- A line segment is a **chord** if both of its endpoints are points on a circle.

Line segments *BD* and *EF* in Figure 7-1 are chords. In fact, *AB* is also a chord, but it's a special kind of chord called a diameter:

- A line segment is a **diameter** if both of its endpoints are points on a circle and one of the points on the line segment is also the center of the circle.

The diameter is the longest chord that can be drawn in a circle. Its length is twice that of a radius, and every diameter contains two radii.

- A line is a **secant** if the line passes through a circle at two distinct points.

Line $\overleftrightarrow{EF}$ in Figure 7-1 is a secant.

note: Every secant (line) contains a chord (line segment), and every chord is on a secant.

- A line is a **tangent** if it touches (intersects) a circle at only one point. The point at which a tangent intersects a circle is called the **point of tangency**, or the **point of contact**.

Line $\overleftrightarrow{HI}$, which intersects the circle in Figure 7-1 at *G*, is a tangent with point of tangency *G*.

note: You can think of a tangent as a secant that has been steadily moved until the secant's two crossing points merge into one—the point of tangency. [See also Example 7-10.]

C. Arcs and angles

We might think of a piece of a circle that extends from one endpoint of a chord to another endpoint as an *arc*. Consider, for example, Figure 7-2, in which the chord *BD* divides the circle *O* into two unequal pieces. We call the larger piece the *major arc* $\overset{\frown}{BD}$, and we call the smaller piece the *minor arc* $\overset{\frown}{BD}$. Sometimes, of course, the chord is a diameter like *AB* in Figure 7-2. In this case, the arcs are equal and are known as *semicircles*.

But what do we mean exactly when we speak of arcs? This question leads us to some formal definitions of arc terminology—and these definitions have to do with angles, specifically central angles:

- An angle is a **central angle** of a circle if its vertex is the center of the circle and its sides are radii.

An **arc** is a set of points of the circle that are cut off by a central angle:

- A **minor arc** is a set of points of the circle that are on, or in the interior of, a central angle.
  A **major arc** is a set of points of the circle that are on, or in the exterior of, a central angle.
- A **semicircle** is a set of points of the circle that are on, or on one side of, a line containing a diameter.

Now, going back to Figure 7-2 and the arc $\overset{\frown}{BD}$, we can see that the major arc $\overset{\frown}{BD}$ is defined by the sides and exterior of the central angle $\angle BOD$. And we can see that the minor arc $\overset{\frown}{BD}$ is defined by the sides and interior of $\angle BOD$—whose chord is *BD*.

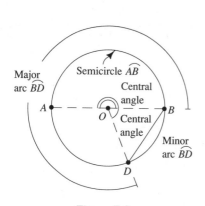

Figure 7-2

note: In general, when we speak of the chord of an arc, we mean the minor arc only.

Similarly, we can see that the two semicircles $\overparen{AB}$ are defined by the central (straight) angle $\angle AOB$.

D. Angles

There are many names for the special angles that can be formed by the chords, secants, and/or tangents of a circle—and there are several ways of classifying these angles. But, in terms of their measurement (which we'll discuss in more detail in Section 7-4), we can group these angles into three classes, depending on where their vertices are: *on* the circle, *interior to* the circle, or *exterior to* the circle.

1. Angles with vertices on the circle

- If an angle has its vertex on the circle such that each of its sides contains another point of the circle, it is called an **inscribed angle** of the circle.

In general, then, an inscribed angle is composed of two secants whose defining rays intersect at a point on the circle, like $\angle ABC$ in Figure 7-3.

note: We can also say that $\angle ABC$ is *inscribed in the major arc* $\overparen{AC}$.

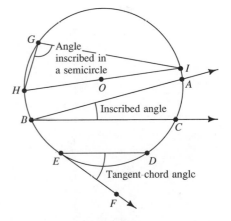

Figure 7-3

Then there's the special case in which the defining rays of the angle intersect the circle at opposite ends of a diameter, as in $\angle HGI$ in Figure 7-3; this angle is said to be *inscribed in a semicircle.*

Finally, there's the oddball case in which a tangent and a secant intersect at a point on the circle, as in $\angle DEF$ in Figure 7-3. This angle, whose vertex is the intersection of a point of tangency and a secant (or chord contained in a secant) with the circle, is known as a **tangent-chord angle.**

note 1: The tangent-chord angle doesn't really fit the definition of an inscribed angle—but it behaves like one in terms of its measurement. It helps to think of the tangent-chord angle as a degenerate inscribed angle, in which one of the defining rays is rotated down to form a tangent, with the vertex at the point of contact.

note 2: The tangent-chord angle is also called the "tangent-secant angle." But since it is possible to have a tangent-secant angle whose vertex is *exterior* to the circle (see Section 7-D3), we'll use the term "tangent-chord" to indicate an angle whose vertex is *on* the circle.

2. Angles with vertices interior to the circle

There are two cases of angles with vertices inside the circle: a general case, in which the vertex can be anywhere in the interior of the circle, and a special case, in which the vertex of the angle is the center of the circle. In the general case, an interior angle is composed of two chords that intersect anywhere in the circle. This general interior angle is illustrated by $\angle FCG$ in Figure 7-4 and is known as a **chord-chord angle.** (Note that all the angles supplementary to and congruent to this angle formed by the chords DG and EF are also chord-chord angles.) In the special case, the chords that make up the angle are intersecting diameters, so that the vertex of the angle is the center of the circle and its defining rays are radii. This angle is, of course, a *central angle*, as illustrated by $\angle AOB$ in Figure 7-4.

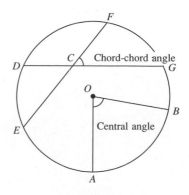

Figure 7-4

3. Angles with vertices outside the circle

Angles whose vertices are outside the circle can be separated into three cases: a general case and two special ones. In the general case, the defining rays of the angle are both secants, which intersect somewhere in the exterior of the circle. This general exterior angle is illustrated by ∠*BCD* in Figure 7-5 and is called a **secant-secant angle**. If, however, one of the rays is rotated so that it becomes a tangent, the angle becomes a **tangent-secant angle**, as illustrated by ∠*IJK* in Figure 7-5. Then, if both of the rays are rotated to form tangents, as in ∠*FGH*, the angle becomes a **tangent-tangent angle**.

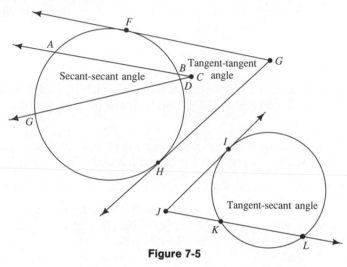

Figure 7-5

note: Each of the angles described in Sections 7-D1 through 7-D3 above can be said to intercept an arc. We say that an *angle intercepts an arc* if the endpoints of the arc lie on the defining rays of the angle, if all the other points in the arc lie in the interior of the angle, and if one point on each ray of the angle is also an endpoint of the arc. Thus, both rays of the angle that intercepts an arc must be some line or line segment that touches or intersects the circle at least once. If one of its rays is not a chord or secant or tangent, an angle cannot be said to intercept an arc (and all bets are off with respect to measurement).

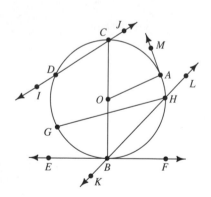

Figure 7-6

EXAMPLE 7-1:   Identify the lines, rays, and line segments associated with the circle shown in Figure 7-6.

Solution: Lines $\overleftrightarrow{IJ}$ and $\overleftrightarrow{KL}$ are secants.
Line $\overleftrightarrow{EF}$ and ray $\overrightarrow{AM}$ are tangents.
Line segments *OA*, *OB*, and *OC* are radii.
Line segment *BC* is a chord that is a diameter.
Line segments *DC* (of the line $\overleftrightarrow{IJ}$), *BH* (of line $\overleftrightarrow{KL}$), and *GH* are chords.

EXAMPLE 7-2:   Identify the angles formed by the rays and line segments associated with the circle shown in Figure 7-7.

Solution: ∠*COD*, ∠*DOE*, ∠*EOB*, ∠*BOC*, and ∠*COE* are central angles.
∠*CDB* and ∠*DBG* are inscribed angles.
∠*BAE* is a tangent-secant angle.
∠*ABD* is a tangent-chord angle.
∠*BHI*, ∠*BHF*, ∠*GHI*, and ∠*GHF* are chord-chord angles.

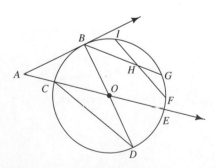

Figure 7-7

7-2. Circumference
A. What a circumference is and how it's calculated

It seems reasonable to think of the circumference of a circle as the "length" of the circle itself. That is, if we think of a circle as a piece of string that we could cut and straighten into a line segment, then the circumference should be the length of the line segment. This makes intuitive sense—and, in fact, it works, every time. But there are a few problems with this common-sensical approach. In the first place, measuring the "length" of one circle (made of string, yet!) doesn't tell us anything about other circles. In the second—and more important—place, we can't properly talk about the "length" of a circle because the idea of length applies only to line segments by definition! (Length is the distance between the two endpoints of a line segment; see Section 1-2.) Now, how do we resolve the dilemma of a measuring method that works and a definition that doesn't?

We resolve this dilemma by "sneaking up" on the circle by way of inscribed polygons. We know that any regular n-gon can be inscribed in a circle so that the radius of the n-gon is equal to the radius of the circle. (See Section 6-2.) We also know that the higher n is (i.e., the more sides the polygon has), the closer the n-gon comes to the circle. (See Section 6-3.) Thus, we can say that the perimeter of an n-gon is an *approximation* to the circumference of a circle—and the approximation gets closer as n gets higher. This "sneaking up," or approximation, process gives us a new way to define the circumference:

$$p \to C$$

where p is the perimeter of an inscribed polygon and C is the circumference of the circle. This definiton, called a *limit*, is read "p approaches C as a limit"; that is,

- The **circumference of a circle** is the limit of the perimeters of the regular polygons inscribed in the circle.

The approximation process is a complicated one, but it gives us a formula we can use for finding the circumference C of a circle with radius r:

CIRCUMFERENCE OF A CIRCLE
$$C = 2\pi r$$

This formula involves a number π (pi), which is defined as the ratio of the circumference to the diameter (d); that is, $\pi = C/d = C/2r$. This ratio is the same for all circles. The value of π is an irrational number $3.14159\ldots$, which is a nonrepeating, nonterminating decimal. Usually, it can be shortened to 3.1416 or 3.14; in crude calculations it can be represented as 22/7, which is a repeating, nonterminating decimal, $3.\overline{142857}\ldots$.

EXAMPLE 7-3: A circle has radius 3 inches. Find the circumference C of the circle.

Solution: Let $r = 3$ inches and approximate π by 22/7. Then

$$C = 2\pi r$$

$$= 2\left(\frac{22}{7}\right)(3 \text{ in.}) = \frac{132}{7} \text{ in.}$$

$$= 18\tfrac{6}{7} \text{ in.}$$

EXAMPLE 7-4: A window frame is to be made in the shape of a semicircle whose inner radius is 4 meters. Find the perimeter of the window.

Solution: The perimeter of the window is the distance around the frame (on the inside). Since the frame is to be semicircular, the perimeter will be the length of one diameter plus one-half the circumference of a circle of radius 4 meters. First,

$$d = 2r = 8 \text{ m}$$

Then

$$C_{\text{semicircle}} = \frac{2\pi r}{2} = \pi r = (4 \text{ m})\left(\frac{22}{7}\right) = \frac{88}{7} \text{ m}$$

Thus

$$p = d + r = 8 \text{ m} + \frac{88}{7} \text{ m} = 20\tfrac{4}{7} \text{ m}$$

B. Where the circumference formula comes from— A trigonometric approach

In order to see where the formula $C = 2\pi r$ comes from, we first have to find out how the perimeter of an *n*-gon is related to its radius. So we begin with a regular *n*-gon cut up into *n* isosceles triangles whose apex angles meet at the center of the *n*-gon (Figure 7-8a). All of these triangles are congruent, so we can choose one of them (Figure 7-8b) to represent what we know about each of them:

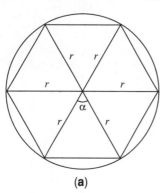

the two equal sides of the triangle form the radius *r* of the *n*-gon;
the altitude of the triangle forms the apothem *h* of the *n*-gon;
the base *b* of the triangle is the side *s* of the *n*-gon.

(See Section 6.2.) Then we also know from Section 6.2 that the apex angle α of each isosceles triangle measures $360°/n$. So

the angle formed by the apothem (altitude) and the radius is $\alpha/2$, which measures $180°/n$; and

the base angles β of the isosceles triangle each measure $\frac{1}{2}\left(180° - \frac{360°}{n}\right) = 90° - \frac{180°}{n}$.

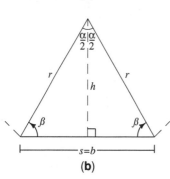

(a)

(b)

Figure 7-8

Now, using the trigonometric ratio for β, we have

$$\cos \beta = \frac{\text{adjacent side}}{\text{hypotenuse}} = \frac{b/2}{r} = \frac{b}{2r}$$

Thus

$$b = 2r \cos \beta = 2r \cos\left(90° - \frac{180°}{n}\right)$$

And since the perimeter *p* of an *n*-gon is the sum of the *n* sides, or $p = ns$, we have

$$p = ns = nb = n\left[2r\cos\left(90° - \frac{180°}{n}\right)\right]$$

$$= 2nr\cos\left(90° - \frac{180°}{n}\right)$$

Finally, we know from Section 3-3C that $\sin \alpha/2 = \cos \beta$, so

$$\sin \frac{180°}{n} = \cos\left(90° - \frac{180°}{n}\right)$$

and

$$p = 2nr \sin \frac{180°}{n}$$

This formula tells precisely how the perimeter of an *n*-gon is related to its radius.

Next, suppose we steadily increase the number *n* of sides in the *n*-gon but hold the radius *r* fixed. We see that the perimeter of the *n*-gon lengthens while the *n*-gon expands to fold into a circle of radius *r*, as in Figure 7-9. Thus we can think of the circle as the limit to which the perimeters are approaching; that is, $p \to C$.

note: This notion of a limit as a steady progression of polygonal lines is really the first extension of our ideas of length to include something that is curved in space.

Then if $p \to C$, the circumference *C* of a circle with radius *r* is whatever number $p = 2nr \sin(180°/n)$ approaches as *n* steadily increases. If we actually perform these calculations (which we won't here), we find that $n \sin(180°/n)$ tends closer and closer to a limit of $3.14159\ldots$, which is the number we call π. Thus we can conclude that $C = 2\pi r = \pi d$.

Figure 7-9

7-3. Area of a Circle
A. How the area of a circle is calculated

To find the area of a circle, we can take an approach similar to the one we use for finding the circumference. Here again, we think of a circle of radius *r* as the limit of a sequence of steadily expanding inscribed *n*-gons.

We know from Section 6-2 that the area of a regular *n*-gon is

$$A(n\text{-gon}) = \frac{nhs}{2} \text{ or } \frac{hp}{2}$$

where *n* is the number of sides, *h* is the length of the apothem, *s* is the length of a side, and $p = ns$ is the perimeter of the *n*-gon. Then, because the apothem *h* of a steadily expanding inscribed *n*-gon must approach the radius *r* of the circle as a limit, $h \to r$, we can think of the radius *r* of the circle as "equal" to *h* and write

$$A\text{ (circle)} = \frac{Cr}{2} = \frac{(2\pi r)r}{2}$$

or

AREA OF A CIRCLE $A\text{ (circle)} = \pi r^2$

B. Where the area formula comes from— A trigonometric approach

We know from Section 7-2B and Figure 7-8b that

$$s = b = 2r \cos \beta$$

and we can see from Figure 7-8b that $\sin \beta = h/r$, so

$$h = r \sin \beta$$

Thus, from the area formula for a regular n-gon we have

$$A(n\text{-gon}) = \frac{nbh}{2} = \frac{n(r \sin \beta)(2r \cos \beta)}{2} = nr^2(\sin \beta)(\cos \beta)$$

And, since $\sin \alpha/2 = \cos \beta$ and $\cos \alpha/2 = \sin \beta$, we have

$$A(n\text{-gon}) = nr^2(\sin \beta)(\cos \beta) = nr^2\left(\cos \frac{\alpha}{2}\right)\left(\sin \frac{\alpha}{2}\right)$$

Then, by the double-angle identity $\sin 2\theta = 2(\cos \theta)(\sin \theta)$, we have

$$\sin 2\left(\frac{\alpha}{2}\right) = 2\left(\cos \frac{\alpha}{2}\right)\left(\sin \frac{\alpha}{2}\right)$$

$$\sin \alpha = 2\left(\cos \frac{\alpha}{2}\right)\left(\sin \frac{\alpha}{2}\right)$$

$$\frac{\sin \alpha}{2} = \left(\cos \frac{\alpha}{2}\right)\left(\sin \frac{\alpha}{2}\right)$$

So

$$A(n\text{-gon}) = nr^2\left(\cos \frac{\alpha}{2}\right)\left(\sin \frac{\alpha}{2}\right) = nr^2\left(\frac{\sin \alpha}{2}\right)$$

And since $\alpha = 360°/n$,

$$A(n\text{-gon}) = nr^2 \frac{\sin(360°/n)}{2}$$

Then, since the circle is the limit the inscribed n-gon is approaching as n increases, the area of a circle of radius r depends on the limit of the factor $n \sin(360°/n)$, which can be shown by calculation with increasing values of n to be exactly 2π. Thus we can write

$$A(\text{circle}) = r^2\left(\frac{n \sin(360°/n)}{2}\right) = r^2\left(\frac{2\pi}{2}\right) = \pi r^2$$

EXAMPLE 7-5: Find the area A of a circle whose radius is 12.3 centimeters.

Solution: Let $\pi = 3.14$

$$A(\text{circle}) = \pi r^2$$
$$= \pi(12.3 \text{ cm})^2$$
$$= 475.29 \text{ cm}^2$$

note: If you use the 22/7 approximation for π, you get $A(\text{circle}) = 475.48$. But if you use a calculator with a special key for π, you should find that the value it gives for the area is close to the first one given here.

EXAMPLE 7-6: Given a circle whose area A is 12.32 square inches and whose circumference C is 12.44 inches, find the radius r of the circle.

Solution

$$\frac{A}{C} = \frac{\pi r^2}{2\pi r} = \frac{r}{2}$$

so

$$r = \frac{2A}{C} = \frac{2(12.32 \text{ in}^2)}{12.44 \text{ in.}} = 1.98 \text{ in.}$$

7-4. Arc Measurement

A protractor, which we use to measure angles, is really just a semicircle that has been divided into 180 parts. Each of these parts is called a degree. Similarly, any circle can be viewed as two semicircles and can be divided into 360 degrees. Thus, instead of measuring arcs (which are parts of circles) in terms of length, we can also measure arcs in terms of degrees, as shown in Figure 7-10. That is,

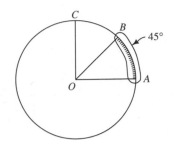

Figure 7-10

- When we measure an arc in degrees, we are measuring it in terms of a percentage—or ratio—of the whole circle of which the arc is a part.

Thus, when we talk about a 45° arc, for instance, we mean that this arc constitutes 45/360 of its circle.

We can quickly convert arc measure into length units if we know the radius of the circle. To do this, we need only calculate the circumference of the circle, multiply the circumference by the degree measure of the arc, and then divide by 360 degrees:

CONVERSION OF DEGREE MEASURE TO LENGTH MEASURE

$$\frac{\text{arc measure in length}}{2\pi r} = \frac{\text{arc measure in degrees}}{360°}$$

Since arcs can be measured in degrees, we can rephrase the definitions of major arcs, minor arcs, and semicircles as follows:

- A **major arc** has a measure greater than 180°.
- A **minor arc** has a measure less than 180°.
- A **semicircle** has a measure of 180°.

EXAMPLE 7-7: An arc of a circle whose radius is 3.20 centimeters measures 4.02 centimeters in length. Find the degree measure of the arc.

Solution: Let x be the measure of the arc in degrees; then

$$\frac{x}{360°} = \frac{4.02 \text{ cm}}{2\pi r}$$

$$x = \frac{360°(4.02 \text{ cm})}{2\pi(3.20 \text{ cm})}$$

$$x = 72°$$

EXAMPLE 7-8: Find the number of degrees in an arc of a circle whose diameter is 5 feet if the length of the arc is (**a**) one-third of the circumference and (**b**) one-twentieth of the circumference of the circle.

Solution: The diameter of 5 feet (radius 2.5 feet) is not relevant here—only the ratios are important.

(**a**) If the arc is one-third of the circumference, then the arc is 360°/3 = 120°.
(**b**) If the arc is one-twentieth of the circumference, then the arc is 360°/20 = 18°.

The measurement of arcs both in length and in degrees bring us to one final point about arcs. Arc measurements are *additive*. Thus,

- If B is a point on $\overset{\frown}{AC}$, then the measurement of $\overset{\frown}{ABC}$ is equal to the sum of the measures of $\overset{\frown}{AB}$ and $\overset{\frown}{AC}$.

note: When you have two arcs that have the same two-letter designation—say a major arc $\overset{\frown}{AC}$ and a minor arc $\overset{\frown}{AC}$—there is always a possibility that the one you're interested in will not be clearly indicated. When it isn't clear from context which arc you mean, you can pick an arbitrary point (or use a point that's already designated) on the arc of interest to give it a three-letter designation. In Figure 7-10, for example, $\overset{\frown}{AC}$ denotes both the major and the minor arc $\overset{\frown}{AC}$, but the designation $\overset{\frown}{ABC}$ clearly indicates the minor arc $\overset{\frown}{AC}$.

7-5. Theorems: Angles of a Circle

The measures of angles whose sides intersect circles can be related to the degree measure of the arcs they cut from the circle. Depending on where its vertex falls (on, interior to, or exterior to the circle), each type of angle intercepts an arc whose measure can be predicted by a few simple theorems.

A. The central-angle theorem

The first of these theorems is the simplest of all. In fact, it's so simple that we won't prove it formally here:

- **CENTRAL ANGLE THEOREM:**
  A central angle has the same degree measure as the arc it cuts from the circle.

  note: The proof of this theorem is really just a repetition of the definition of degree measure for arcs of a circle.

B. Angles with vertices on the circle

The second theorem has to do with angles whose vertices fall on the circle—including the two possible cases of the inscribed angle.

- **INSCRIBED-ANGLE THEOREM:**
  The degree measure of an angle inscribed in a circle is one-half the degree measure of the arc it cuts from the circle.

EXAMPLE 7-9:   Prove the inscribed-angle theorem.

Solution: There are two possible cases to consider here: (**a**) the case in which the center of the circle lies either in the interior of the angle or on one of its defining rays, and (**b**) the case in which the center of the circle lies exterior to the inscribed angle.

(**a**) *Proof:*   Draw an angle $\angle ABC$ inscribed in a circle, so the circle's center is in the angle's interior, as in Figure 7-11.

Step 1: Draw radii OA, OC, and OB, which are congruent by definition.

Step 2: $\angle AOC$ is a central angle and therefore equal in degree measure to $\overset{\frown}{AC}$ by the central-angle theorem.

Step 3: $\angle BOA + \angle BOC + \angle AOC = 360°$ by construction.

Step 4: Since the measures of all angles in a triangle sum to 180°,

$$\angle 1 + \angle 2 + \angle BOA = 180° \quad \text{and} \quad \angle 3 + \angle 4 + \angle BOC = 180°$$

Step 5: Since $OA \cong OB$ and $OB \cong OC$, triangles $\triangle AOB$ and $\triangle BOC$ are both isosceles; thus $\angle 1 \cong \angle 2$ and $\angle 3 \cong \angle 4$.

Step 6: The inscribed angle $\angle ABC = \angle 1 + \angle 3$ by construction.

Step 7: Adding the two equations in Step 4, we get

$$\angle 1 + \angle 2 + \angle 3 + \angle 4 + \angle BOA + \angle BOC = 360°$$

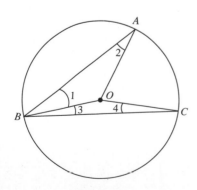

Figure 7-11

then, using Steps 5 and 6, we have

$$\angle 1 + \angle 1 + \angle 3 + \angle 3 + \angle BOA + \angle BOC = 360°$$

$$\angle 1 + \angle 3 + \angle 1 + \angle 3 + \angle BOA + \angle BOC = 360°$$

$$2\angle ABC + \angle BOA + \angle BOC = 360°$$

Step 8: From Steps 3 and 7 we have $\angle BOA + \angle BOC + \angle AOC = 360°$ and $2\angle ABC + \angle BOA + \angle BOC = 360°$, so

$$2\angle ABC + \angle BOA + \angle BOC = \angle BOA + \angle BOC + \angle AOC$$

which reduces to

$$2\angle ABC = \angle AOC$$

or

$$\angle ABC = \frac{\angle AOC}{2}$$

Step 9: $\angle AOC/2 = \overset{\frown}{AB}/2$ by the principle of additivity and the central-angle theorem, so

$$\angle ABC = \frac{\overset{\frown}{AB}}{2}$$

(b) *Proof:* Draw an angle $\angle ABC$ inscribed in a circle, so the circle's center falls exterior to the angle, as in Figure 7-12.

Step 1: Draw a chord BD such that the center of the circle O is interior to $\angle CBD$.

Step 2: $\angle ABD = \overset{\frown}{AD}/2$ and $\angle CBD = \overset{\frown}{CD}/2$ by the results of part **(a)**.

Step 3: $\angle ABC + \angle CBD = \angle ABD$ by construction.

Step 4: $\angle ABC + \overset{\frown}{CD}/2 = \overset{\frown}{AD}/2$ by Steps 2 and 3.

Step 5: Thus

$$\angle ABC = \frac{\overset{\frown}{AD} - \overset{\frown}{CD}}{2} = \frac{\overset{\frown}{AC}}{2}$$

by the principle of additivity. QED

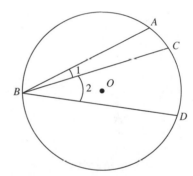

Figure 7-12

- *SEMICIRCLE COROLLARY:*
  Any angle inscribed in a semicircle is a right angle.

The proof of this theorem is nicely obvious: Any angle inscribed in a semicircle must have a measure of $180°/2 = 90°$, which is, of course, a right angle.

Then there's one more type of angle whose vertex falls on the circle—the tangent-chord angle. Although this type of angle cannot be precisely described as "inscribed," we can state its measurement theorem as a corollary to the inscribed-angle theorem.

- *TANGENT-CHORD COROLLARY:*
  The degree measure of a tangent-chord angle is one-half the degree measure of the arc it cuts from the circle.

This corollary can be proved directly from the inscribed-angle theorem after we've established one more fact—the fact that

- **tangent line is perpendicular to a radius of a circle at the point of contact.**

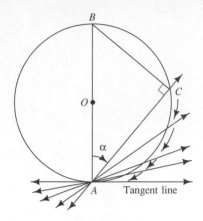

Figure 7-13

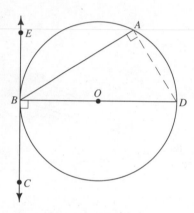

Figure 7-14

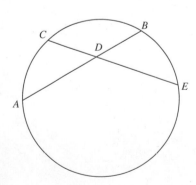

Figure 7-15

EXAMPLE 7-10: Show that a radius drawn to a tangent at the point of contact must be perpendicular to the tangent.

Solution: Consider diameter AB and secant $\overleftrightarrow{AC}$ as shown in Figure 7-13. Then with a scribe mark of a ruler fixed on point A, rotate $\overleftrightarrow{AC}$ down until the line touches the circle at just one point (A). The line that results from this downward rotation is, by definition, a tangent. Thus you can see from Figure 7-13 that the secant $\overleftrightarrow{AC}$ is steadily approaching the tangent as point C moves toward point A.

Now $\angle\alpha$ is $\overparen{BC}/2 < (180°/2 = 90°)$. But as C moves toward A, the arc $\overparen{BC}$ steadily approaches a semicircle—which means that $\angle\alpha$ is steadily approaching $90°$. When the point of tangency is actually achieved, $\angle\alpha$ is $90°$. Thus, radius OA on the diameter AB is perpendicular to the tangent line.

EXAMPLE 7-11: Prove the tangent-chord corollary.

Solution: There are two possible cases to consider here: (**a**) the case in which the center of the circle is in the interior of the angle, and (**b**) the case in which the center of the circle is exterior to the angle.

(**a**) ***Proof:*** Draw $\angle EBA$ such that one of its defining rays ($\overrightarrow{BE}$) is tangent to the circle O at B and the other ray intersects the circle at A, as in Figure 7-14.

Step 1: Draw diameter BOD and chord AD.

Step 2: $\angle EBD$ and $\angle DBC$ are right angles, since a radius and a tangent are perpendicular at the point of tangency.

Step 3: $\overparen{BAD} = 180°$ by the central-angle theorem, but $\overparen{BAD} = \overparen{BA} + \overparen{AD}$ by the principle of additivity; so $\overparen{BA} + \overparen{AD} = 180°$.

Step 4: $\angle EBA + \angle ABD = \angle EBD$, but $\angle EBD = 90°$ by Step 2; so $\angle EBA + \angle ABD = 90°$.

Step 5: $\angle ABD = \overparen{AD}/2$ by the inscribed-angle theorem, so $\angle EBA = 90° - (180° - \overparen{BA})/2$, or $\angle EBA = \overparen{BA}/2$.

(**b**) ***Proof:*** Since $\angle EBA$ and $\angle ABC$ are supplementary and since major arc $\overparen{BA}$ + minor arc $\overparen{BA} = 360°$, the proof follows directly from the results of part (**a**).   QED

Now we can restate the inscribed-angle theorem to include the special case of the tangent-chord angle:

- **The measure of any arc of a circle intercepted by an angle whose vertex is on the circle is two times the degree measure of the angle.**

C. Angles with vertices interior to the circle

If two chords intersect to form interior angles, as shown in Figure 7-15, two pairs of chord-chord angles are formed. $\angle CDA$ and $\angle BDE$ form one such pair, while $\angle CDB$ and $\angle ADE$ form another. And since these are vertical angles, $\angle CDA \cong \angle BDE$ and $\angle CDB \cong \angle ADE$. If we were to take the measure of both angles in a pair, we would see that the sum of their measures is the sum of their arcs' measures. Thus, we can find the measure of any one angle in a pair by taking half the sum of the degree measures of the pair's two arcs. In Figure 7-15, for example, $\angle CDA$ has a degree measure equal to $(\overparen{AC} + \overparen{BE})/2$. This fact can be stated as a theorem.

- *CHORD-CHORD ANGLE THEOREM:*
  A chord-chord angle measures one-half the sum of the degrees in the arcs intercepted by the chords that form the sides of the angle.

EXAMPLE 7-12: Prove the chord-chord angle theorem.

Proof: Draw a circle with two intersecting chords, AB and CE, as in Figure 7-16.

Step 1: Draw chord AE, as shown in Figure 7-16.

Step 2: $\angle BAE = \overarc{BE}/2$ and $\angle CEA = \overarc{CA}/2$, by the inscribed-angle theorem.

Step 3: $\angle ADE + \angle BAE + \angle CEA = 180°$ since the sum of the angles of a triangle is always $180°$.

Step 4: $\angle ADE + \angle BDE = 180°$ since $\angle ADE$ and $\angle BDE$ are supplementary.

Step 5: Thus

$$\angle ADE + \angle BDE = \angle ADE + \angle BAE + \angle CEA$$

$$\angle BDE = \angle BAE + \angle CEA$$

$$= \frac{\overarc{BE}}{2} + \frac{\overarc{CA}}{2}$$

$$= \frac{\overarc{BE} + \overarc{CA}}{2} \qquad \text{QED}$$

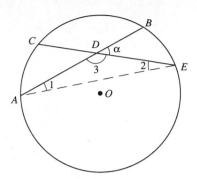

Figure 7-16

note: The central-angle theorem is just a special case of this theorem, in which both of the arcs have the same degree measure.

D. Angles with vertices exterior to the circle

If the vertex of an angle ($\angle ABC$) falls exterior to the circle, we have the three configurations shown in Figure 7-17:

- the *secant-secant angle* (Figure 7-17a), where $\angle ABC$ is formed by two secants that cut arcs $\overarc{AC}$ and $\overarc{DE}$ from the circle;
- the *tangent-secant angle* (Figure 7-17b), where one ray of the angle is a tangent such that points E and C are the same ($C = E$);
- the *tangent-tangent angle* (Figure 7-17c), where both rays of the angle are tangents such that points C and E are the same and points A and D are the same ($C = E$; $A = D$).

Since we can think of a tangent as a secant that's rotated until its two points of intersection with a circle are merged into one point, we can state the measurement theorem for all of the above configurations as one exterior-angle theorem:

- *EXTERIOR-ANGLE THEOREM:*
  The degree measure of an exterior angle of a circle is one-half the non-negative difference of the degree measures of the two arcs intercepted by the sides of the angle.

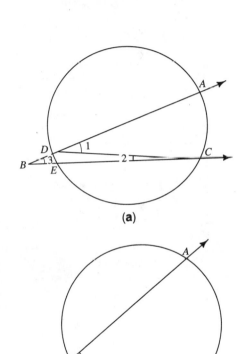

(a)

(b)

EXAMPLE 7-13: Prove the exterior-angle theorem.

Proof: Consider any of the configurations shown in Figure 7-17.

Step 1: Draw chord CD and label the angles for which this chord forms one side as $\angle 1$ and $\angle 2$, as shown in Figures 7-17a, 7-17b, and 7-17c.

Step 2: In each case, $\angle 1 = \angle 2 + \angle ABC$, since the exterior angle of a triangle must equal the sum of the other two interior angles; hence $\angle ABC = \angle 1 - \angle 2$.

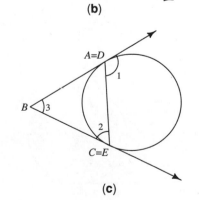

(c)

Figure 7-17

Step 3: Since $\angle 1$ and $\angle 2$ are inscribed angles, $\angle 1 = \overarc{AC}/2$ and $\angle 2 = \overarc{DE}/2$, where AC is the larger arc and DE is the smaller arc.

Step 4: Thus

$$\angle ABC = \frac{\overarc{AC}}{2} - \frac{\overarc{DE}}{2} = \frac{\text{larger arc} - \text{smaller arc}}{2} \qquad \text{QED}$$

note: In the case of the tangent-tangent arc, we have

$$\angle ABC = \frac{\text{major arc} - \text{minor arc}}{2}$$

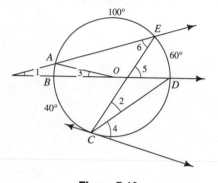

Figure 7-18

EXAMPLE 7-14: Find the degree measure of angles 1 through 6 in Figure 7-18.

Solution: Start by finding the measures of all the arcs on the circle: Since BD is a diameter, both arcs $\overarc{BD}$ are semicircles which measure 180°. Then

$$\overarc{AB} = 180° - (100° + 60°) = 20° \qquad \overarc{BC} = 40° \quad \text{(given)}$$
$$\overarc{AE} = 100° \quad \text{(given)} \qquad\qquad \overarc{CD} = 180° - 40° = 140°$$
$$\overarc{ED} = 60° \quad \text{(given)}$$

Now use the measurement theorems to find the measure of each angle:

$\angle 1$ is a secant-secant angle, so

$$1 = \frac{\overarc{ED} - \overarc{AB}}{2} = \frac{60° - 20°}{2} = 20°$$

$\angle 2$ is an inscribed angle, so

$$\angle 2 = \frac{\overarc{ED}}{2} = \frac{60°}{2} = 30°$$

$\angle 3$ is a central angle, so

$$\angle 3 = \overarc{AB} = 20°$$

$\angle 4$ is a tangent-chord angle, so

$$\angle 4 = \frac{\overarc{CD}}{2} = \frac{140°}{2} = 70°$$

$\angle 5$ is a chord-chord angle, so

$$\angle 5 = \frac{\overarc{DE} + \overarc{BC}}{2} = \frac{40° + 60°}{2} = 50°$$

$\angle 6$ is an inscribed angle, so

$$\angle 6 = \frac{\overarc{AC}}{2} = \frac{\overarc{AB} + \overarc{BC}}{2} = \frac{20° + 40°}{2} = 30°$$

EXAMPLE 7-15: Given that $\overrightarrow{BA}$ and $\overrightarrow{BC}$ are both tangents to the circle shown in Figure 7-19, find the measures of $\angle 3$ and $\angle 1$ if $\angle 2$ has a measure of 53°.

Solution: Minor arc $\overarc{AC}$ cut from the circle by tangent-chord angle $\angle 2$ is $2(53°) = 106°$. Thus major arc $\overarc{AC}$ is $360° - 106° = 254°$. Then, since $\angle 3$ is also a tangent-chord angle,

$$\angle 3 = \frac{254°}{2} = 127°$$

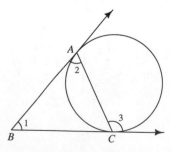

Figure 7-19

and since $\angle 1$ is a tangent-tangent angle,

$$\angle 1 = \frac{\text{major arc } \overset{\frown}{AC} - \text{minor arc } \overset{\frown}{AC}}{2} = \frac{254° - 106°}{2} = \frac{148°}{2} = 74°$$

note: Since $\angle 3$ is also an exterior angle of $\triangle ABC$, $\angle 1 + \angle 2 = \angle 3$. Thus, once you know the measure of $\angle 3$, you can write

$$\angle 1 = \angle 3 - \angle 2 = 127° - 53° = 74°$$

7-6. Theorems: Lines of a Circle

In Example 7-10 we showed that a radius drawn to a tangent line at the point of contact is perpendicular to the tangent. This is one of several important theorems that describe the relationships between line segments that intersect with, inside, or outside a circle. Three more of these theorems and their proofs are given below.

- **A diameter (radius) drawn perpendicular to a chord bisects both the chord and the arc determined by the chord.**

The converse of this theorem is also true; that is

- **If a diameter bisects a chord and its arc, then the diameter and the chord are perpendicular.**

EXAMPLE 7-16: Prove that a diameter drawn perpendicular to a chord bisects both the chord and its arc.

Proof: Assume a configuration like that shown in Figure 7-20, where diameter *DE* intersects chord *AB* at a right angle.

Step 1: Draw radii *OA* and *OB*.

Step 2: $r = OA \cong OB$ by definition.

Step 3: $OC \cong OC$.

Step 4: Since $\angle ACO$ and $\angle BCO$ are right angles by construction, triangles $\triangle ACO$ and $\triangle BCO$ are congruent by hypotenuse-leg.

Step 5: $AC \cong CB$ as corresponding sides of congruent triangles.

Step 6: $AB = AC + CB$, so chord *AB* is bisected.

Step 7: $\angle AOC \cong \angle BOC$ as corresponding angles of congruent triangles.

Step 8: Since $\angle AOC$ and $\angle BOC$ are central angles equal in measure, they must cut out equal arcs, so $\overset{\frown}{AD}/2 = \overset{\frown}{DB}/2$, or $\overset{\frown}{AD} = \overset{\frown}{DB}$.

Step 9: $\overset{\frown}{AB} = \overset{\frown}{AD} + \overset{\frown}{DB}$ by the principle of additivity, so $\overset{\frown}{AB}$ is bisected.

note: The converse of this theorem can also be proved. Since the proof is similar to the one given here, the details of proving the converse are left to you—as an exercise.

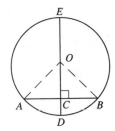

Figure 7-20

- **If two rays are drawn tangent to a circle from the same point *P*, then the lengths of the line segments from point *P* to the points of tangency must be equal in measure.**

EXAMPLE 7-17: Prove that two line segments drawn tangent to a circle from the same point *P* are equal in length.

Proof: Assume a configuration like that shown in Figure 7-21, where *P* is the point from which the tangents are drawn and *A* and *B* are the points of tangency.

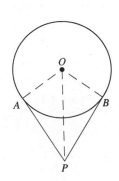

Figure 7-21

Step 1: Draw radii OA and OB.

Step 2: Draw line OP.

Step 3: $\angle OAP$ and $\angle OBP$ are both right angles since tangents and radii must meet at right angles.

Step 4: $\triangle OPA$ and $\triangle OPB$ are right triangles.

Step 5: $r = OA \cong OB$ by definition, so OA and OB are equal corresponding parts of $\triangle OPA$ and $\triangle OPB$.

Step 6: Right triangles $\triangle OPA$ and $\triangle OPB$ share a common side, $OP \cong OP$, as their hypotenuses so $\triangle OPA \cong \triangle OPB$.

Step 7: $AP \cong BP$ as corresponding sides of congruent triangles, so these sides must be equal in measure.

- **If two chords are of equal length, then they are equidistant from the center of the circle.**

EXAMPLE 7-18:   Prove that two chords of equal length are equidistant from the center of a circle.

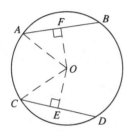

Figure 7-22

Proof: Consider a configuration like that of Figure 7-22, where AB and CD are chords of equal length.

Step 1: Draw OE perpendicular to CD and OF perpendicular to AB. Since the distance between a point and a line is the length of the perpendicular line segment between them, the distance from O to CD is the length of OE and the distance from O to AB is the length of OF.

Step 2: Draw radii OA and OC.

Step 3: $\triangle AFO$ and $\triangle CEO$ are right triangles by construction.

Step 4: $AB \cong CD$ (given).

Step 5: $AF \cong FB$ and $CE \cong ED$, since a radius drawn perpendicular to a chord bisects the chord.

Step 6: $AF = CE$, since these line segments are the results of the bisection of equal chords.

Step 7: $r = OA \cong OC$ by definition.

Step 8: $\triangle AFO \cong \triangle CEO$ by hypotenuse-leg.

Step 9: $OF \cong OE$ as corresponding sides of congruent triangles, so OF and OE are of equal measure.   QED

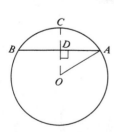

Figure 7-23

EXAMPLE 7-19:   The circle shown in Figure 7-23 has chord AB of 8 centimeters and radius OA of 5 centimeters. Find the distance of the chord from the center of the circle.

Solution: The distance from O to AB is the length of the perpendicular line segment OD. And since OD is on a radius OC, which is the perpendicular bisector of chord AB, D is the midpoint of AB. Thus

$$DA = \frac{AB}{2} = \frac{8 \text{ cm}}{2} = 4 \text{ cm}$$

You also know that

$$AO = r = 5 \text{ cm}$$

Then, since $\triangle OAD$ is a right triangle, you can find $d = OD$ by using the Pythagorean Theorem:

$$d^2 + (4 \text{ cm})^2 = (5 \text{ cm})^2$$
$$d = \sqrt{25 \text{ cm}^2 - 16 \text{ cm}^2}$$
$$= 3 \text{ cm}$$

EXAMPLE 7-20: A ray $\overrightarrow{PA}$ from point P to a circle is tangent to the circle at A. If PA is 12 inches long and the circle's radius is 5 inches, what is the distance from P to the circle?

Solution: First draw a figure that meets the given conditions: Draw a ray from an exterior point P such that the ray is tangent to a circle at A, and draw a radius AO. Note that $PA = 12$ inches and $AO = r = 5$ inches. Then let PO be a line segment that pierces the circle at B, as shown in Figure 7-24. This line segment PO is the shortest distance between two points, so the length of the line segment PB must be the length you want. Next, you know that $\angle PAO$ must be a right angle, since this is the angle where a tangent and a radius meet; so $\triangle PAO$ is a right triangle, with a hypotenuse OP. Then, by the Pythagorean Theorem,

$$(OP)^2 = (PA)^2 + (AO)^2$$
$$= (12 \text{ in.})^2 + (5 \text{ in.})^2$$
$$= 169 \text{ in}^2$$

So

$$OP = \sqrt{169 \text{ in}^2} = 13 \text{ in.}$$

and

$$PB = OP - OB = 13 \text{ in.} - 5 \text{ in.} = 8 \text{ in.}$$

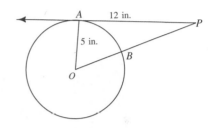

Figure 7-24

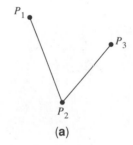

(a)

7-7. Analytic Expression of a Circle

A. Three points determine a circle

We have seen that a circle is completely described by its center and its radial distance. We have also seen that a circle contains no line segments—which means that **no three points on a circle are collinear**. These facts can be rephrased into one more important geometric fact:

- **Any three points on a circle completely specify that circle.**

B. Construction of a circle from three given points

Given three noncollinear points P_1, P_2, and P_3.

Step 1: Draw two line segments determined by these points, say $P_1 P_2$ and $P_2 P_3$, as shown in Figure 7-25a.

Step 2: Draw the perpendicular bisectors of $P_1 P_2$ and $P_2 P_3$, as shown in Figure 7-25b.

Step 3: Stop and think:

    (a) Since $P_1 P_2$ and $P_2 P_3$ are not parallel, their perpendicular bisectors must intersect at some point, O.

    (b) If P_1, P_2, and P_3 lie on a circle, then line segments $P_1 P_2$ and $P_2 P_3$ must be chords.

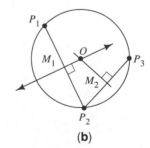

(b)

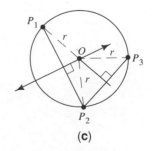

(c)

Figure 7-25

(c) If P_1P_2 and P_2P_3 are chords, then both their perpendicular bisectors must pass through the center of the circle.

(d) Since the only point both perpendicular bisectors pass through is O, point O must be the center of the circle.

(e) If O is the center of the circle, then $P_1O = P_2O = P_3O = r$; that is, the distance between any point P and the center O must be equal to the distance between any other point P and the center—so this distance is a radius by definition.

Step 4: Draw the circle with radius r determined by the three given points, as shown in Figure 7-25c.

C. The equation of the circle

The procedure described in Section 7-7B has an analytic analog—an equation that can be written in two forms. Let's see how this equation is developed.

Let the center O of a circle have coordinates (h, k) and let any point P on the circle have coordinates (x, y). If the circle has radius r—which is the distance between two points $O(h, k)$ and $P(x, y)$—then (x, y) must satisfy the distance relation (Section 1-7).

EQUATION OF A CIRCLE (coordinate form)
$$r = \sqrt{(x - h)^2 + (y - k)^2}$$
$$r^2 = (x - h)^2 + (y - k)^2$$

If we multiply this expression out, we get

$$x^2 + y^2 - 2hx - 2ky + h^2 + k^2 - r^2 = 0$$

Then if we set $A = -2h$, $B = -2k$, and $C = h^2 + k^2 - r^2$, we get the following general equation:

GENERAL EQUATION OF A CIRCLE
$$x^2 + y^2 + Ax + By + C = 0$$

Thus if the coordinates for three noncollinear points are given as (x_1, y_1), (x_2, y_2), and (x_3, y_3), we can substitute them into the general equation with A, B, and C as unknowns. This procedure yields three equations in A, B, and C, which can, in general, be solved for unique values of A, B, and C. Then we can calculate the coordinates of the circle's center and find the radius:

$$h = -A/2$$
$$k = -B/2$$
$$r = \sqrt{h^2 + k^2 - C}$$

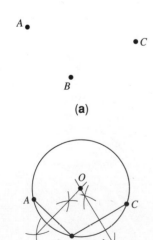

(a)

(b)

Figure 7-26

EXAMPLE 7-21: Use a ruler and compass to draw the circle that passes through the points A, B, and C shown in Figure 7-26a.

Solution

Step 1: Draw the line segments AB and BC.

Step 2: Use the procedure described in Example 1-4 to construct the perpendicular bisectors of $\overleftrightarrow{AB}$ and $\overleftrightarrow{BC}$.

Step 3: Extend the perpendicular bisectors until they meet at point O, as shown in Figure 7-26b.

Step 4: With the point of your compass at O, extend the legs of the compass to a width equal to OA. Use this length as a radius and draw the circle determined by the three given points, as shown in Figure 7-26b.

EXAMPLE 7-22: A circle in the Cartesian plane has center $O(h, k) = (-2, 7)$ and passes through $P(x, y) = (1, 3)$. **(a)** Write the equation of this circle in its general form; **(b)** find the radius of this circle.

Solution

(a) The general form of the equation of a circle is

$$x^2 + y^2 + Ax + By + C = 0$$

When the circle is centered at $(h, k) = (-2, 7)$,

$$A = -2h = -2(-2) = 4$$
$$B = -2k = -2(7) = -14$$

and you can find the value of C by substituting the given point's coordinates, $x = 1$ and $y = 3$, into the general equation along with the values of A and B you just calculated. Thus

$$1^2 + 3^2 + 4(1) + (-14)3 + C = 0$$
$$-28 + C = 0$$
$$C = 28$$

So the general form of the equation is

$$x^2 + y^2 + 4x + -14y - 28 = 0$$

(b)
$$r = \sqrt{h^2 + k^2 - C}$$
$$= \sqrt{(-2)^2 + 7^2 - 28} = \sqrt{53 - 28} = \sqrt{25}$$
$$= 5$$

SUMMARY

1. The circumference of a circle is $2\pi r$ or πd.
2. The area of a circle is πr^2.
3. Arcs of a circle can be measured in degrees:

$$\frac{\text{arc in degrees}}{360°} = \frac{\text{arc in distance units}}{(2\pi r)}$$

4. A central angle has the same number of degrees as the arc that it cuts from the circle.
5. An inscribed angle (including the tangent-chord special case) has a measure equal to one-half the number of degrees in the arc it cuts from the circle.
6. An exterior angle has a measure equal to one-half the difference of the arcs that the angle's sides cut from the circle.
7. An interior chord-chord angle has a measure equal to one-half the sum of the two arcs that the extended sides cut from the circle.
8. An angle inscribed in a semicircle is a right angle.
9. A radius drawn to a tangent at the point of contact is perpendicular to the tangent.
10. A radius drawn perpendicular to a chord bisects the chord and the chord's arc.
11. Two rays drawn tangent to a circle from the same point P through their points of tangency determine tangent line segments of equal length.
12. Two chords of equal length in a circle are an equal distance from the circle's center.
13. Any three noncollinear points determine a circle.

14. The coordinate form of the general equation of a circle of radius r with center (h, k) is

$$(x - h)^2 + (y - k)^2 = r^2$$

15. The general equation

$$x^2 + y^2 + Ax + By + C = 0$$

represents a circle in the Cartesian plane.

RAISE YOUR GRADES

Can you . . . ?

☑ identify and name the different lines and line segments associated with a circle
☑ recognize an inscribed angle and a tangent-chord angle
☑ recognize a central angle
☑ recognize a secant-secant angle, a tangent-secant angle, or a tangent-tangent angle
☑ find the area and circumference of a circle, given a radius
☑ convert the measure of an arc in degrees into a measure in length units, given a radius
☑ find the degree measure of an inscribed angle from that of the arc cut out of the circle
☑ find the degree measure of a exterior angle from that of the arcs cut out of a circle by its sides
☑ find the degree measure of a chord-chord angle from the arcs cut out of a circle by its sides
☑ prove a theorem or give a counterexample relating to chords and tangent lines to circles
☑ construct a circle, given three noncollinear points
☑ find the Cartesian equation of a circle, given the coordinates of three points that are not collinear

SOLVED PROBLEMS

PROBLEM 7-1   Find the circumference of a circle whose radius is 10 feet.

Solution $C = 2\pi r$. If π is approximated as $\frac{22}{7}$ and $r = 10$, then

$$C = 2(\tfrac{22}{7})(10 \text{ ft}) = 62\tfrac{6}{7} \text{ feet}$$

PROBLEM 7-2   Find the circumference of a circle inscribed in an equilateral triangle whose side has a measure of $12\sqrt{3}$.

Solution   If a circle is inscribed in an equilateral triangle, each side of the triangle is tangent to the circle at its midpoint. Draw the prescribed circle in its triangle, then construct radii from the center of the circle to the points of contact, as shown in Figure 7-27. Next, construct the perpendicular bisectors for each side of the triangle. Note that these perpendicular bisectors are also angle bisectors and that these bisectors intersect at one common point—the center of the triangle, which is also the center of the circle. Finally, observe that these bisectors cut the triangle up into three congruent isosceles triangles while each radius cuts each of these isosceles triangles into two congruent right triangles; so there are six congruent right triangles in the original triangle.

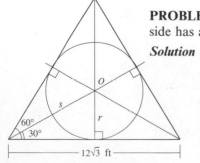

Figure 7-27

Now that you've done all that, you're ready to find a way to calculate the radius of the circle, from which you can find the circumference. To find the radius, then, focus on one of the right triangles you've just constructed and think about the relationship that exists among the sides. You know first that the length of the side that falls on a tangent to the circle must be one-half the length of the equilateral triangle's side, or $12\sqrt{3}/2 = 6\sqrt{3}$. Next you know that each angle of an equilateral triangle has a measure of 60°, so the angle bisector must form an angle whose measure is 30°. Then you can see that the radius r of the circle is the side opposite the 30° angle of a 30°-60°-90° right triangle (see Section 4-7B.2), where the hypotenuse has length s. Thus the side opposite the 60° angle has length

$$\frac{s\sqrt{3}}{2} = 6\sqrt{3}$$

So, since the side opposite a 30° angle of a 30°-60°-90° triangle has length $s/2$, you can find the radius from the equation above:

$$r = \frac{s}{2} = \frac{6\sqrt{3}}{\sqrt{3}} = 6$$

(Or you can take the long way around and show that $s = 2(6)\sqrt{3}/\sqrt{3} = 12$, so $r = s/2 = 12/2 = 6$.) Then, at last, the circumference C is

$$C = 2\pi r = 2\pi 6 = 12\pi$$

PROBLEM 7-3   Find the area A of a circle in terms of π for a circle whose radius r is 3/2 meters.

Solution
$$A = \pi r^2$$
$$= \pi(\tfrac{3}{2}\text{ m})^2$$
$$= \tfrac{9}{4}\pi \text{ m}^2$$

PROBLEM 7-4   Given the results of Problem 7-2 and Figure 7-27, find the area inside the equilateral triangle but outside the circle.

Solution   You know from the results of Problem 7-2 that the radius of the circle is $r = 6$. Therefore, the area of the circle is

$$A(\text{circle}) = \pi r^2 = \pi 6^2 = 36\pi$$

Then since any side of an equilateral triangle can be a base, you know that the base of the triangle is $b = 12\sqrt{3}$. Therefore, the height of the triangle is

$$h = \frac{b\sqrt{3}}{2} = \frac{(12\sqrt{3})(\sqrt{3})}{2} = (6\sqrt{3})(\sqrt{3}) = 18 \quad \text{(See Section 7-4C.)}$$

and the area of the triangle is

$$A(\text{triangle}) = \frac{bh}{2} = \frac{(12\sqrt{3})(18)}{2} = 108\sqrt{3}$$

So the area you want is

$$A(\text{triangle}) - A(\text{circle}) = 108\sqrt{3} - 36\pi = 73.96$$

PROBLEM 7-5   Find the area of the sector of the circle—the pie-shaped wedge—shown in Figure 7-28.

note: Any portion of the interior of a circle that is cut out by a central angle is known as a **sector of the circle**.

Solution   You're given the measure of the central angle of the sector (43°) and the length of the radius (2 ft). And you know that the degree measure of an arc is simply a percentage,

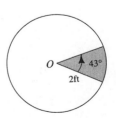

Figure 7-28

or ratio, of the arc to the whole circle. Thus you can see that the same kind of ratio applies to areas; that is,

$$\frac{\text{angle of sector}}{\text{angle of circle}} = \frac{\text{area of sector}}{\text{area of circle}} = \frac{43}{360}$$

In this case, then, the area of the circle is $\pi(2 \text{ ft})^2 = \pi(4 \text{ ft}^2)$, and the angle of the sector is 43°. So

$$A(\text{sector}) = \frac{\pi(4 \text{ ft}^2)(43°)}{360°} = 1.5 \text{ ft}^2$$

PROBLEM 7-6   Find the measures of the angles $\angle 1$ through $\angle 4$ of the circle shown in Figure 7-29.

Solution   You'll need to use the arc measurement theorems here. First, find the measures of all the arcs; then go after the angles.

$$100° = \frac{\widehat{AB} + \widehat{DE}}{2}   \text{[chord-chord angle]}$$

Then, since $\widehat{AB} = 40°$ (given) and $\widehat{DE} = \widehat{DG} + \widehat{GE} = 20° + \widehat{GE}$ (given),

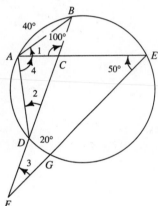

Figure 7-29

$$100° = \frac{40° + 20° + \widehat{GE}}{2}$$

or

$$200° = 60° + \widehat{GE}$$

so

$$\widehat{GE} = 200° - 60° = 140°$$

Step 2: $\angle AEG = 50°$ (given), so

$$50° = \frac{\widehat{AG}}{2}   \text{[inscribed angle]}$$

Then, since $\widehat{AG} = \widehat{AD} + \widehat{DG}$ (given) and $\widehat{DG} = 20°$ (given),

$$50° = \frac{\widehat{AG}}{2} = \frac{\widehat{AD} + \widehat{DG}}{2} = \frac{\widehat{AD} + 20°}{2}$$

or

$$100° = \widehat{AD} + 20°$$

so

$$\widehat{AD} = 100° - 20° = 80°$$

Step 3: Since the complete circle is 360°,

$$\widehat{BE} = 360° - (\widehat{BA} + \widehat{AD} + \widehat{DG} + \widehat{GE})$$
$$= 360° - 40° - 80° - 20° - 140°$$
$$= 80°$$

Step 4: Now that you've found the measures of all the arcs, you can find the measures of the angles. $\angle 1$, $\angle 2$, and $\angle 4$ are all inscribed angles, so

$$\angle 1 = \frac{\widehat{BE}}{2} = 40°   \qquad \angle 2 = \frac{\widehat{AB}}{2} = 20°   \qquad \angle 4 = \frac{\widehat{DG} + \widehat{GE}}{2} = 80°$$

Finally, $\angle 3$ is a secant-secant angle, so

$$\angle 3 = \frac{\widehat{BE} - \widehat{DG}}{2} = \frac{80° - 20°}{2} = 30°$$

PROBLEM 7-7 Prove that when two chords AC and DE intersect at a point B inside a circle as in Figure 7-30 the lengths of the segments that the two chords are divided into stand in the ratio $AB/BE = DB/BC$.

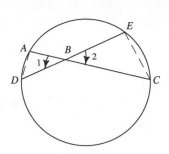

Solution

Step 1: Draw chords AD and EC.

Step 2: Since $\angle 1$ and $\angle 2$ are vertical angles, $\angle 1 \cong \angle 2$.

Step 3: $\angle CAD$ and $\angle DEC$ both cut out arc $\overarc{DC}$. Since both of these angles are inscribed angles, each has a measure of $\overarc{DC}/2$ and $\angle CAD \cong \angle DEC$.

Step 4: Similarly, $\angle ADE \cong \angle ECA$, since both of these angles cut out arc $\overarc{AE}$.

Step 5: Triangles $\triangle BAD$ and $\triangle BEC$ are similar since their three corresponding angles are congruent.

Step 6: Since the corresponding sides of similar triangles always stand in the same ratio, $AB/BE = DB/BC$.

Figure 7-30

PROBLEM 7-8 Prove that when two secants $\overleftrightarrow{PB}$ and $\overleftrightarrow{PD}$ cut a circle as shown in Figure 7-31 the segments of the secants stand in the ratio $PA/PC = PD/PB$.

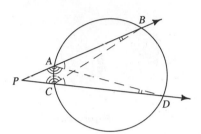

Solution

Step 1: Construct $\triangle PAD$ and $\triangle PCB$ by drawing chords AD and CB.

Step 2: Since $\angle BCD$ and $\angle DAB$ are inscribed angles that intercept the same arc $\overarc{BD}$, $\angle BCD \cong \angle DAB$.

Step 3: $\angle PCB$ and $\angle PAD$ are supplements of $\angle BCD$ and $\angle DAB$, respectively, so $\angle PCB \cong \angle PAD$.

Step 4: $\angle ADC$ and $\angle CBA$ are both inscribed angles that intercept the same arc, so $\angle ADC \cong \angle CBA$.

Figure 7-31

Step 5: $\angle BPC \cong \angle DPA$, since these angles are really the same angle.

Step 6: $\triangle PAD$ is similar to $\triangle PCB$ since all three of their corresponding angles are congruent.

Step 7: Since the corresponding sides of similar triangles stand in the same ratio, $PA/PC = PD/PB$.

PROBLEM 7-9 Find the general form of the equation of the circle that passes through the three points $(0,0)$, $(0,1)$, and $(2,0)$.

Solution The general equation of a circle is

$$x^2 + y^2 + Ax + By + C = 0$$

So each of the three given points must satisfy this equation. Thus, at $(x, y) = (0,0)$

$$0^2 + 0^2 + A(0) + B(0) + C = 0$$

$$0 + 0 + 0 + 0 + C = 0$$

$$C = 0$$

Then at $(x, y) = (0, 1)$ where $C = 0$,

$$0^2 + 1^2 + A(0) + B(1) + 0 = 0$$

$$0 + 1 + 0 + B + 0 = 0$$

$$B = -1$$

And at $(x, y) = (2, 0)$ where $B = -1$ and $C = 0$,

$$2^2 + 0^2 + A(2) + (-1)(0) + 0 = 0$$

$$4 + 0 + 2A - 0 + 0 = 0$$

$$2A = -4$$

$$A = -2$$

Thus the equation of the circle determined by the three points $(0, 0)$, $(0, 1)$, and $(2, 0)$ is

$$x^2 + y^2 + (-2)x + (-1)y + 0 = 0$$
$$x^2 + y^2 - 2x - y = 0$$

Review Exercises

(All dimensions below are assumed to be in some appropriate set of units!)

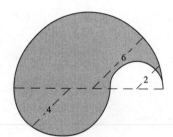

Figure 7-32

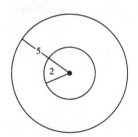

Figure 7-33

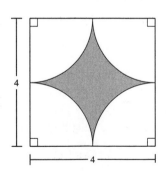

Figure 7-34

EXERCISE 7-1 Find the diameter of a circle of radius 2.

EXERCISE 7-2 Find the circumference of a circle of radius 3.

EXERCISE 7-3 If the area of a semicircular region is 8π, what is the *entire* perimeter of the region?

EXERCISE 7-4 Find the perimeter of the region shown in Figure 7-32.

EXERCISE 7-5 Find the area of the shaded region in Figure 7-32.

EXERCISE 7-6 Find the area between the two **concentric circles** (circles having the same center) shown in Figure 7-33.

EXERCISE 7-7 Four quarter-circles of radius 2 are cut from a square of side 4 as shown in Figure 7-34. Find the area of the (shaded) portion of the square that remains.

EXERCISE 7-8 A central angle of $30°$ cuts a sector (a pie-shaped wedge) from a circle of radius 2.5. Find the area of this sector as shown in Figure 7-35.

EXERCISE 7-9 The central angle of $30°$ in Figure 7-35 cuts a $30°$ arc from the circle. Find the length of this arc.

EXERCISE 7-10 Assume that the sector of Exercise 7-8 is removed from the circle. Find the perimeter of the region that remains.

EXERCISE 7-11 Find the measure of $\angle x$ in Figure 7-36.

EXERCISE 7-12 Find the measures of $\angle x$ and $\angle y$ in Figure 7-37.

EXERCISE 7-13 Find the measure of $\angle x$ in Figure 7-38.

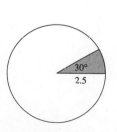

Figure 7-35

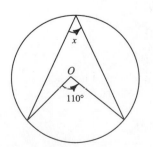

Figure 7-36

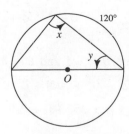

Figure 7-37

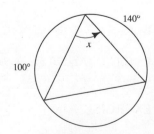

Figure 7-38

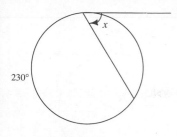

Figure 7-39

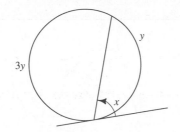

Figure 7-40

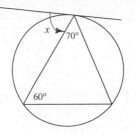

Figure 7-41

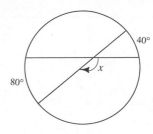

Figure 7-42

EXERCISE 7-14   Find the measure of ∠x in Figure 7-39.

EXERCISE 7-15   Find the measures of arc y and ∠x in Figure 7-40.

EXERCISE 7-16   Find the measure of ∠x in Figure 7-41.

EXERCISE 7-17   Find the measure of ∠x in Figure 7-42.

EXERCISE 7-18   Find the measures of ∠x and ∠y in Figure 7-43.

EXERCISE 7-19   Find the measure of the arc x in Figure 7-44.

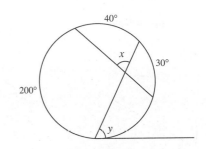

Figure 7-43

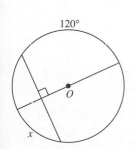

Figure 7-44

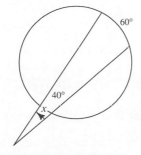

Figure 7-45

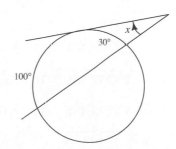

Figure 7-46

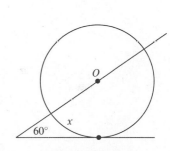

Figure 7-47

EXERCISE 7-20   Find the measure of ∠x in Figure 7-45.

EXERCISE 7-21   Find the measure of ∠x in Figure 7-46.

EXERCISE 7-22   Find the measure of arc x in Figure 7-47.

EXERCISE 7-23   Find the measures of ∠x and ∠y in Figure 7-48.

EXERCISE 7-24   Find the measure of arc x in Figure 7-49.

EXERCISE 7-25   Three straight lines are tangents to the circle shown in Figure 7-50. Find the length x.

EXERCISE 7-26   Find the lengths x and y shown in Figure 7-51.

EXERCISE 7-27   A secant and a tangent to a circle are drawn as shown in Figure 7-52. Prove that the segments satisfy the relation $AD/AB = AC/AD$.

EXERCISE 7-28   Find length x in Figure 7-53.

EXERCISE 7-29   Find length x in Figure 7-54.

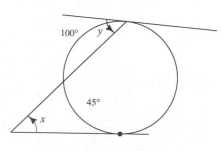

Figure 7-48

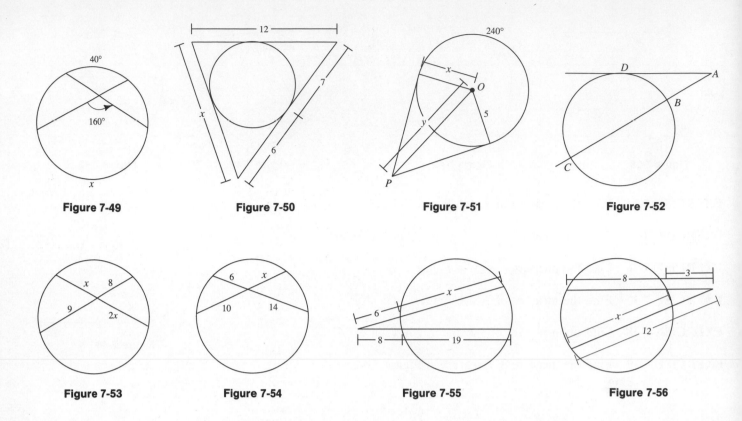

Figure 7-49   Figure 7-50   Figure 7-51   Figure 7-52

Figure 7-53   Figure 7-54   Figure 7-55   Figure 7-56

EXERCISE 7-30 Find length x in Figure 7-55.

EXERCISE 7-31 Find length x in Figure 7-56.

EXERCISE 7-32 Prove that two parallel lines that intersect a circle cut out equal arcs.

EXERCISE 7-33 Prove that chords of equal length cut arcs of equal length from a circle.

EXERCISE 7-34 Find the Cartesian equation of a circle that is centered at $(1, 2)$ and has radius 3.

EXERCISE 7-35 Find the center of the circle whose Cartesian equation is $x^2 + y^2 + 5x - 6y = 0$.

EXERCISE 7-36 Find the equation of the circle that passes through the points $(2, 2)$ $(0, 4)$, and $(2, 0)$.

EXERCISE 7-37 Use a ruler and compass to construct the circle of Exercise 7-36.

Answers to Review Exercises

7-1 4

7-2 $6\pi \approx 18.85$

7-3 $r = 4$, perimeter $= 4\pi + 8 \approx 20.57$

7-4 perimeter $= (6\pi + 4\pi + 2\pi) = 12\pi \approx 37.70$

7-5 area $= (36\pi + 16\pi - 4\pi)/2 = 24\pi \approx 75.40$

7-6 area $= (25 - 4)\pi = 21\pi \approx 65.97$

7-7 $16 - 4\pi \approx 3.43$

7-8 area $= (6.25/12)\pi \approx 1.64$

7-9 length $= \frac{5}{12}\pi \approx 1.31$

7-10 perimeter $= (5 - \frac{5}{12})\pi + 5 \approx 19.40$

7-11 55°

7-12 $x = 90°$, $y = 30°$

7-13 $x = 60°$

7-14 $x = 65°$

7-15 $y = 90°$, $x = 45°$

7-16 $x = 50°$

7-17 $x = 120°$

7-18 $x = 65°$, $y = 60°$

7-19 $x = 60°$

7-20 $x = 10°$

7-21 $x = 35°$

7-22 $x = 30°$

7-23 $y = 50°$, $x = 85°$

7-24 $x = 280°$

7-25 $x = 6 + 5 = 11$

7-26 $x = 5$, $y = 10$

7-27 Draw segments CD and DB. Then show that $\triangle ADC$ and $\triangle ABD$ are similar.

7-28 $x = 6$

7-29 $x = 8.4$

7-30 $x = 30$

7-31 $x = 10$

7-32 Draw a transversal so that the alternate interior angles are inscribed angles with the arcs in question.

7-33 Draw radii to the ends of the chords. The two triangles formed are congruent, and the central angles correspond to the arcs.

7-34 $(x - 1)^2 + (y - 2)^2 = 9$

7-35 $(-\frac{5}{2}, 3)$

7-36 $x^2 + y^2 + 2x - 2y - 8 = 0$

7-37 See Figure 7-57

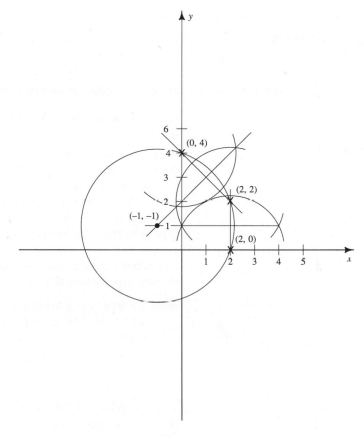

Figure 7-57

3-SPACE

THIS CHAPTER IS ABOUT

☑ **Space and Dimensions**
☑ **Points, Lines, and Planes in 3-Space**
☑ **Directions of Planes and Lines**
☑ **Angles between Lines and Planes**
☑ **Cartesian Space**
☑ **Spheres**

8-1. Space and Dimensions

We do not live in the flat world of the Euclidean plane. Instead, we live in a three-dimensional world, or **3-space**, in which we can move up and down *off* the plane, as well as on it. We can think of this capacity for motion in space as a **dimension**. Lines, then, have only one dimension. We can move only back and forth on a line l_1. The Euclidean plane has two dimensions. We can move back and forth on line l_1 in the Euclidean plane, and we can move on a line l_2 perpendicular to line l_1. All motions in the Euclidean plane can be expressed as combinations of these two kinds of motion, which we identify as the first and second dimensions; so the Euclidean plane is sometimes called 2-space. Finally, any motion that is not on one plane takes us into the third dimension.

note: In order to visualize three-dimensional space, you might close this book and look at it from the end, as shown in Figure 8-1. Each page in the closed book is, in effect, a Euclidean plane. Now you can imagine that the space occupied by this book is made up of lots of planes stacked one on top of the other, like the pages in a book.

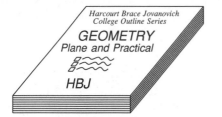

Harcourt Brace Jovanovich
College Outline Series
GEOMETRY
Plane and Practical
HBJ

Figure 8-1

8-2. Points, Lines, and Planes in 3-Space

A. Points and lines

We know that a point in 2-space is a location only—i.e., it has no dimensions (see Section 1-1). A point in 3-space is also just a location, but it is a location that is not confined to a given plane. For example, a point P_1 on page 17 of this book may lie 5 inches down from the top of the page and 5 inches from the right edge while a point P_2 that lies in exactly the same position on page 2 will not coincide with P_1 because it is on a different plane. We also know that any two points in 2-space determine a line. This is true in 3-space as well. That is, once we have identified any two points in 3-space, we can lay a ruler or a piece of taut string between them to determine a line.

B. Planes

The **plane postulate**

• **Any three distinct points that are not collinear determine a plane in 3-space.**

Consider any two points P_1 and P_2 in 3-space. These two points determine a line l. Now consider any point P_3 that is not on line l. We can draw as many lines as we want to from point P_3 to any point on line l to get a picture that resembles a Japanese fan, as in Figure 8-2. All these lines could be drawn on a single flat piece of paper, so they are all on one plane. Thus we can say that the set of points on all these lines is a plane R in 3-space.

A plane, however, is not limited to a single piece of paper. Like lines, planes are infinitely extended. And planes can occur in any orientation in our three-dimensional world.

note: When we are dealing with points that lie completely on one plane, that is, with **coplanar points**, we will usually think of that plane as rotated to lie on our piece of paper and draw pictures as if we were in the flat Euclidean 2-space discussed in Chapters 1 through 7. At other times we will make diagrams that show abbreviated planes drawn at some perspective angle, as in Figure 8-3.

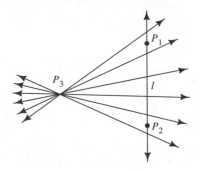

Figure 8-2

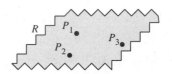

EXAMPLE 8-1: Describe how you could illustrate a line through two points in 3-space.

Solution: Choose two points in the room you're in. Pick points that are identified by some rigid object, say a paperweight, the top of a chair, or the corner of a door. Then get a ball of string. Tie or tape one end of the string at one of the points you've chosen. Next, pull the string taut and tie or tape the string to the other point, as shown in Figure 8-4.

EXAMPLE 8-2: Describe how you could illustrate a plane formed by three points using simple, everyday objects.

Solution: Find three sharpened pencils and support them so their three points stick up in the air. (You could do this by inserting the eraser ends of the pencils into some soft surface like modeling clay or potting soil.) Then think of the three pencil points as the three points that determine a plane. Take a stiff piece of cardboard or a pane of glass and lay it across the top of the pencil points. The flat cardboard or glass should rest stably on the pencil points, as illustrated in Figure 8-5. This piece of cardboard or pane of glass is part of the plane determined by the three pencil points.

Figure 8-3

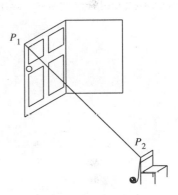

Figure 8-4

C. Other ways to determine a plane

We know that three noncollinear points determine a plane, and we can use this knowledge to find three other ways to determine a plane:

- **A line l and a point P not on line l determine a plane.**
- **Two intersecting lines l_1 and l_2 determine a plane.**
- **Two parallel lines l_1 and l_2 determine a plane.**

(See Figure 8-6.)

A point and a line: The discussion in Section 8-2B and Figure 8-2 show us that a line l and a point P not on the line determine a plane. It follows that if three noncollinear points determine a plane, then a point and a line must determine a plane because any two of the three noncollinear points form a line, as shown in Figure 8-6a.

Intersecting lines: If two lines l_1 and l_2 cross each other, they must intersect at some point P. Thus if we let P play the role of point P_3 and pick any point P_1 on line l_1 and any point P_2 on line l_2 as in Figure 8-6b, we can see that the two intersecting lines l_1 and l_2 are part of the plane formed by the three points P_1, P_2, and P_3.

Parallel lines: Finally, it is possible for two lines l_1 and l_2 to be coplanar even if they don't intersect. That is, all the points on two nonintersecting

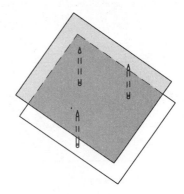

Figure 8-5

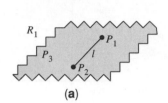

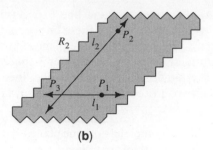

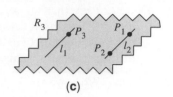

(a)   **(b)**   **(c)**

Figure 8-6

lines in 3-space can lie in some common plane, as shown in Figure 8-6c. Thus if we pick any point P_3 on line l_1 and then pick any two points P_1 and P_2 on line l_2, we have the plane generated by the three points P_1, P_2, and P_3.

D. Parallel lines and skew lines

We know that two parallel lines in 3-space determine a plane. For example, the left- and right-hand edges of this paper are parallel lines that lie on the plane of this page. Since these lines lie in one flat plane and do not intersect, we can say that

- Two *coplanar* lines that do not intersect in 3-space are called **parallel lines**.

But in 3-space it is also possible to have two lines that never intersect and yet are not parallel:

- Two lines that do not intersect in 3-space but are *not coplanar* are called **skew lines**.

Two skew lines l_1 and l_2 are shown in Figure 8-7.

note: It's hard to picture skew lines. You might get a better image if you think of the top edge of this book's back cover as l_1 and the right edge of the front cover as l_2. When the book is closed, these are skew lines: They do not intersect, they are not coplanar, and they certainly are not parallel.

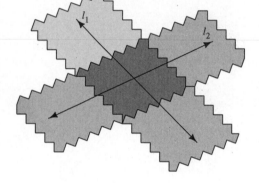

Figure 8-7

8-3. Directions of Planes and Lines

A. Parallel planes

Just as two lines l_1 and l_2 are parallel lines in 2-space if they do not intersect when they are infinitely extended in the same direction.

- Two planes R_1 and R_2 are parallel planes in 3-space if they do not intersect when they are infinitely extended in the same direction.

For example, when this book is closed, all its pages are in planes that are parallel to each other. Figure 8-8 shows abbreviated versions of parallel planes, where the broken lines in plane R_2 indicate pieces of that plane that would normally be hidden from our view by plane R_1. Finally, we can make an important observation about parallel planes:

- If two planes R_1 and R_2 are parallel to a third plane R_3, then each of these planes is parallel to the other.

B. Intersecting planes

If two planes R_1 and R_2 intersect, R_1 and R_2 must have at least one point P in common. Then if we draw a line l through P in R_1 and rotate l about P, eventually l must fall in R_2, as shown in Figure 8-9. Now we can see

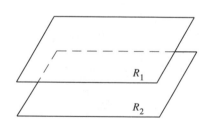

Figure 8-8

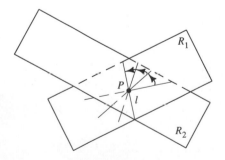

Figure 8-9

that, just as two distinct lines can intersect at only one point,

- **Two distinct planes can intersect at only one line**.

This line is called the **line of intersection**, or simply the **intersection** of two planes. (See Solved Problem 8-2 for additional observations on the intersection of two planes.)

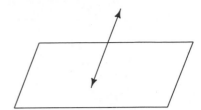

C. Lines parallel to planes

If a line *l* does not intersect a plane *R*, then line *l* is said to be parallel to the plane *R*. Figure 8-10 shows a line parallel to a plane.

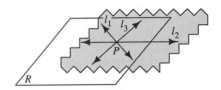

Figure 8-10

We have to be careful, however, that we don't read too much into the fact that a given line is parallel to a plane. There can be many lines that are parallel to one plane but are not parallel to each other, as Figure 8-11 shows. In Figure 8-11, lines l_1, l_2, and l_3 each pass through *P* and are parallel to plane *R*. Thus we can say that

- **Any lines that lie in a plane R_1 parallel to plane R_2 will be parallel to plane R_2, but these lines will not necessarily be parallel to each other.**

Similarly, if two planes R_1 and R_2 are both parallel to line *l*, as shown in Figure 8-12, R_1 and R_2 need not be parallel to each other. The only thing we can say for certain about this situation is that

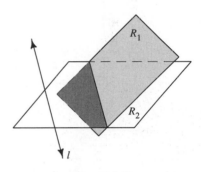

Figure 8-11

- **If R_1 and R_2 are both parallel to line *l* and if R_1 and R_2 intersect, then the line of intersection between R_1 and R_2 is parallel to line *l*.**

Finally, if line *l* is *not* parallel to a plane *R*, then either *l* must lie completely in *R*, as shown in Figure 8-13a, or *l* must meet *R* in only one point, as shown in Figure 8-13b. This point at which a line and a plane meet is called the **foot** of the line.

Figure 8-12

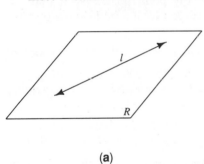

(a)

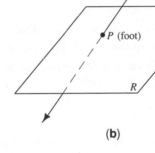

(b)

Figure 8-13

D. Perpendiculars

Suppose we draw a line segment from a point *P* not on plane *R* to a point *Q* in *R*. Then, if we draw a line *l* through the foot *Q* such that *l* is completely in *R* as in Figure 8-14, the line segment *PQ* makes an angle with *l* in the plane that is defined by *PQ* and *l*. If this angle is 90° for every possible line *l* that can be drawn through *Q* in *R*, then *PQ* is said to be **perpendicular to the plane**. (When we draw such a line, we speak of "dropping a perpendicular" from *P* to *R*.)

note: Imagine *R* as the floor of the room you're in. You could then tie a small weight to the end of a string and gently lower the weight with the string from a point in space until the weight touches the floor *R*. This string-plus-weight arrangement is called a **plumb line**, and the line obtained with a plumb line is exactly perpendicular to the plane of a floor. Plumb lines are frequently used by carpenters and paper hangers to guarantee that walls and wallpaper sheets are perfectly perpendicular to the floor.

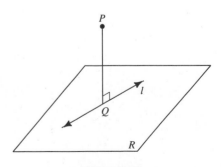

Figure 8-14

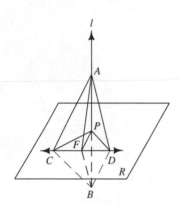

Figure 8-15

EXAMPLE 8-3: Plane R is perpendicular to line l_1 at point P. Show that if line l_2 is also perpendicular to line l_1 at P, then l_2 must lie in plane R.

Solution

Step 1: Since l_1 and l_2 are intersecting lines, they define a plane S whose line of intersection with plane R is l_3, as shown in Figure 8-15.

Step 2: Since l_3 lies in R, l_3 must be perpendicular to l_1 at P by definition.

Step 3: Now l_2 and l_3 must both lie in S *and* be perpendicular to l_1—which is also in S—at P.

Step 4: But since only one line can be perpendicular to another line at a given point in a plane (see Section 1-4), it isn't possible to have two distinct lines l_2 and l_3 that are both perpendicular to l_1 at P in S. Thus l_2 and l_3 must be the same line.

Step 5: Since l_3 lies in R, l_2 must also lie in R.   QED

EXAMPLE 8-4: Show that all lines drawn perpendicular to line l at point P must lie in the same plane.

Solution: This one's tricky and requires a lot of construction, which is shown in Figure 8-16. We can use this figure to show several things. First, we'll show that the plane determined by any two lines drawn perpendicular to l through P is a perpendicular plane. Then we can use Example 8-3 to show that all other lines that are perpendicular to l at P must lie in that plane.

Step 1: Let CP and DP be segments of any two lines that are perpendicular to l at P. These two lines define a plane R.

Step 2: Draw line $\overleftrightarrow{CD}$ in R, and let F be any point on $\overleftrightarrow{CD}$ such that $\overleftrightarrow{FP}$ represents any other line through P in R. (We'll need to show that $\overleftrightarrow{FP}$ is perpendicular to line l.)

Step 3: Pick a point A on l, then pick a point B on l on the opposite side of R so that $AP \cong BP$.

Step 4: Draw line segments AC, BC, AD, BD, AF, and BF, thereby creating lots of triangles.

Step 5: Since CP and DP are perpendicular to l (by construction), triangles $\triangle APC$ and $\triangle BPC$ are right triangles, as are $\triangle APD$ and $\triangle BPD$.

> *note:* This is where we're going. We want to prove that $\triangle APF \cong \triangle BPF$. If these two triangles are shown to be congruent, then $\angle APF \cong \angle BPF$. And since the congruent angles $\angle APF$ and $\angle BPF$ sum to a straight angle, each of these angles must be a right angle.

Step 6: Since $AP \cong BP$ and $CP \cong CP$, right triangle $\triangle APC \cong$ right triangle $\triangle BPC$ by leg-leg. Similarly, right triangle $\triangle APD \cong$ right triangle $\triangle BPD$ by the same reasoning.

Step 7: $AC \cong BC$ and $AD \cong BD$ as corresponding parts of congruent triangles.

Step 8: Since $CD \cong CD$, $\triangle ACD \cong \triangle BCD$ by SSS.

Step 9: Thus $\angle ACF \cong \angle BCF$ as corresponding angles of the congruent triangles $\triangle ACD$ and $\triangle BCD$.

Step 10: Since $CF \cong CF$, $\triangle ACF = \triangle BCF$ by SAS.

Step 11: $AF \cong BF$ as corresponding sides of congruent triangles.

Step 12: Since $PF \cong PF$, $\triangle APF \cong \triangle BPF$ by SSS (finally!).

Step 13: $\angle APF \cong \angle BPF$ as corresponding angles of congruent triangles; and $\angle APB$ is a straight angle, so FP is perpendicular to l.   QED

EXAMPLE 8-5: Line l_1 in Figure 8-17a is perpendicular to plane R at P. Line l_2 lies in R. Line segment PD to l_2 is constructed perpendicular to l_2 at D in R. Show that if A is any other point on l_1, then AD is also perpendicular to l_2.

Solution

Step 1: Choose two points B and C on l_2 such that $BD \cong DC$. Draw line segments AB, AC, PB, and PC, as in Figure 8-17b.

Step 2: Since $PD \cong PD$ and $BD \cong DC$, right triangle $\triangle PDC \cong$ right triangle $\triangle PDB$ by leg-leg.

Step 3: $PB \cong PC$ as corresponding parts of congruent triangles.

Step 4: AP on l_1 is perpendicular to PB and PC in R by definition, since l_1 is perpendicular to R (given).

Step 5: Right triangles $\triangle APB$ and $\triangle ACP$ have legs $AP \cong AP$ and $PB \cong PC$, so $\triangle APB \cong \triangle ACP$.

Step 6: $AB \cong AC$ as corresponding parts of congruent triangles.

Step 7: Since $AD \cong AD$, $BD \cong CD$, and $AB \cong AC$, $\triangle ABD \cong \triangle ACD$ by SSS.

Step 8: Thus $\angle ADC \cong \angle ADB$ as corresponding parts of congruent triangles.

Step 9: $\angle ADC$ and $\angle ADB$ are congruent supplements, so each must measure $90°$; therefore, AD is perpendicular to l_2.

EXAMPLE 8-6: Show that if two lines l_1 and l_2 are perpendicular to the same plane R, then l_1 and l_2 must also be coplanar.

Solution

Step 1: Let line segments AB and CD be segments of two lines l_1 and l_2 perpendicular to plane R at points B and D, respectively, as shown in Figure 8-18.

Step 2: Draw line segment BD in R.

Step 3: Draw line l_3 in R perpendicular to BD at point D.

Step 4: Since CD is perpendicular to plane R, CD is perpendicular to l_3 by definition.

Step 5: From Example 8-4, the plane S formed by intersecting line segments CD and BD is also perpendicular to l_3.

Step 6: By the results of Example 8-5, AD must also be perpendicular to l_3; thus, AD must lie in plane S.

Step 7: Since A and B are both in plane S, line l_1 must also lie in plane S. Therefore, l_1 and l_2 are coplanar.

E. Projections

As we've seen in Examples 8-3 through 8-6, we can use 2-space figures, which we can easily draw, to analyze 3-space figures, which we can't really draw without distortion. But there's another way to relate 3-space figures to 2-space figures. If, for instance, we have a three-dimensional object whose points and lines are solid, we might place a bright light in front of the object and a white sheet behind it, so that the object casts a shadow on the sheet. Then if we move the light, we can see that the shadow changes shape. This shadow is a 2-space figure on a plane (the sheet), and we can tell something about the 3-space object from its 2-space shadow. This shadow is, in fact, a *projection* of the object onto the sheet.

In mathematics, we think of a projection of a 3-space figure onto a plane as what we would see if we were to look down upon the figure from

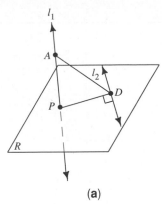

(a)

Figure 8-16

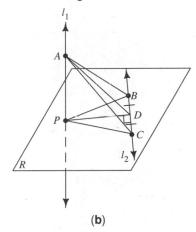

(b)

Figure 8-17

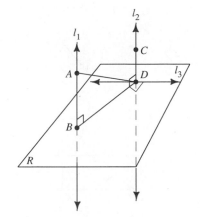

Figure 8-18

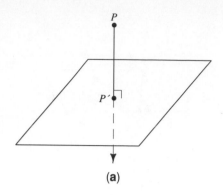

(a)

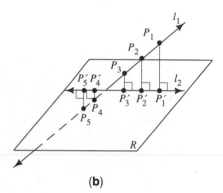

(b)

Figure 8-19

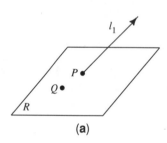

(a)

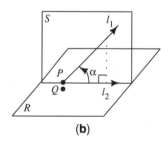

(b)

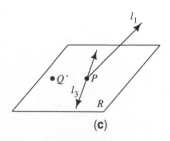

(c)

Figure 8-20

a vantage point high above the figure. But this point of view isn't exactly a formal definition, so let's clean this idea up and bring it into our geometric model of space. That is, let's look at the formal definition of the projection of a point in space onto a plane and the projection of a line in space onto a plane.

If we take a point P not in plane R and drop a perpendicular from P to R, as in Figure 8-19a, we get a point P' in R.

- The point P', which is the foot of the perpendicular from the point P to the plane R, is the **projection of the point** P onto R.

Then, if we have a line l_1 not in the plane R, we can project l_1 onto R by projecting every point P in l_1 onto R; that is,

- The *set* of all points P' of the plane R that are projections of the points P in line l_1 is the **projection of the line** onto the plane.

Usually, the projection of a line l_1 onto a plane R is another line l_2, as shown in Figure 8-19b, where each point P on line l_1 has a corresponding point P' on the line l_2. Thus the set of points P', which make up the line l_2, is the projection of l_1 when l_1 is not perpendicular to R. If, however, a line l is perpendicular to a plane R, the set of points that make up its projection contains only one point. That is, the projection of a line that is perpendicular to a plane is a point.

EXAMPLE 8-7: Given that l_1 is not perpendicular to R as shown in Figure 8-19b, show that the projection of line l_1 onto plane R is also a line.

Solution

Step 1: Pick any two points, say P_1 and P_2, on line l_1 and draw lines through these points perpendicular to plane R.

Step 2: By Example 8-6, we know that if two lines are perpendicular to the same plane R then they must be coplanar. Thus every line from l_1 perpendicular to R must lie in some common plane S.

Step 3: The intersection of plane S with plane R is a line l_2, and this line l_2 is the projection of l_1 onto R.   QED

8-4. Angles between Lines and Planes
A. The angle between a line and a plane

Lines and planes that intersect clearly have different directions. We need some way of measuring the direction of one relative to the other—that is, we need some way of finding the angle between a line and a plane. But first, we have to determine which angle we mean when we speak about the "angle between a line and a plane."

Consider a line l_1 that meets plane R at point P, as in Figure 8-20a. Obviously, line l_1 forms many angles with R, any of which might be measured with a protractor. The angle we're interested in, however, is the *smallest* angle that l_1 forms with some line in R that passes through P. And this line, l_2, is the line of intersection between plane R and plane S, where plane S is perpendicular to plane R, as shown in Figure 8-20b. That is, l_2 is the projection of l_1 onto plane R, and the acute angle formed between l_1 and l_2 is the angle between the line l_1 and the plane R.

note: Think of yourself as standing at point Q in Figure 8-20b. From this point what you're really trying to do is to look along R, perpendicular to l_1, and measure the angle that l_1 forms with the horizon" l_2. But suppose you are standing on point Q', as in Figure 8-20c, so that a line l_3—which is *not* the projection of l_1 onto R—is the horizon.

The angle that l_1 forms with l_3 is larger than the acute angle between l_1 and l_2; in fact, the angle between l_1 and l_3 could have any value between "the angle" above and 90°. That is, the plane determined by l_1 and l_3 is *not* perpendicular to plane R.

B. Dihedral angles

Suppose two planes R_1 and R_2 intersect in line l as in Figure 8-21a. In this figure the two half-planes R_1 and R_2 each form a *side* of a **dihedral angle** with a common *edge l*. Now we can describe the particular dihedral angle shown in Figure 8-21a by first identifying the common edge, and then identifying a point on each side. Thus we can write $\angle \overset{\frown}{A\text{-}PQ\text{-}B}$ for the dihedral angle, where A is any point on half-plane R_1, PQ is the common edge, and B is any point on half-plane R_2.

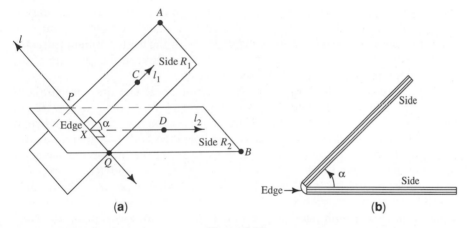

Figure 8-21

Next we need a way to measure $\angle A\text{-}PQ\text{-}B$. We measure dihedral angles by the angle of intersection formed in a third plane perpendicular to the line of intersection of the two original planes. This third plane is called the **angle plane** of the dihedral angle, which may be formed as follows. First, we construct a line l_1 in R_1 perpendicular to any point X on edge l. Then we construct a line l_2 in R_2 perpendicular to l at X. The plane formed by the intersection of l_1 and l_2 is therefore perpendicular to the line of intersection $\overleftrightarrow{PQ}$ by construction, and the measure of the acute **plane angle** α formed by these two lines is the measure of the dihedral angle $\angle \overset{\frown}{A\text{-}PQ\text{-}B}$. Thus, $\angle CXD$ is the intersection of lines $\overleftrightarrow{CX}$ and $\overleftrightarrow{DX}$ (l_1 and l_2), and the measure of $\angle CXD$ is the measure of the dihedral angle $\angle \overset{\frown}{A\text{-}PQ\text{-}B}$ in Figure 8-21a.

note: You can think of a dihedral angle as the acute angle formed by the pages of a partially opened book, where page 10, say, is one side of the angle, page 11 is the other side, and the center gutter is the line of intersection or edge. The measure of the dihedral angle formed by the pages of the book is the measure of the acute plane angle α you would see if you were to look at the book, end-on, from a distance— as shown in Figure 8-21b.

When two planes intersect, four dihedral angles are formed around the common edge. The measure of each of these angles is determined by the plane angle whose sides are lines perpendicular to the common edge. If the measure of each of these plane angles is 90°, then the four dihedral angles are congruent and the planes that make up the dihedral angles are **perpendicular planes**.

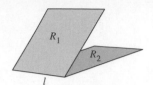

Figure 8-22

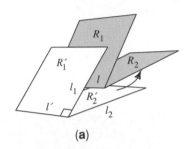

(a)

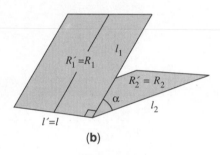

(b)

Figure 8-23

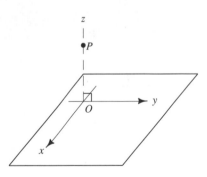

Figure 8-24

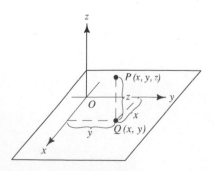

Figure 8-25

EXAMPLE 8-8:   Use a piece of paper to illustrate a dihedral angle. Identify the sides and the edge of the dihedral angle.

Solution: Fold any piece of paper in half, firmly creasing the fold. Unfold the paper slightly. The crease of the fold is the line of intersection between two planes, so it is the edge l of the dihedral angle. The two halves of the paper on either side of the crease represent the two intersecting half-planes, so you can say that each half of the paper forms a side R of the dihedral angle, as shown in Figure 8-22.

EXAMPLE 8-9:   Describe how you could measure the dihedral angle you constructed in Example 8-8.

Solution: If you knew for sure that the lines formed by the edges of the folded paper were straight lines perpendicular to the line of intersection l, the easiest way to measure the dihedral angle in Figure 8-22 would be to place your protractor against the outside edges of the paper, with the center of the protractor at the endpoint of the crease l. Then you could simply read the measure of the angle by comparing its sides to the scribe marks.

But what if you don't know for certain that these end lines are straight? In that case you have to make your own plane angle. To do this, draw a straight line l' on a second piece of paper. Then construct another line perpendicular to l'. Cut the paper along the perpendicular line; then fold the paper along line l' so that the perpendicular is intersected by l'. Label the intersected segments of the perpendicular l_1 and l_2, which are on planes R_1' and R_2', respectively, as shown in Figure 8-23a. Now slide the newly folded paper over the original intersecting planes R_1 and R_2 in Figure 8-22, so that line l' coincides with l, plane R_1' coincides with plane R_1, and plane R_2' coincides with plane R_2. Now you can use your protractor to measure the angle α made by the intersection of lines l_1 and l_2, as shown in Figure 8-23b.

note: It doesn't matter *where* you slide this second dihedral angle on the first, as long as the lines of intersection are collinear and the respective planes are coplanar. The measure of the angle will be the same. From this fact, you may infer that every plane angle that can be constructed on a dihedral angle is congruent to every other plane angle on the same dihedral angle.

8-5. Cartesian Space
A. Coordinates

The Cartesian plane described in Section 1-7 gives us a way to represent points on a plane analytically by using an ordered pair of numbers (x, y). We can do something similar to represent points in 3-space, only instead of an ordered pair, we use a triple of numbers (x, y, z).

We start with a plane like the one shown in Figure 8-24. Then we draw a set of x and y axes on the plane such that these axes meet at the point O. Thus we have the xy plane. Next, we pick a point P in space such that the *projection* of P on the plane is O. Then, if we draw the line PO and label this line z, we have the graph or model of 3-space, where the three axes x, y, and z can be used to locate any point in 3-space.

Let's look now at Figure 8-25. Point P corresponds to three numbers: a pair of numbers (x, y), which are the coordinates of P's projection Q onto the xy plane, and a third number z, which is the length of the line segment joining the point P to Q. The length of PQ is P's distance from the xy plane. But, just as a point in 2-space can lie on either side of O on the x or y axes, a point in 3-space can lie above or below the xy plane. We can eliminate this ambiguity by tagging distances of points below the plane with a minus sign. When the distance above or below the xy plane is tagged, we call it the **z coordinate**. Finally, we put an upward-pointing arrow on the z axis to indicate that the z coordinates of points above the xy plane are positive, while the z coordinates of points below the xy plane are negative.

Figure 8-26 shows us how points in 3-space are fully specified by the three numbers (x, y, z). For example, a point with coordinates $(1, 1, 1)$ projects onto $(1, 1)$ in the xy plane and is a distance of 1 unit above the plane, while the point with coordinates $(1, 1, -1)$ also projects onto $(1, 1)$ in the xy plane but is 1 unit below the plane.

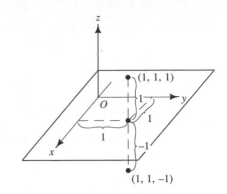

Figure 8-26

note: You know that the x and y axes determine the xy plane—so it's easy to see that the x and z axes determine the xz plane and the y and z axes determine the yz plane. The xz and yz planes are both perpendicular to the xy plane and to each other. This means that we could have used either one of these to replace the xy plane in the discussion above, with the other perpendicular axis playing the role of the z axis. For example, we could accurately have said that point P with coordinates (x, y, z) projects onto (x, z) in the xz plane and lies y units away from the xz plane, or P projects onto (y, z) in the yz plane and lies x units away from the yz plane. Finally, the sign always indicates the location of a point with respect to the common intersection O, and we use the arrowhead on each axis to indicate the positive direction.

EXAMPLE 8-10: (a) Make an accurate perspective drawing—a graph—of three-dimensional Cartesian space. (b) Plot the point P with coordinates $(1, 3, 2)$.

Solution

(a) *Step 1:* Draw a horizontal line segment on your page and place an arrowhead pointing to the right on its right endpoint. Mark one-unit intervals on this line. This line segment is the y axis.

Step 2: Draw a line segment perpendicular to the y axis on the page and put an arrowhead pointing in the upward direction at its upper endpoint. Label the intersection of this line segment with the y axis O and mark one-unit intervals along this segment's length. This line segment is the z axis.

Step 3: Draw a third line through the intersection O of the y and z axes so that this line segment slants down to the left in a southwest direction, as shown in Figure 8-27a. Mark one-unit intervals on this line segment and place an arrowhead pointing in the southwest direction on the endpoint. This is the x axis.

(b) *Step 1:* To plot a point with x coordinate 1, place a ruler on the x axis at the one-unit mark and draw a line parallel to the y axis.

Step 2: To plot a point with y coordinate 3, place a ruler on the y axis at the three-unit mark and draw a line parallel to the x axis. Note that the two lines you've just drawn intersect at point Q with coordinates $(1, 3)$ in the xy plane. The point Q is the projection of the required point $P(1, 3, 2)$ onto the xy plane.

Step 3: Draw a line through Q parallel to the z axis and measure two units along this line from Q. The point on this line at the two-unit distance is $P(1, 3, 2)$, as shown in Figure 8-27b.

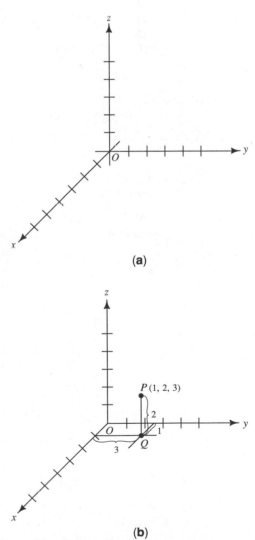

(a)

(b)

Figure 8-27

EXAMPLE 8-11: A point P_1 has coordinates $(1, 3, 4)$. A plane R is passed through this point parallel to the xy plane. A second point P_2 with coordinates $(2, 7, 9)$ is projected onto R. Find the coordinates of this projection Q.

Solution

Step 1: Since R is parallel to the xy plane (given) and $\overleftrightarrow{P_2Q}$ is perpendicular to R (by definition), $\overleftrightarrow{P_2Q}$ must also be perpendicular to the xy plane.

Step 2: Thus points P_2 and Q both have the same x and y coordinates, $x = 2$ and $y = 7$.

Step 3: Point P_1 has z coordinate $z = 4$. This is 4 units above the xy plane. Since P_1 is in R, and R is parallel to the xy plane, then any point in R must also have z coordinate $z = 4$.

Step 4: Thus Q has coordinates $(2, 7, 4)$.

B. Distance

If we know the coordinates of two points in 3-space, we can easily find the distance between them by the Pythagorean theorem. First, we let P_1 and P_2 have respective coordinates (x_1, y_1, z_1) and (x_2, y_2, z_2). Then we draw a plane perpendicular to the z axis (parallel to the xy plane) at height z_1 as in Figure 8-28. Next, we project P_2 onto this plane at Q; this point, Q, therefore has coordinates (x_2, y_2, z_1). Since P_1 and Q lie in a plane parallel to the xy plane, we can, for all practical purposes, ignore the z_1 coordinate here. Now we can treat the points as if they were in the xy plane at (x_1, y_1) and (x_2, y_2) and use the distance formula (1-1) for finding $d(P_1, Q)$. Thus

$$d(P_1, Q) = \sqrt{(x_2 - x_1)^2 + (y_2 - y_1)^2}$$

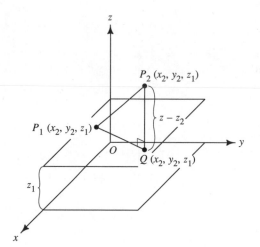

Figure 8-28

Now P_1Q and P_2Q define a plane containing right triangle $\triangle P_1 P_2 Q$, where $P_1 P_2$ is the hypotenuse and $P_1 Q$ and $P_2 Q$ are the legs. Thus

$$d(P_2, Q) = |z_2 - z_1| = \sqrt{(z_2 - z_1)^2}$$

Then by the Pythagorean Theorem

$$d(P_1, P_2)^2 = d(P_1, Q)^2 + d(P_2, Q)^2$$
$$= (x_2 - x_1)^2 + (y_2 - y_1)^2 + (z_2 - z_1)^2$$

Thus

$$d(P_1, P_2) = \sqrt{(x_2 - x_1)^2 + (y_2 - y_1)^2 + (z_2 - z_1)^2} \qquad \textbf{(8-1)}$$

This is the three-dimensional version of the distance formula (1-1).

EXAMPLE 8-12: Find the distance between P_1 and P_2 if their respective coordinates are $(1, 3, 2)$ and $(3, 4, -7)$.

Solution

$$d(P_1, P_2) = \sqrt{(3 - 1)^2 + (4 - 3)^2 + (-7 - 2)^2}$$
$$\sqrt{2^2 + 1^2 + (-9)^2} = \sqrt{4 + 1 + 81} = \sqrt{86} \approx 9.27$$

EXAMPLE 8-13: Find the midpoint M of the line segment joining (x_1, y_1, z_1) to (x_2, y_2, z_2).

Solution

Step 1: Let M have coordinates (x, y, z).

Step 2: $x = x_1 + \dfrac{x_2 - x_1}{2} = \dfrac{x_1 + x_2}{2}$

Step 3: $y = y_1 + \dfrac{y_2 - y_1}{2} = \dfrac{y_1 + y_2}{2}$

Step 4: $z = z_1 + \dfrac{z_2 - z_1}{2} = \dfrac{z_1 + z_2}{2}$

So M has the coordinates $\left(\dfrac{x_1 + x_2}{2}, \dfrac{y_1 + y_2}{2}, \dfrac{z_1 + z_2}{2}\right)$.

8-6. Spheres

We know what a circle is (see Section 7-1): It is the set of all points in the plane that fall a fixed distance (the radius) from a given point (the center). And just as the circle is a basic figure in 2-space, the sphere is a basic figure in 3-space.

If we had a circle with a rigid diameter running through it, we could rotate that circle up out of the plane around the diameter, as shown in Figure 8-29. The result of a complete rotation of this circle would be a sphere in 3-space. A **sphere**, then, is a three-dimensional circle, all of whose points lie a fixed distance r (the radius) from a given point C (the center) in space.

Given the coordinates of the center C of a sphere, (x_0, y_0, z_0), and the coordinates of a point P on the sphere, (x, y, z), we can use the distance formula (8-1) to write

$$r = \sqrt{(x - x_0)^2 + (y - y_0)^2 + (z - z_0)^2}$$

So the equation of a sphere is

EQUATION OF A SPHERE $r^2 = (x - x_0)^2 + (y - y_0)^2 + (z - z_0)^2$ **(8-2)**

Figure 8-29

EXAMPLE 8-14: Find the equation of a sphere centered at $(0, 0, 1)$ whose radius is 2.

Solution: Use the equation of a sphere (8-2):

$$r^2 = (x - x_0)^2 + (y - y_0)^2 + (z - z_0)^2$$
$$2^2 = (x - 0)^2 + (y - 0)^2 + (z - 1)^2$$
$$4 = x^2 + y^2 + (z - 1)^2$$

If a plane is passed through a sphere, the result is a circle, as shown in Figure 8-30. If the plane passes through the center of the sphere, the circle is known as a great circle, because this circle is the circle of the largest diameter that can be cut from a given sphere. Any other circle that can be cut from the same sphere will have a diameter smaller than that of the great circle.

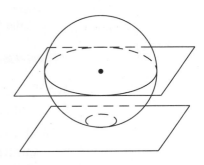

Figure 8-30

SUMMARY

1. Three noncollinear points determine a plane. Also,

   - A line and a point not on the line determine a plane.
   - Two intersecting lines determine a plane.
   - Two parallel lines determine a plane.

2. Two distinct planes can intersect in only one line.
3. Parallel planes are planes that do not intersect.
4. Two lines that do not intersect are parallel if and only if they are coplanar.
5. Skew lines are not parallel and do not intersect.
6. A line is perpendicular to a plane if and only if it is perpendicular to every line in the plane that passes through its foot.
7. All lines that are perpendicular to a common line at one point form a plane.

8. If two lines l_1 and l_2 are perpendicular to a third line l_3 at a point, the plane formed by l_1 and l_2 is perpendicular to line l_3.
9. Lines perpendicular to parallel planes are parallel.
10. If line l is perpendicular to plane R at point P, then any other line also perpendicular to l at P must lie in R.
11. The projection of a point P onto plane R is a point P' where a line through P perpendicular to R intersects R.
12. The projection of a line onto a plane is the set of points P' which are the projections of all the points in the line.

  • The projection of a line onto a plane perpendicular to the line is a point.
  • The projection of a line onto a plane not perpendicular to the line is a line.

13. The angle line l makes with plane R is the angle between l and its projection onto R.
14. A dihedral angle is an angle between two planes.
15. The measure of the dihedral angle of planes R_1 and R_2 is the measure of the angle between each of the intersections of R_1 and R_2 with a third plane that is perpendicular to the line of intersection of R_1 and R_2.
16. Cartesian 3-space represents each point with three numbers that are the point's signed distances from three mutually perpendicular plane.

  • The point (x, y, z) projects onto the point (x, y) in the xy plane and is z units above it.
  • The point (x, y, z) projects onto the point (x, z) in the xz plane and is y units above it.
  • The point (x, y, z) projects onto the point (y, z) in the yz plane and is x units above it.

17. The distance d between $P_1(x_1, y_1, z_1)$ and $P_2(x_2, y_2, z_2)$ is
$$d(P_1, P_2) = \sqrt{(x_1 - x_2)^2 + (y_1 - y_2)^2 + (z_1 - z_2)^2}$$

18. A sphere is the set of points that fall a fixed distance (the radius) from a given point (the center).
19. The equation of a sphere is
$$r^2 = (x - x_0)^2 + (y - y_0)^2 + (z - z_0)^2$$
where r is the radius of the sphere, (x_0, y_0, z_0) are the coordinates of the center, and (x, y, z) are the coordinates of a point on the sphere.

RAISE YOUR GRADES
Can you ... ?

☑ find a plane determined by three points
☑ accurately plot points on a Cartesian 3-space system of axes
☑ find the distance between two points in 3-space
☑ find the intersection of two planes
☑ tell when a line is perpendicular to a plane
☑ tell when two planes are parallel
☑ recognize skew lines
☑ tell when lines in 3-space are parallel
☑ find the measure of the angle between two planes
☑ find the angle between a line and a plane
☑ find the midpoint of a line in 3-space
☑ write the equation of a sphere, given the radius and center
☑ tell when two planes are perpendicular
☑ find the projection of a point on a plane
☑ find the projection of a line on a plane
☑ describe how to drop a perpendicular from a point to a plane
☑ describe how to construct a plane perpendicular to a line at a point

SOLVED PROBLEMS

PROBLEM 8-1 Demonstrate (**a**) two parallel lines in 3-space and (**b**) two skew lines in 3-space.

Solution Hold a pad of paper at any angle in 3-space to represent a plane.

    (**a**) Place one pencil along the top edge of the pad and another along the bottom edge. These pencils are on parallel lines.

    (**b**) Place a pencil on the face of the pad. Now place a second pencil on the back of the pad so that the pencils appear to cross when you look down from above. (They don't actually cross since the pad's thickness separates them.) These pencils are on skew lines.

PROBLEM 8-2 We know that when two planes R_1 and R_2 intersect, they have at least one line in common (see Section 8-3B). Is it possible, then, that there could be a point Q in both R_1 and R_2 that is *not* on l?

Solution

Step 1: Let l be a line in both R_1 and R_2.

Step 2: Let Q be a point in both R_1 and R_2 but not on l.

Step 3: Since a line and a point not on a line determine a plane, l and Q completely determine a plane S.

Step 4: Since Q and l lie in R_1, S must be R_1.

Step 5: Since Q and l lie in R_2, S must also be R_2.

Conclusion If such a point Q exists, then R_1 and R_2 are one and the same plane. If R_1 and R_2 are not the same plane, then R_1 and R_2 can only meet in line l.

PROBLEM 8-3 Given that planes R_1 and R_2 are parallel, show that if line l is perpendicular to R_1, then l must also be perpendicular to R_2.

Solution

Step 1: Let l intersect R_1 and R_2 at P_1 and P_2, respectively.

Step 2: Draw any line l_2 through P_2 in plane R_2, as in Figure 8-31.

Step 3: Lines l and l_2 define a plane S that intersects R_1 in line l_1.

Step 4: Lines l_1 and l_2 are coplanar lines that do not intersect since R_1 and R_2 do not intersect. Hence l_1 and l_2 are parallel lines in S.

Step 5: Since line l is perpendicular to l_1 in plane S, l must also be perpendicular to l_2 (see Section 1-6).

Step 6: Line l is perpendicular to R_2 since it is perpendicular to any line in R_2 that intersects l.

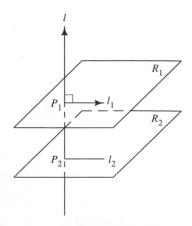

Figure 8-31

PROBLEM 8-4 Describe a way to project a point on the corner of your desk onto the plane of the floor.

Solution Use a plumb line. That is, put a weight like a fishermen's sinker on a string. Place the string on the point that you want to project. Slowly lower the weight to the floor. The projection is the point where the weight touches the floor.

PROBLEM 8-5 Plot the point $P(1, 3, 2)$ by using the projection first (**a**) on the xz plane and then (**b**) on the yz plane rather than the xy plane.

Solution

(a) *Step 1:* Draw the three axes as in Example 8-10.

   Step 2: Draw a line through the one-unit mark on the x axis parallel to the z axis.

   Step 3: Draw a line through the two-unit mark on the z axis parallel to the x axis. These lines intersect at Q on the xz plane.

   Step 4: Draw a line through Q parallel to the y-axis and measure off three units along this line and mark the point. This is the desired point $P(1, 3, 2)$, as shown in Figure 8-32a.

(b) See Figure 8-32b.

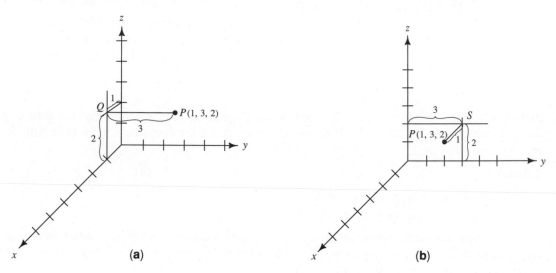

Figure 8-32

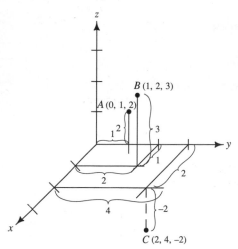

Figure 8-33

PROBLEM 8-6 Make an accurate perspective drawing of 3-dimensional Cartesian space, and plot the points $A(0, 1, 2)$, $B(1, 2, 3)$, and $C(2, 4, -2)$.

Solution See Figure 8-33.

PROBLEM 8-7 A line l passes through the points O and P whose coordinates are $(0, 0, 0)$ and $(1, 1, 3)$, respectively. (a) Draw the projection l_1 of this line in the xy plane. (b) Draw the projection l_2 of this line in the xz plane. (c) Draw all three lines, l, l_1, and l_2 in one Cartesian system.

Solution

(a) Points $(0, 0, 0)$ and $(1, 1, 3)$ project onto $(0, 0)$ and $(1, 1)$, respectively, in the xy plane. The projection l is drawn in Figure 8-34a.

(b) Points $(0, 0, 0)$ and $(1, 1, 3)$ project onto $(0, 0)$ and $(1, 3)$, respectively, in the xz plane. The projection is drawn in Figure 8-34b.

(c) The line and the two projections are drawn in perspective in Figure 8-34c.

Notice that l_1 is determined by $(0, 0, 0)$ and $(1, 1, 0)$ while l_2 is determined by $(0, 0, 0)$ and $(1, 0, 3)$.

PROBLEM 8-8 A line passes through points P_1 and P_2 whose coordinates are $(1, 3, 2)$ and $(2, 4, 2)$. Does the line intersect the xy plane? Why or why not?

Solution This line does not intersect the xy plane. Both points have 2 as their z coordinate. Thus both points lie at a distance of 2 units above the xy plane. Since the two points are the same distance from the xy plane, any ruler placed through these two points must also lie 2 units above the xy plane. So the line doesn't *intersect* the xy plane—it's *parallel* to it.

PROBLEM 8-9 A line l passes through the points $(0, 1, 0)$ and $(1, 2, 1)$. Find the angle α between l and the xy plane.

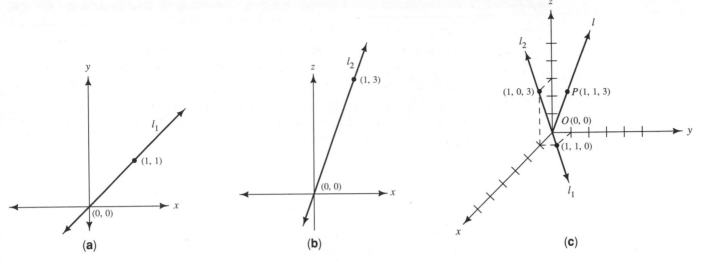

Figure 8-34

Solution

Step 1: Project line l onto the xy plane. To do this, let B be $(1,2,1)$ and C be the projection of B, so that C has coordinates $(1,2,0)$. Then point A with coordinates $(0,1,0)$ is its own projection into the xy plane. So $\overleftrightarrow{AC}$ in Figure 8-35 is the projection of l onto the xy plane.

Step 2: Since C is the projection of B, $\angle C$ is a right angle and $\triangle ACB$ is a right triangle with hypotenuse AB.

Step 3:
$$d(A,C) = \sqrt{(1-0)^2 + (2-1)^2 + (0-0)^2} = \sqrt{2}$$

and

$$d(B,C) = \sqrt{(1-1)^2 + (2-2)^2 + (1-0)^2} = 1$$

So leg AC is $\sqrt{2}$ units long and leg BC is 1 unit long.

Step 4:
$$\tan \alpha = \frac{BC}{AC} = \frac{1}{\sqrt{2}} \cong 0.707$$

Figure 8-35

Thus, α is approximately $35.26°$.

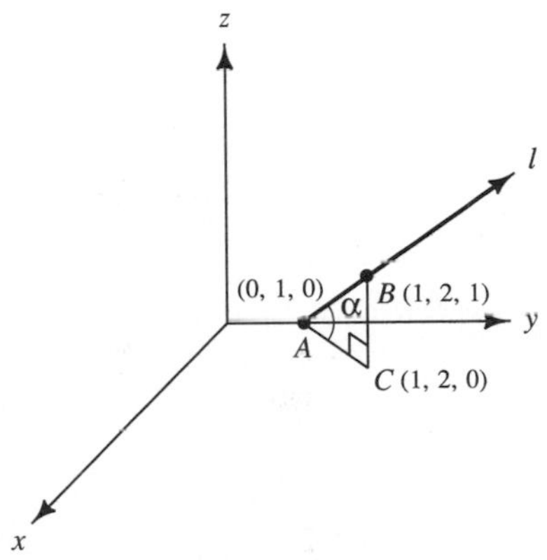

PROBLEM 8-10 Plane R is parallel to the xy plane and 3 units above it. Plane R slices through a sphere of radius 4 centered at $(1,1,2)$ and cuts out a circle C. Find the radius of C.

Solution

Step 1: The equation of points (x,y,z) on the sphere with center $(x_0, y_0, z_0) = (1,1,2)$ is

$$(x-1)^2 + (y-1)^2 + (z-2)^2 = 4^2$$

Step 2: All the points on plane R have z coordinate 3. Thus the xy coordinates of points in C satisfy

$$(x-1)^2 + (y-1)^2 + (3-2)^2 = 16$$

or

$$(x-1)^2 + (y-1)^2 = 15$$

Step 3: R is basically a copy of the xy plane, and this is just like the equation of a circle in the xy plane centered at $(1,1)$ with radius $\sqrt{15} \cong 3.87$.

Reviews Exercises

EXERCISE 8-1   Are the following statements true or false? (Circle T or F.)

(a)	Any two lines always determine a plane.	T	F
(b)	Three lines may intersect at a point so that each one is perpendicular to the other two.	T	F
(c)	Two planes may intersect in a line segment.	T	F
(d)	There is exactly one plane containing any three points.	T	F
(e)	A line l that is perpendicular to each of two coplanar lines l_1 and l_2 is perpendicular to the plane containing the lines l_1 and l_2.	T	F
(f)	Two lines parallel to a plane are parallel to each other.	T	F
(g)	Any four points completely determine a plane.	T	F
(h)	Two lines that do not intersect are parallel.	T	F
(i)	At any point on a line, there is exactly one plane perpendicular to the line.	T	F
(j)	The projection of a line onto a plane is always a line.	T	F
(k)	You can draw only one line l_1 perpendicular to a given line l at a specified point on line l.	T	F
(l)	A line and a point not on the line completely determine a plane.	T	F
(m)	A line perpendicular to one of two parallel planes is perpendicular to the other plane.	T	F
(n)	If a line l_1 intersects a plane at a point P and is not perpendicular to the plane, then you can draw a line l_2 in the plane perpendicular to line l_1 at point P.	T	F
(o)	Three lines that intersect pairwise such that there is no common intersection point for all three must be coplanar.	T	F
(p)	Three planes may intersect pairwise in such a way that their lines of intersection do not meet.	T	F
(q)	If two planes are parallel, then every line in the first plane is parallel to every line in the second plane.	T	F
(r)	The projection of a line segment has the same length as the original line segment.	T	F
(s)	The projections of two perpendicular lines are always perpendicular to each other.	T	F
(t)	The projections of two parallel lines onto a plane can intersect.	T	F
(u)	The projections of two skew lines onto a plane must intersect.	T	F

EXERCISE 8-2   Show that any plane R that intersects two parallel planes R_1 and R_2 must intersect R_1 and R_2 in two parallel lines l_1 and l_2.

EXERCISE 8-3   Show that any two distinct planes R_1 and R_2 that are perpendicular to the same line l must be parallel. [*Hint:* Assume they are not parallel.)

EXERCISE 8-4   Use Exercise 8-3 to show that two distinct planes that are parallel to a common third plane are parallel to each other.

EXERCISE 8-5   Show that if two intersecting planes are both perpendicular to a third plane, then their line of intersection is perpendicular to the third plane.

EXERCISE 8-6 Two planes R and S intersect in line l. Line segment PQ extends from point P to line l in plane S and is perpendicular to at point Q. If the projection of PQ onto plane R has half the length of PQ, find the measure of the dihedral angle between R and S.

EXERCISE 8-7 Point M is the midpoint of the line segment joining points P_1 and P_2. This segment P_1P_2 is projected onto plane R to produce another line segment $P'_1P'_2$. Show that the projection M' of M is the midpoint of the projection $P'_1P'_2$.

EXERCISE 8-8 Plane R contains two points A and B. P is a point whose projection is point Q in R. Show that if $AQ \cong BQ$, then $PA \cong PB$.

EXERCISE 8-9 Line l_1 is parallel to plane R. Line l_2 is perpendicular to both l_1 and R. Show that l_1's projection l_3 in R is perpendicular to l_2 also.

EXERCISE 8-10 Show that if l_1 and l_2 intersect and are both parallel to plane R, then the plane determined by l_1 and l_2 is also parallel to R.

EXERCISE 8-11 Find the distance between the following pairs of points whose Cartesian coordinates are:

(a) $(0, 1, 3)$ and $(1, 4, 2)$ (c) $(1, 4, -3)$ and $(2, 4, -5)$

(b) $(-1, 3, 0)$ and $(1, 3, 2)$ (d) $(0, 0, 0)$ and $(0, -1, -3)$

EXERCISE 8-12 Find the coordinates of the midpoints of the line segments joining the points of Exercise 8-11.

EXERCISE 8-13 Point P has coordinates $(1, 2, 2)$ and Q has coordinates $(4, 5, 8)$. Point M lies on the segment PQ and is twice as far from P as it is from Q. Find M's coordinates.

EXERCISE 8-14 For each of the following points find its projection on (a) the xy plane, (b) the xz plane, and (c) the yz plane:

     (i) $(1, 2, 4)$ (ii) $(7, 2, -1)$ (iii) $(3, -1, 5)$

EXERCISE 8-15 For each of the following points find how far it lies from (a) the xy plane, (b) the xz plane, and (c) the yz plane:

     (i) $(1, 2, 4)$ (ii) $(7, 2, -1)$ (iii) $(3, -1, 5)$

EXERCISE 8-16 Find the angle between the line through $(0, 2, 0)$ and $(5, 4, 6)$ and (a) the xy plane, (b) the xz plane, (c) the yz plane.

Answers to Review Exercises

8-1 (a) False—two skew lines don't even lie in the same plane.
    (b) True
    (c) False—planes extend infinitely, so their intersection must be a line, not a line segment.
    (d) False—three points could be collinear. (If the points were not collinear, they would determine a plane.)
    (e) False—line l could lie perpendicular to l_1 and l_2 in their plane if the two lines l_1 and l_2 are parallel. If l_1 and l_2 intersect, l is perpendicular to their plane.
    (f) False—the lines could be skew lines.
    (g) False—four points could lie on two skew lines, which don't determine a plane.
    (h) False—two nonintersecting lines could be skew lines.

(i) True

(j) False—if the line is perpendicular to the plane, the projection is a single point.

(k) False—an entire plane of lines can be perpendicular to a given line at a specified point.

(l) True

(m) True

(n) True

(o) True

(p) True

(q) False—the lines could be skew.

(r) False

(s) False

(t) True—projections of two parallel lines can meet if the plane of the parallel lines is perpendicular to the plane of projection. Otherwise the projections would be parallel.

(u) False—their projections could also be parallel lines, or a point and a line not containing that point.

8-2 *Proof by contradiction:* The two lines of intersection, l_1 and l_2, are both in the intersecting plane R; so l_1 and l_2 must be coplanar. If l_1 and l_2 meet at some point P, then the two parallel planes R_1 and R_2 must also meet at P, which is impossible. So l_1 and l_2 must be parallel.

8-3 See Figure 8-36. Let A and B be the points where line l intersects planes R_1 and R_2, respectively. If R_1 and R_2 are not parallel, then they must meet at some point P. Thus the line from P to A in R_1 and the line from P to B in R_2 must both be perpendicular to line l, and PA and PB must define a plane S that contains l. But in a plane, there can be only one perpendicular from a point P to a line l. Therefore, A and B must be the same point. Now, every point on R_1 or R_2 lies on a line through A which is perpendicular to l. But, according to the results of Example 8-4, all the lines perpendicular to l at point A must be coplanar. So all the points on R_1 and R_2 are coplanar, imply-

ing that R_1 and R_2 are the same plane. Contradiction!

8-4 Suppose R_1 and R_2 are both parallel to R_3. Pick a point on R_1 and run a line l perpendicular to R_1 through it. Line l is perpendicular to R_3 by Solved Problem 8-3. Since l is perpendicular to R_3 and R_2 is parallel to R_3, l is perpendicular to R_2. Thus R_1 and R_2 are both perpendicular to l. By Exercise 8-3, R_1 and R_2 are parallel.

8-5 Let planes R_1 and R_2 both be perpendicular to plane R_3. R_1 intersects R_3 in l_1, and R_2 intersects R_3 in l_2. Let l be the line of intersection of R_1 and R_2. Then l_1 and l_2 are perpendicular to l. Thus, by the discussion at the beginning of Example 8-4, the plane R_3 determined by l_1 and l_2 is also perpendicular to l.

8-6 See Figure 8-37. Line segment PQ in plane S has length x, so its projection on plane R has length $x/2$. Thus the angle α between R and S has a measure of $60°$.

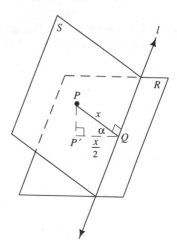

Figure 8-37

8-7 See Figure 8-38. Draw line segment $P_1 P_2^*$ parallel to projection $P_1' P_2'$. $P_1' M' \cong P_1 M^*$ and $P_2' M' \cong P_2^* M^*$. $\triangle P_1 M M^*$ is similar to $\triangle P_1 P_2 P_2^*$ and $\dfrac{P_1 P_2}{P_2 M} = 2$.

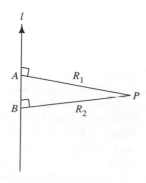

Figure 8-36

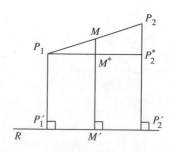

Figure 8-38

Thus $\dfrac{P_1P_2^*}{P_1M^*} = 2$ and $P_1M^* \cong M^*P_2^*$ or $P_1'M' \cong M'P_2'$.

8-8 See Figure 8-39. $AQ \cong BQ$, $PQ \cong PQ$; and since Q is the projection of P, $\angle PQA \cong \angle PQB$ as right angles. Therefore, $\triangle PQA \cong \triangle PQB$, and $PA \cong PB$ as corresponding parts of congruent triangles.

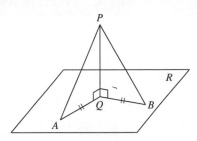

Figure 8-39

8-9 See Figure 8-40. Lines l_1 and l_2 determine a plane, which intersects R at l_3. Line l_3 is l_1's

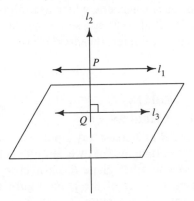

Figure 8-40

projection in R. Since l_1 and l_3 are coplanar and l_1 is parallel to R, which contains l_3, l_1 is parallel to l_3. Thus l_3 must be perpendicular to l_2 since l_1 is perpendicular to l_2.

8-10 Let l_1 and l_2 intersect at point P. Drop a perpendicular line l from P to plane R. Since l_1 and l_2 are each parallel to R, l_1 and l_2 must be perpendicular to l. (If this were not the case, then one of the lines, say l_1 for example, and its projection would not be parallel in the plane of l and l_1. Then l_1 and its projection in R would meet—but this would mean that R and l_1 were not parallel. This would also be the case for l_2.) Therefore, the plane of l_1 and l_2 is perpendicular to l at P; and by Exercise 8-3, this plane is parallel to R.

8-11 (a) $\sqrt{11}$ (b) $2\sqrt{2}$ (c) $\sqrt{5}$ (d) $\sqrt{10}$

8-12 (a) $(0.5, 2.5, 2.5)$ (c) $(1.5, 4, -4)$
(b) $(0, 3, 1)$ (d) $(0, -0.5, -1.5)$

8-13 $(3, 4, 6)$

8-14 (*i*) (a) $(1, 2)$ (b) $(1, 4)$ (c) $(2, 4)$
(*ii*) (a) $(7, 2)$ (b) $(7, -1)$ (c) $(2, -1)$
(*iii*) (a) $(3, -1)$ (b) $(3, 5)$ (c) $(-1, 5)$

8-15 (*i*) (a) 4 units above (b) 2 units above
(c) 1 unit above
(*ii*) (a) 1 unit below (b) 2 units above
(c) 7 units above
(*iii*) (a) 5 units above (b) 1 unit below
(c) 3 units above

8-16 (a) approx. $48.1°$ (c) approx. $38.3°$
(b) approx. $14.4°$

9 SOLID FIGURES

THIS CHAPTER IS ABOUT

☑ **Prisms and Circular Cylinders**
☑ **Volume**
☑ **Pyramids and Cones**
☑ **Surface Area**
☑ **Frustra**
☑ **Spheres**

This chapter is about *solid figures*—the three-dimensional counterparts of two-dimensional plane figures. Here we'll discuss **polyhedra**, which are solids whose surfaces consist of intersecting planes. We'll also discuss some specific curved-surface analogs of polyhedra—the circular cylinder, the circular cone, and the sphere.

9-1. Prisms and Circular Cylinders
A. Generating prisms and circular cylinders

Let's consider some figures in a plane R in 3-space, say a polygon—such as a triangle or a rectangle—and a circle. Imagine that we can lift each figure steadily upward off the plane R to some other plane R' parallel to R. The lifting motion must be such that each point of the plane figure steadily sweeps out a straight line PP' from its starting point P to its resting point P'—but the straight line PP' need *not* be perpendicular to planes R and R'. Figure 9-1 shows the results of this lifting process for (**a**) an equilateral triangle, (**b**) a rectangle, (**c**) a circle, and (**d**) an obtuse triangle.

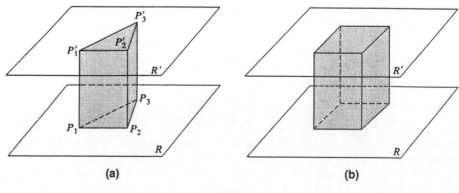

(a)   (b)

Figure 9-1

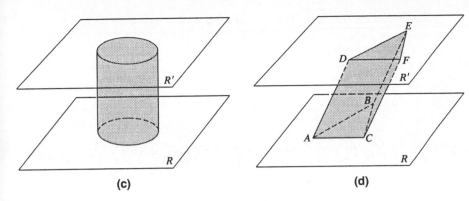

Figure 9-1. Continued

When a plane figure—that is, a set of points P in a plane R—is lifted, point for point, to another plane R' parallel to plane R, it sweeps out another set of points PP' whose *union* is a **geometric solid**: If the plane figure is a polygon, the result of this lifting process is called a **prism**—and if the plane figure is a circle, the result of the lifting process is called a **circular cylinder**.

B. Characteristics of prisms and cylinders

1. Prisms

When we generate a prism from a polygon, we call the original figure in plane R the **lower base** and the displaced copy of this figure in parallel plane R' the **upper base**. These two bases are parallel and congruent. The line segments that join a vertex of the lower base to the corresponding vertex in the upper base are called the **lateral edges** of the prism. The construction of the prism requires that the lateral edges form a set of parallel lines in 3-space. For example, in the prism shown in Figure 9-2 $\triangle ABC$ is the lower base and $\triangle DEF$ is the upper base, while the line segments AD, BE, and CF are the lateral edges and AD is parallel to BE and CF. Thus, as Figure 9-2 shows, if two lateral edges of a prism such as AD and CF join consecutive vertices D, F and A, C of the bases, the lateral edges and the line segments DF and AC of the base all lie in one plane and form a parallelogram. This parallelogram is called a **lateral face** (side) of the prism. Thus we can say that

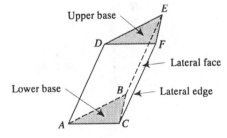

Figure 9-2

- A **prism** is a polyhedron whose bases are congruent polygons in parallel planes and whose lateral faces are parallelograms.

2. Circular cylinders

Just as circles are the curved analogs of polygons, cylinders are the curved-space analogs of prisms. They have no lateral faces or edges, but are characterized by circular bases in parallel planes. Thus

- A **circular cylinder** is a solid figure whose bases are congruent circles in parallel planes.

C. Classifying prisms and cylinders

Prisms and cylinders fall into two fundamental categories, which can be determined by drawing a perpendicular from the plane R of the lower base to the plane R' of the upper base. This perpendicular, whose length is the distance from plane R to plane R', is called the **altitude** or **height** (h). Then we pick any point P on the lower base of the figure and draw another line from this point P to its exact counterpart P' on the upper base. If this line PP'

is *parallel* to the altitude, then the figure is a **right prism** or a **right cylinder**, as shown in Figure 9-3a. If this line *PP'* is *not parallel* to the altitude, the figure is **oblique**, as shown in Figure 9-3b.

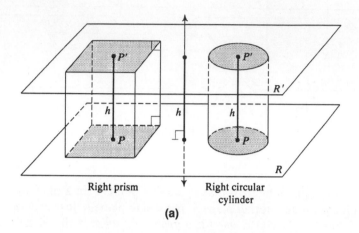

Right prism   Right circular cylinder

(a)

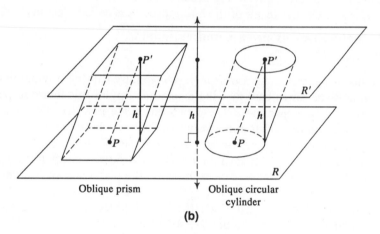

Oblique prism   Oblique circular cylinder

(b)

Figure 9-3

note: In a right prism or cylinder, the length of *PP'* is equal to the distance *h*, but in an oblique prism or cylinder the length of *PP'* is greater than *h*.

As Figure 9-3a shows, the lateral edges of a right prism are parallel to the altitude, which means that they are also perpendicular to the planes of the bases. Thus we can see that the lateral faces of a right prism are not only parallelograms, they are rectangles.

We also classify prisms and cylinders by the shapes of their bases or cross sections. If we pass a plane parallel to the planes of the two bases through a prism, the result is a **cross section** of the prism.

- The cross section of a cylinder or prism is always parallel and congruent to the two bases of the cylinder or prism.

Thus we can say that a prism whose cross sections (or bases) are triangles is a **triangular prism**, a prism whose cross sections are rectangles is a **rectangular prism**, a prism whose cross sections are trapezoids is a **trapezoidal prism**, and so on. Finally, we have a special name for a prism whose cross sections are parallelograms; we call such a prism a **parallelepiped**.

note: We already know the name of the figure whose cross section is a circle—it's a circular cylinder.

EXAMPLE 9-1:   Classify the solid figures shown in Figure 9-4.

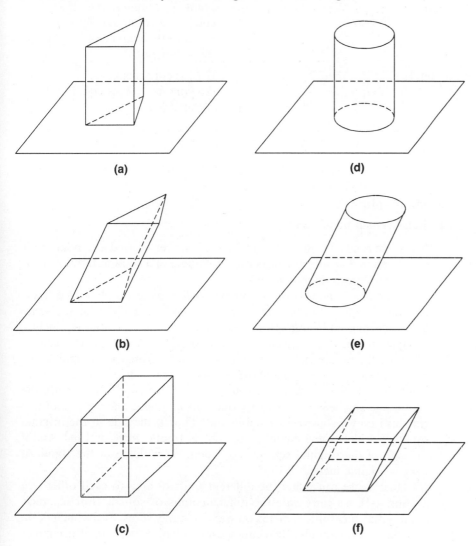

Figure 9-4

Solution

(a) Triangular right prism
(b) Triangular oblique prism
(c) Rectangular right prism
    or right parallelepiped

(d) Right circular cylinder
(e) Oblique circular cylinder
(f) Oblique parallelepiped

9-2. Volume
A. Properties of volume

Just as plane figures enclose regions in the plane, prisms and other 3-space figures enclose *space*. And just as the measure of the region enclosed by a 2-space figure is its area, the measure of the space enclosed by a 3-space figure is its *volume*.

Volume, then, is the direct three-dimensional analog of area and can be defined abstractly in terms of the following properties:

• If T is a set of points in 3-space, its volume V is an assignment of a number $V(T)$ to T such that

(a) $V(T) \geq 0$ Volume is never negative

(b) If $T_1 \cap T_2 = \varnothing$, then
$$V(T_1 + T_2) = V(T_1) + V(T_2)$$

If T_1 and T_2 are two sets of points in 3-space that have no points in common, then the combined volume of $T_1 + T_2$ is equal to the volume of T_1 plus the volume of T_2.

(c) If $T_1 \cong T_2$, then
$$V(T_1) = V(T_2)$$

If T_1 is congruent to T_2, then the volume of T_1 is equal to the volume of T_2.

note: T_1 is said to be congruent to T_2 if the figure represented by T_1 is the exact size and shape as the figure represented by T_2.

Notice that the properties of volume are just like those of area (see Section 4-1), except they are applied to 3-space sets rather than 2-space sets.

B. Finding volumes

1. Standard unit of volume

Just as we need a standard unit of area in 2-space, we need a standard unit of volume in 3-space. We measure the volume of a solid in terms of how many standard units it contains.

To find a standard unit of volume, we first picture a plane in 3-space on which a square has been drawn, so that each side of the square is one unit of distance in length. (The units can be any unit of distance—inches, centimeters, meters, ...) Now we form a right prism from this square by lifting the square exactly one unit straight up from the original plane. The resulting prism is a standard unit of volume called a **cube**, as illustrated in Figure 9-5. Then, since each side of this cube—the length, the width, and the height—is exactly one unit in length, the volume of this standard unit is known as 1 cubic unit. Thus if the side of the original square is 1 inch, the volume of the cube is 1 cubic inch, or 1 in³. And if the side of the original square is 1 meter, the volume of the standard cube is 1 cubic meter, or 1 m³.

Just as we can relate the different units of area to each other (see Section 4-1), we can relate the different units of volume. For instance, a cube that measures 1 meter on each side measures 10 decimeters on each side, so it contains 1000 cubes, each of which measures 1 decimeter on a side. That is,

$$1 \text{ m}^3 = (10 \text{ dm})(10 \text{ dm})(10 \text{ dm}) = 1000 \text{ dm}^3$$

(See Figure 9-6.) Similarly, a cube that measures 1 decimeter on each

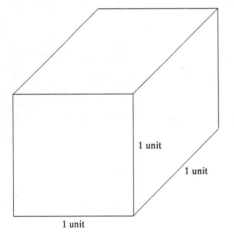

1 unit

1 unit

1 unit

Figure 9-5

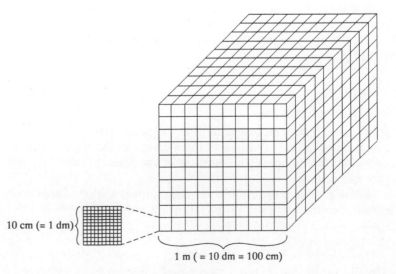

10 cm (= 1 dm)

1 m (= 10 dm = 100 cm)

Figure 9-6

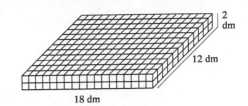

Figure 9-7

side measures 10 centimeters on each side, so it contains 1000 cubes, each of which measures 1 centimeter on a side. That is,

$$1 \text{ dm}^3 = (10 \text{ cm})(10 \text{ cm})(10 \text{ cm}) = 1000 \text{ cm}^3$$

Thus

$$1 \text{ m}^3 = 1000 \text{ dm}^3 = (1000)(1000 \text{ cm}^3) = 1{,}000{,}000 \text{ cm}^3$$

Now consider a right rectangular prism whose base measures 1.2 meters by 1.8 meters and whose height is 0.2 meters. Since there are 10 decimeters in a meter, this rectangular prism must therefore have a base that measures 18 decimeters by 12 decimeters and a height of 2 decimeters, as shown in Figure 9-7. The volume of this prism must therefore be

$$(18 \text{ dm})(12 \text{ dm})(2 \text{ dm}) = 432 \text{ dm}^3$$

That is, this prism contains 432 one-decimeter cubes. We can also say that since 1 cubic meter equals 1000 cubic decimeters, the volume of this prism is 0.432 cubic meters.

2. Volume of a right rectangular prism

As we have just seen, finding the volume of a right rectangular prism is simple. It can be reduced to a formula, where a is the length of the base, b is the width of the base, and c is the altitude:

VOLUME OF A RIGHT RECTANGULAR PRISM	$V = abc$

That is, the volume of a right rectangular prism is the length of the base a times the width of the base b times the altitude c. (See Figure 9-8.)

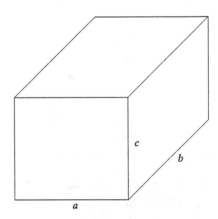

Figure 9-8

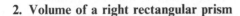

EXAMPLE 9-2: Find the volume of a right rectangular solid whose base is 23 centimeters by 14 centimeters and whose height is 6 decimeters.

Solution: First convert all the measurements into the same units. If you choose to work in centimeters, the prism is 23 cm by 14 cm by 10(6) = 60 cm. Then the volume is

$$V = abc$$
$$= (23 \text{ cm})(14 \text{ cm})(60 \text{ cm}) = 19{,}320 \text{ cm}^3$$

Or, in decimeters,

$$(2.3 \text{ dm})(1.4 \text{ dm})(6 \text{ dm}) = 19.320 \text{ dm}^3$$

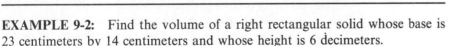

3. Volumes of other right prisms and cylinders

We can use volume properties **(b)** and **(c)** (Section 9-2A) to find the volume of any right prism or right cylinder. Take a right prism whose bases are right triangles, for example. If a right prism has a right triangle as its base, where the legs of the triangle measure a and b and the altitude of the prism measures c, then this triangular prism is really half of a right rectangular prism that measures a by b by c, as shown in Figure 9-9. Now, since one half of a right rectangular prism is congruent to the other half, the two halves must have equal volumes, by property **(c)**. And by property **(b)**, the volumes of the two halves must sum to abc, which is

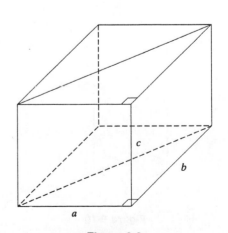

Figure 9-9

the volume of the right rectangular prism. Thus, the volume of a right triangular prism is

VOLUME OF A RIGHT TRIANGULAR PRISM $$V = \frac{abc}{2}$$

We could go on like this for general triangles, polygons, and even circles—but we won't. Instead, a moment's thought tells us the principle involved. What we're really doing in each case is computing the volume of a right prism or cylinder by finding the area A of a base and then multiplying that area by the altitude h of the solid. That is,

VOLUME OF ANY RIGHT PRISM (OR CYLINDER) $$V = Ah$$

or

- The volume of a right prism (or cylinder) is the product of the altitude and the base area.

note 1: Recall that the area of any triangle may be calculated by the formula $A = bh/2$, where b is the base of the triangle and h is the altitude or height of the triangle (see Section 4-2B). Also recall that the area of a circle may be calculated by the formula $A = \pi r^2$, where $\pi = 3.1416\ldots$ and r is the radius of the circle (see Section 7-3B).

note 2: When you have a prism with a base whose area cannot be calculated with a simple formula, you can use a modification of volume property (**b**) similar to our modification of area property (**b**) (Section 4-1C). Thus, if set T lies completely in a plane, then $V(T) = 0$. So, if figures T_1 and T_2 have points in common, but overlap only in a plane figure (either a face or an edge), you may still use property (**b**) to say that $V(T_1 + T_2) = V(T_1) + V(T_2)$. This really means that volume measures enclosed space only. The faces and edges of a 3-space figure are not enclosed space, but rather are boundary points between enclosed and unenclosed space. As such, they don't enter into the calculation of volume.

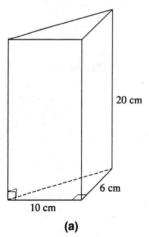

(a)

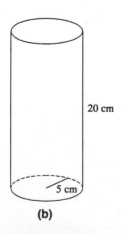

(b)

Figure 9-10

EXAMPLE 9-3: Find the volume of (**a**) the right triangular prism and (**b**) the right circular cylinder shown in Figure 9-10.

Solution

(**a**) The base of the prism is a right triangle with legs of 10 cm and 6 cm. The area of the base is therefore

$$A = \frac{(10\ \text{cm})(6\ \text{cm})}{2} = 30\ \text{cm}^2$$

The height of the prism is 20 cm. The volume of the prism is therefore

$$V = Ah = A(20\ \text{cm}) = (30\ \text{cm}^2)(20\ \text{cm}) = 600\ \text{cm}^3$$

(**b**) $$V = Ah = \pi r^2 h = \pi(5\ \text{cm})^2(20\ \text{cm}) = 1570.8\ \text{cm}^3$$

EXAMPLE 9-4: Find the volume of the right prism shown in Figure 9-11.

Solution: The height of this prism is 5 cm. The base of the prism is a polygon known as a Swiss cross, which is formed by removing four squares from a larger square. In this prism, four squares of sides 2 cm have been removed from a square of side 6 cm. Thus the area of the base of this polygon is

$$A = (6 \text{ cm})(6 \text{ cm}) - 4(2 \text{ cm})(2 \text{ cm})$$
$$= 36 \text{ cm}^2 - 16 \text{ cm}^2$$
$$= 20 \text{ cm}^2$$

And the volume of the prism is

$$V = Ah = (20 \text{ cm}^2)(5 \text{ cm}) = 100 \text{ cm}^3$$

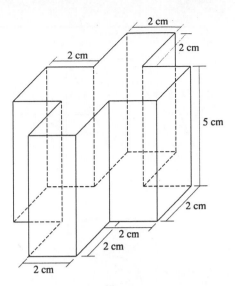

Figure 9-11

4. Oblique prisms and cylinders: Cavalieri's Principle

Let's imagine we have a right prism made up of cross-sectional slices, i.e., many smaller right prisms stacked one on top of the other, as shown in Figure 9-12a. Then, if we slide this stack of small prisms so that it's tilted as shown in Figure 9-12b, we get a figure that looks a little like an oblique prism except for the steplike edges. Next, suppose we could subdivide each of these prisms into thinner and thinner slices, thereby making the edge steps smaller and smaller. After a number of these subdivisions, the tilted stack of prism-slices begins to look more and more like an oblique prism. In fact, we can say that this process of subdividing and sliding a right prism is a way of *approaching* an oblique prism. That is, an oblique prism may be seen as the *limit* of the process of sliding very, very small cross sections of a right prism to one side.

What does this observation tell us about the volume of an oblique prism? To answer this question, let's make some more observations. First, we can see that at each stage of the sliding process, every thin slice has a constant volume, regardless of whether the slice is in the right-prism stack or in the oblique-prism stack. We also know, from volume property **(b)**, that the volume of a solid can be found by summing the volume of all its parts—or in this case, of all its slices. Finally, we can see that if we push the slices in the oblique prism back into a right prism, we get a right prism whose base area is the same as that of the oblique prism and whose altitude is the same length as the *altitude* of the oblique prism.

Now we can draw a simple conclusion:

• **The volume of any prism is the product of its altitude and the area of its base.**

VOLUME OF ANY PRISM $V = Ah$

This discussion of the volume of an oblique prism is an application of **Cavalieri's Principle**, which may be stated as follows:

• **Given two solids and a plane, such that every plane parallel to the given plane intersects either both solids or neither solid, if every plane that intersects both solids cuts out equal-area cross sections of both solids, then the two given solids must have equal volumes.**

note: From Cavalieri's Principle, you can see that the formula for finding volume of any prism also applies to circular cylinders. Thus the volume of a cylinder is $V = Ah = \pi r^2 h$.

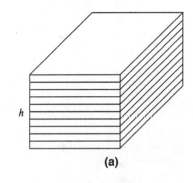

(a)

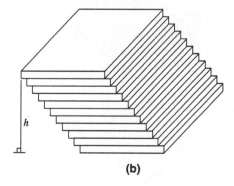

(b)

Figure 9-12

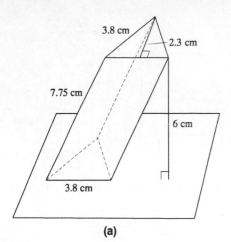

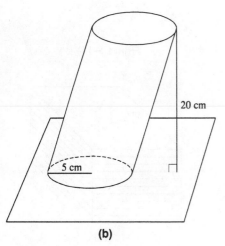

(a)

(b)

Figure 9-13

EXAMPLE 9-5: Find the volume of (**a**) the oblique prism and (**b**) the oblique circular cylinder shown in Figure 9-13.

Solution

(**a**) The base of this prism is an isosceles triangle, two of whose sides measure 3.8 cm each. An altitude from one of the vertices to a side of this triangle is 2.3 cm. Thus the area of the prism's base is

$$A = \frac{(2.3 \text{ cm})(3.8 \text{ cm})}{2} = 4.37 \text{ cm}^2$$

The altitude of the prism is 6 cm. The volume of the prism is therefore

$$V = Ah = (4.37 \text{ cm}^2)(6 \text{ cm}) = 26.22 \text{ cm}^3$$

(**b**) $$V = Ah = \pi r^2 h = \pi(5 \text{ cm}^2)(20 \text{ cm}) = 1570.8 \text{ cm}^3$$

9-3. Pyramids and Cones

A. Characteristics of pyramids and cones

1. General pyramid

Let's begin with a polygon in plane *R*. If we pick a point *V* off the plane *R* and draw a line segment from *V* to each vertex of the polygon, we generate a figure like one of those shown in Figure 9-14a and b. This type of figure is known as a pyramid.

- A **pyramid** is a pointed figure whose base is a polygon and whose vertex is not in the same plane as the polygon. Its *lateral edges* are line segments drawn from the vertex of the pyramid to the vertices of the polygon, and its *lateral faces* are triangles formed from two lateral edges of the pyramid and a side of the base polygon.

In Figure 9-14a, for example, the line segments *VA*, *VB*, *VC*, *VD*, and *VE* are lateral edges, and the triangles △*VAB*, △*VBC*, △*VCD*, △*VDE*, and △*VEA* are lateral faces.

If we drop a perpendicular from the vertex of a pyramid to a point on its base plane, this perpendicular, whose length is the distance from the vertex to the base, is called the **altitude (or height) of the pyramid**. For example, the line segments *h* and *h'* are the altitudes of the pyramids shown in Figure 9-14.

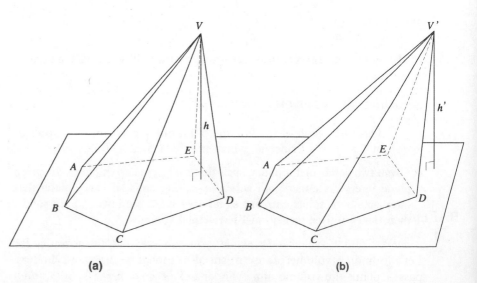

(a)   **(b)**

Figure 9-14

2. Regular pyramids

If a pyramid has a regular (equal-sided) polygon for its base and its vertex is so situated that the altitude of the pyramid meets the center of the base polygon, then that pyramid is known as a **regular pyramid**. In a regular pyramid

- the lateral edges are all congruent line segments, and
- the lateral faces are all congruent isosceles triangles.

Figure 9-15 shows a regular pyramid whose base is a regular (equilateral) triangle $\triangle ABC$. Notice that the line segment VO is the altitude of the pyramid, where O is the center of the base triangle. Also notice that $VA \cong VB \cong VC$ and $\triangle VAB \cong \triangle VBC \cong \triangle VCA$.

Finally, in a regular pyramid, an altitude drawn from the vertex V of the pyramid to the base of any of the face isosceles triangles is called the **slant height of the pyramid**. In Figure 9-15, the line segment VD is the slant height of the pyramid—as well as the altitude of $\triangle VAB$.

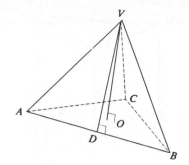

Figure 9-15

EXAMPLE 9-6: Show that the lateral edges of a regular pyramid are all congruent.

Solution

Step 1: Draw a regular pyramid like that shown in Figure 9-16. Draw altitude VO, and from any two consecutive vertices such as A and B of the base, draw radii OA and OB.

Step 2: $\triangle VAO$ and $\triangle VBO$ are right triangles by the definition of the altitude of a pyramid.

Step 3: $\angle VOA \cong \angle VOB$, since they are both right angles.

Step 4: $OA \cong OB$, since they are both radii of a regular polygon.

Step 5: $VO \cong VO$.

Step 6: $\triangle VAO \cong \triangle VBO$ by SAS.

Step 7: $VA \cong VB$ as corresponding parts of congruent triangles.

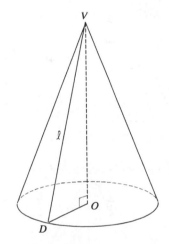

Figure 9-16

---

3. Circular Cones

We know that if we steadily increase the number of sides a regular polygon has, the perimeter of the polygon steadily approaches a circle as a limit (see Section 7-2). Thus, if we steadily increase the number of sides in the base of a regular pyramid, the base will slowly fill out into a circle and the pyramid will become a **right circular cone** like the one shown in Figure 9-17. Notice that, as a result of the limiting process, all the face triangles—which become thinner and thinner as the base polygon gains more sides—collapse to a line segment from the vertex to a point on the base circle in the cone. We call this line segment the **slant height of the cone**. The collection of all possible slant heights forms the surface of the cone.

note: If the base of a regular pyramid is changed to a circle, the result is a right circular cone. But any pyramid can be changed to a cone, including a pyramid whose altitude does not meet the center of its base polygon, so a circular cone may also be oblique.

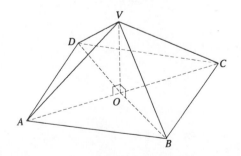

Figure 9-17

B. Finding the volume of a pyramid and cone

1. Volume of a pyramid

Let's look at Figure 9-18a, which shows a triangular right prism. If we pass a plane through points A, G, and B and another plane through points A, G, and F, we get the three pyramids shown in Figure 9-18b.

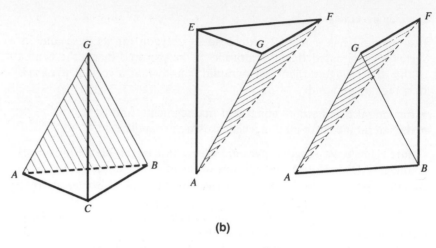

(a)   (b)

Figure 9-18

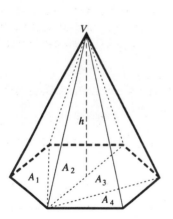

Figure 9-19

These three pyramids, *ACBG*, *EGFA*, and *ABGF* have equal volumes. (Can you see why this is true? See Solved Problems 9-4 and 9-5.) Thus the volume of any one of them by itself is equal to one-third the volume of the original prism. So it's easy to understand that the volume of one of these pyramids, say *ACBG*, is one-third the base area of the prism times the length of its altitude. And since we started with a right prism, we know that the length of its altitude is equal to that of a lateral edge; so the volume of pyramid *ACBG* is equal to the area of △*ACB* times the length of *GC* times 1/3.

Now consider an arbitrary pyramid with vertex *V* and some polygonal base. Such a pyramid is illustrated in Figure 9-19, where the polygonal base is an irregular hexagon. If we draw diagonals in the base polygon of an arbitrary pyramid, we cut that polygon into $n - 2$ triangles, $A_1, A_2, \ldots, A_{n-2}$. Thus, in Figure 9-19, we have triangles A_1, A_2, A_3, and A_4. Each of these triangles is the base of another, smaller pyramid whose vertex is also *V*—that is, each of these smaller triangular pyramids has the same vertex as the original arbitrary pyramid. Then, by the additive volume property (**b**), the volume of the original pyramid must be the sum of the volumes of each of the smaller triangular pyramids.

Next we see that all of the bases of the small triangular pyramids, each of which has the same vertex, are in the same plane as the base of the original pyramid. This means that their altitudes are of equal length; that is, all the triangular pyramids and the original arbitrary pyramid all have altitude *h*. Thus, if *h* is the altitude, *V* the volume, and *A* the area,

$$V(\text{any pyramid}) = V(\text{1st triangular pyramid}) + \cdots$$
$$+ V(\text{last triangular pyramid})$$

$$= \frac{hA(\text{1st triangle base})}{3} + \cdots + \frac{hA(\text{last triangle base})}{3}$$

VOLUME OF ANY PYRAMID
$$= \frac{h[A(\text{1st triangle base}) + \cdots + A(\text{last triangle base})]}{3}$$

$$= \frac{hA(\text{polygon base})}{3}$$

That is

- **For any pyramid, the volume is one-third the product of its altitude and its base area.**

2. Volume of a cone

Consider a regular pyramid whose base is an *n*-sided polygon inscribed in the cone's base circle and whose vertex is the same as the cone's. Both the cone and the pyramid have the same altitude, *h*. As the number of sides *n* increases, the space inside the pyramid approaches the space inside the cone. The volume of the pyramid is $(1/3)hA$(polygon base), and this approaches $(1/3)hA$(circle base), the volume of the cone, as a limit. Thus

VOLUME OF A CONE $$V = \frac{hA}{3} = \frac{h\pi r^2}{3}$$

- **The volume of a cone is one-third the product of its altitude and its base area.**

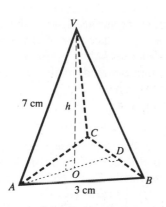

Figure 9-20

EXAMPLE 9-7: Find the volume of the regular triangular pyramid shown in Figure 9-20.

Solution: Find this volume in steps.

Step 1: Find the area of the pyramid's base. Since the base is an equilateral triangle of side $s = 3$ cm, you can calculate the area from the area formula given in Section 4-7C:

$$A = \frac{\sqrt{3}s^2}{4} = \frac{\sqrt{3}(3 \text{ cm})^2}{4} = \frac{\sqrt{3}(9 \text{ cm}^2)}{4} = 3.897 \text{ cm}^2$$

Step 2: Find the length *h* of the altitude of the pyramid.
  (a) Since the altitude of a regular pyramid is a perpendicular from the vertex of a pyramid to the center of its polygonal base, the altitude *VO* forms one of the legs of a right triangle *AOV* whose hypotenuse *AV* is 7 cm and whose other leg is *AO*.
  (b) Leg *AO* is the distance from the vertex of an equilateral triangle to the centroid *O* of the triangle along its altitude (median) *AD*. The length of *AO* is therefore (see Section 4-7C)

$$AO = \frac{\sqrt{3}s}{3} = \frac{\sqrt{3}(3 \text{ cm})}{3} = \sqrt{3} \text{ cm} = 1.732 \text{ cm}$$

  (c) Use the Pythagorean Theorem to find the length of $VO = h$:

$$h = VO = \sqrt{(7 \text{ cm})^2 - (1.732 \text{ cm})^2} = \sqrt{46 \text{ cm}^2} = 6.782 \text{ cm}$$

Step 3: Find one-third the product of the base area of the pyramid and the altitude to get the volume:

$$V = \frac{hA}{3} = \frac{(6.782 \text{ cm})(3.897 \text{ cm}^2)}{3} = 8.81 \text{ cm}^3$$

note: Get into the habit of checking your answers for *reasonability*. It can save a lot of grief from careless errors. The pyramid certainly lies inside a right prism with base 3 by 3 cm square and height of 7 cm. This prism thus has a volume of 63 cm³. Your calculated volume for the pyramid should be a lot less than this number. If it isn't, you probably made a mistake with the calculator buttons.

EXAMPLE 9-8: Find the volume of a cone whose height is 7 inches and whose base has a diameter of 4.6 inches.

Solution: The radius of the base of the cone is one-half the length of the diameter, so the length of the radius $r = 2.3$ inches. Thus the area of the base is

$$A = \pi r^2 = (3.1416\ldots)(2.3 \text{ in.})^2 = 16.62 \text{ in}^2$$

And the volume of the cone is

$$V = \frac{hA}{3} = \frac{(7 \text{ in.})(16.62 \text{ in}^2)}{3} = 38.78 \text{ in}^3$$

9-4. Surface Area

When we look at a solid 3-space figure, we see the outside, or *surface*, of the figure. The **surface** of the solid is the *union* of all its faces—the lateral faces and base faces, all of which are 2-space figures. Thus, if we were to paint a 3-space figure, we would be painting all the faces, and we'd use up a certain amount of paint—the same amount of paint that would be used up if all the faces were cut apart and laid out on a single 2-dimensional plane. The combined area of all the faces and bases of a 3-space solid, then, is the **surface area** of the solid.

There is no general formula for calculating the surface area of any 3-space figure. Instead, we have to find the area of each face separately—view it as an isolated 2-space figure and measure its sides—and then add up all the areas of all the faces and bases.

Some special solids, however, do have formulas.

A. Surface area of a right prism

In a right prism, each lateral edge is perpendicular to the base, so the lateral faces of the prism are all rectangles and the altitude h of the prism is equal to the length of the lateral edge. Then, as shown in Figure 9-21, each rectangular face has as one measurement the length h and, as the other measurement, the length l_i, which is a side of the base polygon. The area of each of the lateral faces is therefore hl_i. Thus for a right prism with an n-sided base the area of all the lateral faces is

$$hl_1 + hl_2 + \cdots + hl_n = h(l_1 + l_2 + \cdots + l_n) = hP$$

where $P = l_1 + l_2 + \cdots + l_n$ is the length of the base perimeter. So we can say that the lateral surface area S_L of a right prism is the product of its height and perimeter, or

LATERAL SURFACE AREA OF A RIGHT PRISM $\qquad S_L = hP$

And the total surface area S_T of a right prism must therefore be the product of its height and perimeter plus the area A of each of its bases, or

TOTAL SURFACE AREA OF A RIGHT PRISM $\qquad S_T = hP + 2A$

B. Surface area of a regular pyramid

In a regular pyramid, as shown in Figure 9-22, the lateral faces are all congruent isosceles triangles whose altitude is actually the slant height l of the pyramid and whose base side s is an edge of the pyramid's base. The area of each of the n lateral faces is then $ls/2$, so the area of all n lateral faces is $nls/2$. So we can say that the lateral surface area of a regular pyramid is one-half the product of its slant height l and the perimeter P of its base, where $P = ns$, or

LATERAL SURFACE AREA OF A REGULAR PYRAMID $\qquad S_L = \dfrac{lP}{2}$

Then if A is the area of the pyramid's base, the total surface area S_T of a regular pyramid is one-half the product of its slant height and the perimeter

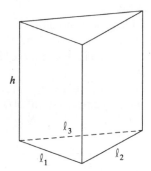

Figure 9-21

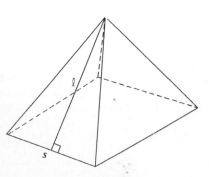

Figure 9-22

of its base plus the area of its base, or

TOTAL SURFACE AREA OF A REGULAR PYRAMID $S_T = \dfrac{lP}{2} + A$

C. Surface area of a right circular cone

We can think of right circular cones as a special case of regular pyramids. That is, the surface area of a cone is what the surface area of a regular pyramid approaches as *n* (the number of lateral faces) increases. Thus, if a right cone's base has radius *r*, its perimeter becomes $2\pi r$ and its lateral surface area becomes

LATERAL SURFACE AREA OF A RIGHT CONE $S_L = \dfrac{lP}{2} = \dfrac{l2\pi r}{2} = l\pi r$

while the area of the right cone's base is πr^2. So the total surface area of a right cone whose base has radius *r* is

TOTAL SURFACE AREA OF A RIGHT CONE $S_T = l\pi r + \pi r^2$

EXAMPLE 9-9: Find the surface area of the right prism shown in Figure 9-11 (Example 9-4), whose base is a Swiss cross with sides of 2 cm and height 5 cm.

Solution: You know from Example 9-4 that the area *A* of the base of the figure is 20 cm². And since there are 12 sides in a Swiss cross, the perimeter *P* of the figure is 12(2 cm) = 24 cm. So the total surface area must be

$$S_T = hP + 2A = (5 \text{ cm})(24 \text{ cm}) + 2(20 \text{ cm}^2) = 160 \text{ cm}^2$$

EXAMPLE 9-10: Find the surface area of the regular pyramid shown in Figure 9-23, whose base is a regular hexagon of radius 5 cm and whose slant height is 7 cm.

Solution: In order to find the surface area of this pyramid, you need to know the measure of the side of its base. But that measure isn't given, so you have to figure it out. First, then, recall that a regular *n*-gon can be cut up into *n* congruent isosceles triangles whose sides are the radii of the *n*-gon and whose apex is the center of the *n*-gon (see Section 6-2). So if you have a regular hexagon, where *n* = 6, you can construct six triangles whose legs are all equal in measure and whose apex angles are all 360°/6 = 60°. This means that each of the angles in these triangles must be 60°, so the triangles are equilateral. Therefore, if the radius of the given hexagon is 5 cm, its side must also be 5 cm.

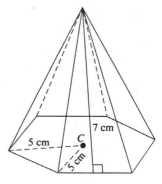

Figure 9-23

If the side of the pyramid's base is 5 cm and the pyramid's slant height (given) is 7 cm, then the lateral surface area is

$$S_L = \frac{lP}{2} = \frac{6(5 \text{ cm})(7 \text{ cm})}{2} = 105 \text{ cm}^2$$

The base area is simply the area of six equilateral triangles of side 5 cm, or

$$A = 6\left(\frac{\sqrt{3}s^2}{4}\right) = \frac{6\sqrt{3}(25 \text{ cm}^2)}{4} = 64.95 \text{ cm}^2$$

So the entire surface area is $S_L + A = 105 \text{ cm}^2 + 64.95 \text{ cm}^2 = 169.95 \text{ cm}^2$

9-5. Frustra

Given a pyramid or a cone, we can pass a plane parallel to the base through the solid, so that the solid is cut into two pieces, as shown in Figure 9-24. Then, if we discard the top, pointed piece—i.e., a small pyramid or a small cone— we are left with a truncated figure called a **frustrum** (pl. *frustra*).

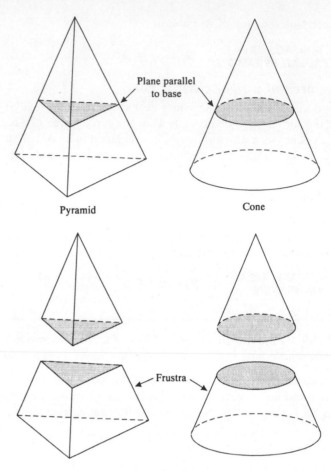

Figure 9-24

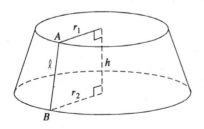

Figure 9-25

Just like any other solid, frustra have volumes and surface areas. And we can generally find these volumes and surface areas by subtracting the volumes and lateral surface areas of the smaller pyramid or cone (i.e., the "discards") from the larger pyramid or cone.

In the case of the frustrum of a right circular cone, as shown in Figure 9-25, we can also use a formula to find the volume and the surface area. Let r_1 be the radius of the upper base of the circular frustrum and let r_2 be the radius of the lower base. Now, if we draw r_1 parallel to r_2, we form a plane that intersects the side of the frustrum along a line segment, which we can label AB. This line segment AB is the slant height of the frustrum, which has length l. The lateral surface area of the circular frustrum is therefore $S_L = \pi l(r_2 + r_1)$, and the total surface area is

TOTAL SURFACE AREA OF A FRUSTRUM OF A RIGHT CIRCULAR CONE
$$S_T = \pi l(r_2 + r_1) + \pi(r_2^2 + r_1^2)$$

The volume of the circular frustrum is

VOLUME OF A FRUSTRUM OF A RIGHT CIRCULAR CONE
$$V = \frac{\pi h}{3}(r_2^2 + r_1^2 + r_1 r_2)$$

where h is the distance between the planes of the upper and lower bases of the frustrum—i.e., the altitude.

9-6. Spheres

A regular polyhedron has *n* faces, each of which has edges of the same length. If we increase the number of faces in a polyhedron, these faces eventually collapse to form a sphere—just as polygons form circles—as a limit.

note: More formally, we can define a sphere as follows:

> **Let *C* be a point and let *r* be any positive number. The set of all points whose distance from *C* is equal to *r* is the sphere with center *C* and radius *r*.**

In other words (i.e., English), a sphere is the surface of a round ball in space.

A. Volume of a sphere

It's mathematically possible to develop this limit concept, from polyhedron to sphere, to determine the volume of a sphere, but such a procedure is outside the "sphere" of this book. So we'll fall back on Cavalieri's Principle instead. But to do this, we have to start with some construction, as follows:

First, we set a sphere with center *C* and radius *r* on a plane *R*, and place a right circular cylinder with base radius *r* and height $2r$ next to the sphere on plane *R*, as shown in Figure 9-26a. Then, letting *V* be the midpoint of the line segment connecting the centers of the cylinder bases (i.e., the *axis* of the cylinder), we draw two cones inside the cylinder. These two cones both have this midpoint *V* as their vertex, while the upper and lower bases of the cylinder are the bases of the upper and lower cones.

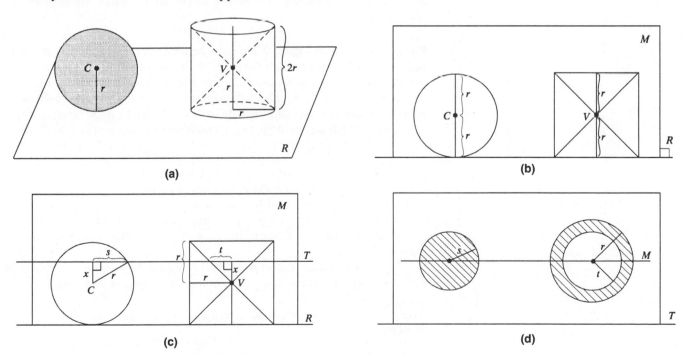

Figure 9-26

Now, if we pass a plane *M* through *C* and *V* such that *M* is perpendicular to plane *R*, we see that plane *M* cuts out a vertical cross section of each solid—a circle of radius *r* from the sphere and a square of side $2r$ from the cylinder, as shown in Figure 9-26b. Note that points *C* and *V* are the same distance (*r*) from *R* in both figures, that both plane figures are $2r$ tall, and that the cross section of the two cones forms the diagonals of a square.

Next, we observe that any plane parallel to R must either intersect both solids or miss both solids (one of the conditions of Cavalieri's Principle). Thus we can pass a plane T parallel to R through both solids, as in Figure 9-26c, so that x is the (perpendicular) distance from the center of each solid to the plane T.

The result of passing plane T through the solids is shown in Figure 9-26d, which is the view we'd have if we were to look directly down onto plane T. Here we see that T cuts out a circle of radius s from the sphere, a circle of radius r from the cylinder, and a circle of radius $t < r$ from one of the cones inside the cylinder. The circles of radius r and radius t have the same center, so they are called *concentric* circles. And the doughnut-shaped region between the concentric circles is known as an **annulus**.

Having done all that, we now need to show that the plane that cuts out a cross-sectional area from each solid does in fact cut out equal-area cross sections in order to apply Cavalieri's Principle. So let's look back at Figure 9-26c to determine some areas.

The circle cut from the sphere by plane T has radius s, and s is one of the legs of a right triangle whose other leg is x and whose hypotenuse is r. By the Pythagorean Theorem, then, $s^2 = r^2 - x^2$, so the area of this circle A_C must be

$$A_C = \pi s^2 = \pi r^2 - \pi x^2$$

Now the circle cut from the cylinder has radius r, so its area is πr^2; and the circle cut from the cone has radius t, so its area is πt^2. Thus the area of the annulus A_A must be the area of the larger circle minus the area of the smaller circle, or

$$A_A = \pi r^2 - \pi t^2$$

But we can see from Figure 9-26c that the cones in the cylinder have a base radius r and height r, which means that the diagonals of the square form isosceles right triangles whose sides are r and whose base angles must be $45°$ each. Thus the diagonals of the square must also form $45°$ angles with plane T. Therefore the right triangle with sides x and t must also have two $45°$ angles, which means that $x = t$. Therefore, we can write

$$\pi r^2 - \pi t^2 = \pi r^2 - \pi x^2 = A_C = A_A$$

Thus the areas of the circle cut from a sphere and of an annulus cut from a cone-within-a-cylinder by a plane that intersects both solids are equal if the radius of the sphere is equal to the radius of the cylinder. And since these cross-sectional areas are equal, the volumes of the sphere and the volume of the cylinder minus the volume of the cones must also be equal by Cavalieri's Principle.

Now, we can determine a formula for finding the volume of a sphere of radius r using the formulas for the volumes of cylinders and cones:

VOLUME OF A SPHERE

Volume of sphere = Volume of cylinder

$$- \text{ Volume of two cones}$$

$$= \pi r^2 h - 2\left(\frac{\pi r^{2r}}{3}\right)$$

$$= \pi r^2 (2r) - 2\left(\frac{\pi r^3}{3}\right) \qquad \text{where } h = 2r$$

$$= 2\pi r^3 - \frac{2\pi r^3}{3}$$

$$= \frac{4\pi r^3}{3}$$

EXAMPLE 9-11: A storage tank is formed in the shape of a circular frustrum whose base radius is 12 feet and top radius is 8 feet. The top and bottom of the tank are 4 feet apart. A hemispherical (half a sphere) dome is placed on top of the frustrum so that the top circle is a great circle of the sphere. Find the volume of the tank.

Solution: The volume of the tank must be the volume of the frustrum plus the volume of the hemisphere. So you find the volume of the frustrum V_F from the formula

$$V_F = \frac{\pi h(r_2^2 + r_1^2 + r_1 r_2)}{3} = \frac{\pi(4 \text{ ft})[(12 \text{ ft})^2 + (8 \text{ ft})^2 + (12 \text{ ft})(8 \text{ ft})]}{3} = 1273.4 \text{ ft}^3$$

Then you find the volume of the hemisphere V_H by using the formula for the volume of a sphere divided by 2:

$$V_H = \frac{1}{2} \cdot \frac{4\pi r^3}{3} = \frac{2\pi r^3}{3} = \frac{2\pi(8 \text{ ft})^3}{3} = 1072.3 \text{ ft}^3$$

So the volume of the tank is

$$V_F + V_H = 1273.4 \text{ ft}^3 + 1072.3 \text{ ft}^3 = 2345.7 \text{ ft}^3$$

B. Surface area of a sphere

Consider two concentric spheres such that one has radius r and the other has radius $r + h$, where h is very small (see Figure 9-27). The volume inside the outer sphere but outside the inner sphere is, in effect, a thin shell for a small h.

note: Imagine that you could cut open a hollow rubber ball, flatten out the resulting piece of rubber, and cut it into smaller pieces whose areas you could calculate. Then if you multiplied the sum of all those areas by the thickness of the rubber (h), you'd have the volume of the rubber "shell." Keep thinking about this hollow ball as you read through the following explanation of surface area.

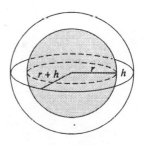

Figure 9-27

Thus, if the surface area of the inner sphere is S, the volume of the shell is the product Sh. But the volume of the shell is also the volume of the larger sphere minus the volume of the smaller sphere, so we have

Volume of shell = Volume of larger sphere − Volume of smaller sphere

$$Sh = \frac{4\pi(r + h)^3}{3} - \frac{4\pi r^3}{3} = \frac{4\pi[(r + h)^3 - r^3]}{3}$$

$$= \frac{4\pi(r^3 + 3r^2 h + 3rh^2 + h^3 - r^3)}{3}$$

$$= \frac{4\pi(3r^2 h + 3rh^2 + h^3)}{3}$$

Then if we divide both sides by h, we get

$$S = \frac{4\pi(3r^2 + 3rh + h^2)}{3}$$

Now the surface area S of a sphere should not depend on the thickness h of the shell. In reality, the product Sh is only a very good approximation to the volume of the shell—and this approximation becomes better and better as h approaches 0. So, if we let $h \to 0$, that is, if we let h be 0 for all practical purposes, the equation becomes

<table>
<tr><td>**SURFACE
AREA OF A
SPHERE**</td><td>$S = \dfrac{4\pi[3r^2 + 3r(0) + 0^2]}{3} = \dfrac{4\pi(3r^2)}{3} = 4\pi r^2$</td></tr>
</table>

- So the surface area of a sphere of radius r is $4\pi r^2$.

EXAMPLE 9-12:   Find the surface area of the tank described in Example 9-11 exclusive of the lower base.

Solution: The area you want is half the surface area of a sphere of radius 8 feet plus the entire lateral surface area of a circular frustrum whose upper and lower bases have radii of 8 and 12 feet, respectively, and whose height is 4 feet.
The surface area of the hemisphere would be half the surface area of the sphere, or

$$S_{\text{hemisphere}} = \frac{1}{2} \cdot 4\pi r^2 = 2\pi r^2 = 2\pi(8 \text{ ft})^2 = 402.1 \text{ ft}^2$$

Since the lateral surface area of a circular frustrum is $\pi l(r_2 + r_1)$, you have to find the frustrum's slant height l. You can do this by treating the slant height as the hypotenuse of a right triangle whose legs are the height ($h = 4$ feet) and the radius of the larger base ($r_2 = 12$ feet) minus the radius of the smaller base ($r_1 = 8$ feet). Thus the slant height can be calculated by the Pythagorean Theorem:

$$\sqrt{l^2} = \sqrt{h^2 + (r_2 - r_1)^2}$$
$$l = \sqrt{(4 \text{ ft})^2 + [(12 \text{ ft}) - (8 \text{ ft})]^2}$$
$$= \sqrt{2(4 \text{ ft})^2} = (4 \text{ ft})\sqrt{2}$$
$$= 5.66 \text{ ft}$$

Therefore the lateral surface area of the frustrum is

$$S_{L(\text{frustrum})} = \pi l(r_2 + r_1) = \pi(5.66 \text{ ft})(20 \text{ ft}) = 355.4 \text{ ft}^2$$

And finally, the surface area of the tank exclusive of the lower base is

$$S_{\text{tank}} = S_{\text{hemisphere}} + S_{L(\text{frustrum})} = (402.1 \text{ ft}^2) + (355.4 \text{ ft}^2) = 757.5 \text{ ft}^2$$

SUMMARY

1. The cross sections of a prism or cylinder, including the upper and lower bases, are parallel and congruent.
2. The lateral edges of a prism are all parallel, and the lateral faces are all parallelograms.
3. The lateral edges and lateral faces of a right prism are perpendicular to the bases.
4. The altitude (or height) of a prism is the (perpendicular) distance from the plane of the upper base to the plane of the lower base.
5. Volume is a measure of the space enclosed by a figure in 3-space.
6. The volume of a prism or cylinder is the product of the base area and the altitude of the solid.
7. The lateral surface area of a right prism is the product of the altitude of the prism and the length of its base perimeter; the total surface area of a right prism is the sum of the lateral surface area and the areas of its two bases.
8. A right rectangular prism has three mutually perpendicular edges at each corner.
9. The volume of a right rectangular prism is the product of the lengths of any three mutually perpendicular edges that meet at a corner.

10. All faces, including the bases, of a parallelepiped are parallelograms.
11. The volume of solid 3-space figures may be determined by Cavalieri's Principle: If a plane parallel to the base plane of two solid figures cuts out equal-area cross sections of those solids, then the two solid figures are equal in volume.
12. A pyramid has a polygonal base and a single vertex that is not in the plane of the base.
13. A regular pyramid has a regular polygon as its base.
14. A circular cone has a circular base and a vertex that is not in the plane of the base.
15. The volume of a pyramid or cone is one-third the product of the base area and the altitude of the solid.
16. The lateral surface area of a regular pyramid or circular cone is one-half the product of the slant height and the length of the base perimeter; the total surface area of a regular pyramid is the sum of the lateral surface area and the area of the base.
17. A frustrum is a truncated pyramid or cone.
18. The volume of a sphere is $4\pi r^3/3$.
19. The surface area of a sphere is $4\pi r^2$.

RAISE YOUR GRADES

Can you . . . ?

☑ distinguish between a right and an oblique prism or cylinder
☑ distinguish between the altitude and lateral edge of a prism
☑ find the volume and surface area of a prism or cylinder
☑ find the volume and surface area of a prism or cone
☑ identify a regular pyramid and find its slant height
☑ identify a circular cone and find its slant height
☑ identify a frustrum and find its volume and surface area
☑ identify a sphere and find its volume and surface area

SOLVED PROBLEMS

PROBLEM 9-1 The right prism shown in Figure 9-28 has a trapezoidal base. Find **(a)** the lateral surface area, **(b)** the total surface area, and **(c)** the volume of this prism.

Solution   One side of the trapezoidal base is its altitude with a measure of 3 cm. The other side of the trapezoid is the hypotenuse of a right triangle with legs of 3 cm and 8 cm − 4 cm = 4 cm. From the Pythagorean Theorem, this side must therefore have a measure of $\sqrt{(3 \text{ cm})^2 + (4 \text{ cm})^2} = \sqrt{25 \text{ cm}^2} = 5$ cm. The perimeter P of the base is thus 4 cm + 3 cm + 8 cm + 5 cm = 20 cm. The base area of a trapezoid is one-half its height times the sum of its two bases, so the base area of this trapezoid is $(\frac{1}{2})(3 \text{ cm})[(4 \text{ cm}) + (8 \text{ cm})] = 18 \text{ cm}^2$.

   (a) The lateral surface area of a prism is the product of its altitude and the perimeter of its base, so the lateral surface area of this prism is

$$(12 \text{ cm})(20 \text{ cm}) = 240 \text{ cm}^2$$

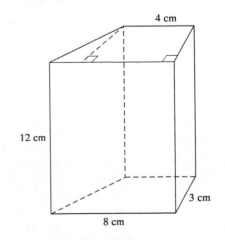

Figure 9-28

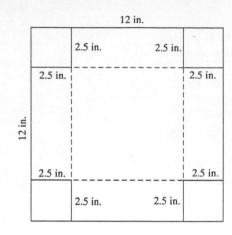

Figure 9-29

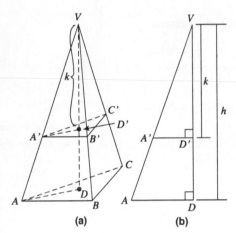

Figure 9-30

(b) The total surface area of a prism is the sum of its lateral surface area and the combined area of its two bases, so the total surface area of this prism is

$$(240 \text{ cm}^2) + 2(18 \text{ cm}^2) = 276 \text{ cm}^2$$

(c) The volume of a prism is the area of the base times the altitude of the prism, so the volume of this prism is

$$(18 \text{ cm}^2)(12 \text{ cm}) = 216 \text{ cm}^3$$

PROBLEM 9-2 You have a square piece of metal measuring 12 inches on a side. You cut out a 2.5-inch square from each corner, as shown in Figure 9-29, and make a topless box by folding along the dotted lines. What is the volume of your box?

Solution The space enclosed by this box is a rectangular right prism—a parallelepiped—measuring 2.5 inches by $[(12 \text{ inches}) - 2(2.5 \text{ inches})] = 7$ inches. The area of its base is $(7 \text{ inches})^2 = 49 \text{ in}^2$. Its volume is the product of its altitude and the area of its base:

$$(2.5 \text{ in.})(49 \text{ in}^2) = 122.5 \text{ in}^3$$

PROBLEM 9-3 Let R be a plane parallel to the base $\triangle ABC$ of pyramid $ABCV$ shown in Figure 9-30a. Plane R cuts out $\triangle A'B'C'$, which is a distance k from V. If the altitude of pyramid $ABCV$ is h, show that the area of $\triangle A'B'C'$ is proportional to the area of $\triangle ABC$ by a factor of $(k/h)^2$; that is,

$$A(\triangle A'B'C') = \left(\frac{k}{h}\right)^2 A(\triangle ABC)$$

Solution If you can show that $A'B'/AB = B'C'/BC = C'A'/CA = k/h$, then $\triangle A'B'C'$ is similar to $\triangle ABC$ and their altitudes must be in the same ratio, k/h. The areas of the two triangles must then have the ratio $(k/h)^2$.

Part I First, show that $VA'/VA = k/h$.

Step 1: Draw the altitude VD of pyramid $ABCV$ to the base $\triangle ABC$. Since VD is by definition perpendicular to $\triangle ABC$ and the planes of $\triangle ABC$ and $\triangle A'B'C'$ are parallel (given), VD must also be perpendicular to $\triangle A'B'C'$.

Step 2: Draw $A'D'$ and AD to form $\triangle VD'A'$ and $\triangle VDA$. Since VA and VD are two line segments, they determine a plane—and since A' is a point on VA and D' is a point on VD, $A'D'$ and AD are on this plane.

> *note:* It helps to draw a vertical cross section of the pyramids like that shown in Figure 9-30b. Here you can *see* that $A'D'$ and AD are on a common plane.

Step 3: $A'D'$ is parallel to AD since these are line segments cut from parallel planes by a common plane.

Step 4: $\angle VA'D' \cong \angle VAD$ as corresponding angles cut from two parallel lines by a transversal.

Step 5: $\angle A'VD' \cong \angle AVD$.

Step 6: As right angles, $\angle VD'A' \cong \angle VDA$.

Step 7: $\triangle VD'A'$ is similar to $\triangle VDA$, since the three corresponding angles of each triangle are equal in measure.

Step 8: $VA'/VA = VD'/VD = k/h$.

Part II: Now show that $\triangle A'B'C'$ is similar to $\triangle ABC$.

Step 1: $\angle A'VB' \cong \angle AVB$

Step 2: $\angle VA'B' \cong \angle VAB$ and $\angle VB'A' \cong \angle VBA$, as corresponding angles cut from two parallel lines by a transversal.

Step 3: Face $\triangle VA'B'$ is similar to face $\triangle VAB$.

Step 4: $A'B'/AB = VB'/VB = VA'/VA = k/h$.

Step 5: $\triangle VB'C'$ is similar to $\triangle VBC$ and $\triangle VC'A'$ is similar to $\triangle VCA$ by the reasoning of Steps 1 through 4.

Step 6: $A'B'/AB = B'C'/BC = C'A'/CA = k/h$.

Step 7: $\triangle A'B'C'$ is similar to $\triangle ABC$ since all the corresponding sides of these triangles stand in the same ratio, k/h.

Thus, if the area of $\triangle ABC$ is its base times its altitude, the area of $\triangle A'B'C'$ is k/h times the altitude of $\triangle ABC$ multiplied by k/h times the base of $\triangle ABC$, or

$$A(\triangle A'B'C') = \left(\frac{k}{h}\right)^2 A(\triangle ABC)$$

note: This problem deals with pyramids whose bases are triangles, but the result holds for any pyramid or cone. (Why?) Thus we can see that

- The cross-sectional area of any pyramid or cone is proportional to the area of any other cross section of that same pyramid or cone.

PROBLEM 9-4 Use Cavalieri's Principle to show that any two pyramids with equal base areas and equal heights must also have equal volumes.

Solution

Step 1: Set two pyramids, 1 and 2, on the same plane R.

Step 2: Since both pyramids have the same height, any plane S parallel to R must intersect either both pyramids or neither pyramid.

Step 3: Let B_1 be the base of Pyramid 1 in R and C_1 its cross section in S. Let B_2 and C_2 be the corresponding items in Pyramid 2.

Step 4: If the vertex of each pyramid is a distance k from S and a distance h from R, then

$$A(C_1) = \left(\frac{k}{h}\right)^2 A(B_1) \qquad \text{and} \qquad A(C_2) = \left(\frac{k}{h}\right)^2 A(B_2)$$

from Problem 9-3.

Step 5: Since $A(B_1) = A(B_2)$, $A(C_1) = A(C_2)$.

Step 6: Since plane S parallel to plane R of the bases cuts out equal-area cross sections from both pyramids, the pyramids must have equal volumes by Cavalieri's Principle.

PROBLEM 9-5 The three pyramids shown in Figure 9-18b were cut from the triangular prism shown in Figure 9-18a. Show that these pyramids have equal volumes.

Solution Consider pyramids $ACBG$ and $EGFA$. If pyramid $ACBG$ has base $\triangle ACB$, it has vertex G and altitude GC. And if pyramid $EGFA$ has base $\triangle EGF$, it has vertex A and altitude AE. Now since these two pyramids were cut from a triangular prism, whose lateral edges and bases are congruent by definition, $GC \cong AE$ and $\triangle ACB \cong \triangle EGF$. Therefore, since both pyramids have equal base areas and altitudes, they must have equal volumes by the result of Problem 9-4.

    Now consider pyramids $ACBG$ and $ABGF$. If pyramid $ACBG$ has base $\triangle GCB$, it has vertex A. And if pyramid $ABGF$ has base $\triangle BFG$, it also has vertex A. Next, observe that both of these bases are in the parallelogram $CBFG$. So, if their bases are in the same plane and they have the same vertex, these two pyramids have equal altitudes. And since $\triangle GCB$ and $\triangle BGF$ have a common side, GB, which is the diagonal of parallelogram $CBFG$, $\triangle GCB \cong \triangle BGF$—which means that the bases of the two pyramids have equal areas. Thus, again, from Problem 9-4, the two pyramids must have equal volumes.

> *note:* Why didn't we just use the volume formula for a pyramid in this proof? Why go to all the trouble of using Problem 9-4? Because the volume formula is *founded* on this proof, that's why. If we had used the formula, we'd have made the classic error of using a result to prove itself.

PROBLEM 9-6   A regular pyramid has a height of 6 feet and a square base whose sides are 6 feet. Find **(a)** the volume and **(b)** the lateral surface area of this pyramid.

Solution   Draw this pyramid as shown in Figure 9-31.

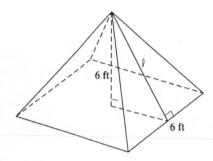

(a) The base area is $(6 \text{ ft})^2 = 36 \text{ ft}^2$. So the volume of the pyramid is

$$V = \frac{Ah}{3} = \frac{(36 \text{ ft}^2)(6 \text{ ft})}{3} = 72 \text{ ft}^3$$

(b) The slant height l of the pyramid is the hypotenuse of a right triangle whose legs are 6 feet and 3 feet long. So, by the Pythagorean Theorem,

$$l = \sqrt{(3 \text{ ft})^2 + (6 \text{ ft})^2} \approx 6.71 \text{ ft}$$

Figure 9-31

The perimeter P of the base is $4(6 \text{ ft}) = 24 \text{ ft}$. So the lateral surface area of the pyramid is

$$\frac{lP}{2} = \frac{(6.71 \text{ ft})(24 \text{ ft})}{2} = 80.5 \text{ ft}^2$$

PROBLEM 9-7   A storage tank is made in the shape of a cylinder whose radius is 9 ft and whose height is 12 ft. Find **(a)** its volume and **(b)** its lateral surface area.

Solution   The base of this cylinder has area $A = \pi r^2 = \pi(9 \text{ ft})^2 = 254.5 \text{ ft}^2$. The perimeter of the base is $P = 2\pi r = 2\pi(9 \text{ ft}) = 56.5 \text{ ft}$.

(a) $$V = hA = (12 \text{ ft})(254.5 \text{ ft}^2) = 3054 \text{ ft}^3$$

(b) $$S = hP = (12 \text{ ft})(56.5 \text{ ft}) = 678 \text{ ft}^2$$

PROBLEM 9-8   A cone has slant height $3\sqrt{5}$ and base perimeter 6π. Find **(a)** its lateral surface area, **(b)** the radius of its base, **(c)** its altitude, and **(d)** its volume.

Solution

(a) The lateral surface area S is

$$S = \frac{lP}{2} = \frac{3\sqrt{5} \times 6\pi}{2} = 20.12$$

(b) The base perimeter measures 6π, so

$$2\pi r = 6\pi$$
$$r = 3$$

(c) The slant height $l = 3\sqrt{5}$ is the hypotenuse of a right triangle formed by the base radius $r = 3$ and the altitude h. Thus, by the Pythagorean Theorem,

$$h = \sqrt{l^2 - r^2} = \sqrt{(3\sqrt{5})^2 - 3^2} = 6$$

(d) The volume of the cone is

$$V = \frac{hA}{3} = \frac{h\pi r^2}{3} = \frac{6(\pi 3^2)}{3} = 18\pi \approx 56.5$$

PROBLEM 9-9 A regular square-based pyramid has a height of 10 cm and a base side of 10 cm. A plane parallel to and 6 cm above the base cuts off a small pyramid from the top and leaves a square-based frustrum. Find (**a**) the lateral surface area, (**b**) the total surface area, and (**c**) the volume of the original pyramid and the frustrum.

Solution Before you can find all of the preceding, you need to find some other things: the areas of the two bases of the frustrum, the perimeters of these two bases, and the slant heights of the original pyramid and the small pyramid.

First, let A_1 be the area of the lower base of the frustrum and A_2 be the area of its upper base. Thus, $A_1 = (10 \text{ cm})^2 = 100 \text{ cm}^2$. Then let h be the height of the original pyramid and k be the height of the smaller pyramid, so $k = h - 6 \text{ cm} = 4 \text{ cm}$. You know from Problem 9-3 that the areas of the two bases of the frustrum must be proportional to each other by a factor of $(k/h)^2$, so

$$A_2 = \left(\frac{k}{h}\right)^2 A_1 = \left(\frac{4}{10}\right)^2 (100 \text{ cm}^2) = 16 \text{ cm}^2$$

Next, the perimeter P_1 of the lower base of the frustrum with side $s_1 = 10 \text{ cm}$ is $4(10 \text{ cm}) = 40 \text{ cm}$, and the perimeter P_2 of the upper base of the frustrum is $4\sqrt{16 \text{ cm}^2} = 16 \text{ cm}$, with side $s_2 = (16 \text{ cm})/4 = 4 \text{ cm}$.

Finally, the slant height l_1 of the original pyramid is the hypotenuse of a right triangle whose legs are the height h and one-half the side s_1 of this pyramid's base. Thus, by the Pythagorean Theorem,

$$l_1 = \sqrt{h^2 + \left(\frac{s_1}{2}\right)^2} = \sqrt{(10 \text{ cm})^2 + (5 \text{ cm})^2} = \sqrt{125 \text{ cm}^2} = 5\sqrt{5} \text{ cm}$$

And the slant height l_2 of the smaller pyramid can also be found by the Pythagorean Theorem, since l_2 is the hypotenuse of a right triangle whose legs are the height k and one-half the side s_2 of the frustrum's upper base:

$$l_2 = \sqrt{k^2 + \left(\frac{s_2}{2}\right)^2} = \sqrt{(4 \text{ cm})^2 + (2 \text{ cm})^2} = \sqrt{20 \text{ cm}^2} = 2\sqrt{5} \text{ cm}$$

(**a**) The lateral surface area of the original pyramid is

$$\frac{l_1 P_1}{2} = \frac{(5\sqrt{5} \text{ cm})(40 \text{ cm})}{2} = 223.6 \text{ cm}$$

Then, if the lateral surface area of the smaller pyramid is $l_2 P_2 / 2 = (2\sqrt{5} \text{ cm})(16 \text{ cm})/2 = 35.8 \text{ cm}^2$, the lateral surface area of the frustrum is

$$\frac{l_1 P_1}{2} - \frac{l_2 P_2}{2} = (223.6 \text{ cm}^2) - (35.8 \text{ cm}^2) = 187.8 \text{ cm}^2$$

(**b**) The total surface area of the original pyramid is its lateral surface area plus the area of its base, or

$$(223.6 \text{ cm}^2) + (100 \text{ cm}^2) = 323.6 \text{ cm}^2$$

And the total surface area of the frustrum is its lateral surface area plus the area of its two bases, or

$$(187.8 \text{ cm}^2) + (100 \text{ cm}^2) + (16 \text{ cm}^2) = 303.8 \text{ cm}^2$$

(**c**) The volume of the original pyramid is one-third the product of its height and the area of its base, or

$$\frac{(10 \text{ cm})(100 \text{ cm}^2)}{3} = 333.3 \text{ cm}^3$$

Then, if the volume of the smaller pyramid is one-third the product of its height and the area of its base $(4 \text{ cm})(16 \text{ cm}^2)/3 = 21.3 \text{ cm}^3$, the volume of the frustrum is the difference

$$\text{Volume of frustrum} = \text{Volume of original pyramid} - \text{Volume of small pyramid}$$
$$= (333.3 \text{ cm}^3) - (21.3 \text{ cm}^3)$$
$$= 312 \text{ cm}^3$$

PROBLEM 9-10   A metal sphere of diameter 10 cm is melted down and recast first into a cylinder and then into a cone. If the base of each new solid has a diameter equal to that of the sphere, what is the altitude of (**a**) the cylinder and (**b**) the cone?

Solution   The sphere has a radius r of 5 cm and hence a volume V of

$$V = \frac{4\pi r^3}{3} = \frac{4\pi (5 \text{ cm})^3}{3}$$

(**a**) If the cylinder cast from a sphere has a radius r and a volume V equal to that of the sphere, you can call the altitude of the cylinder h and write

$$\text{Volume of cylinder} = \text{Volume of sphere}$$

$$\pi r^2 h = \frac{4\pi r^3}{3}$$

so the altitude of the cylinder is

$$h = \frac{4\pi r^3}{3\pi r^2} = \frac{4r}{3} = \frac{4(5 \text{ cm})}{3} = 6\tfrac{2}{3} \text{ cm}$$

(**b**) If the cone cast from a sphere has a radius r and a volume V equal to that of the sphere, you can call the altitude of the cone k and write

$$V_{\text{cone}} = V_{\text{sphere}}$$

$$\frac{\pi r^2 k}{3} = \frac{4\pi r^3}{3}$$

so the altitude of the cone is

$$k = \frac{4\pi r^3}{\pi r^2} = 4r = 4(5 \text{ cm}) = 20 \text{ cm}$$

PROBLEM 9-11   Find the total surface area of (**a**) the sphere, (**b**) the cylinder, and (**c**) the cone in Problem 9-10.

Solution

(**a**) The surface area of the sphere with radius $r = 5$ cm is

$$4\pi r^2 = 4\pi (5 \text{ cm})^2 = 100\pi \text{ cm}^2$$

(**b**) You haven't been given a specific formula for the surface area of a cylinder, but you can easily figure it out. You've seen that a right cylinder is really just a special case of a right prism—one with its edges "collapsed" and circles instead of regular polygons for bases. So it's reasonable to assume that the general formula that applies to regular right prisms also applies to cylinders; that is, the surface area of a cylinder is the product of its altitude (h) and the circumference (P) (perimeter) of its base plus the areas of its two bases. So, here, the area of the base is

$$\pi r^2 = \pi (5 \text{ cm})^2 = 25\pi \text{ cm}^2$$

Then the lateral surface area is

$$hP = 2\pi rh = 2\pi (5 \text{ cm})(6.67 \text{ cm}) = 66.7\pi \text{ cm}^2$$

Thus the total surface area is

$$2(25\pi \text{ cm}^2) + 66.7\pi \text{ cm}^2 = 366.62 \text{ cm}^2$$

(c) To find the total surface area of the cone, you need its slant height *l*; so by the Pythagorean Theorem,

$$l = \sqrt{r^2 + k^2} = \sqrt{(5 \text{ cm})^2 + (20 \text{ cm})^2} = 20.62 \text{ cm}$$

Next, the base perimeter (circumference) *P* of the base is

$$2\pi r = 2(5 \text{ cm})\pi = 10\pi \text{ cm}$$

and the area of the base is

$$\pi r^2 = \pi(5 \text{ cm})^2 = 25\pi \text{ cm}^2$$

The lateral surface area is

$$\frac{lP}{2} = \frac{(10\pi \text{ cm})(20.62 \text{ cm})}{2} = 323.9 \text{ cm}^2$$

So the total surface area is

$$323.9 \text{ cm}^2 + 25\pi \text{ cm}^2 = 402.4 \text{ cm}^2$$

Review Exercises

EXERCISE 9-1   A brick has a face that measures 4 inches by 8 inches. The thickness is 2 inches. Find **(a)** the volume and **(b)** the total surface area of the brick.

EXERCISE 9-2   Find the lateral surface area of a right prism with pentagonal bases whose sides measure 3 cm, 5 cm, 6 cm, 7 cm, and 9 cm, if the bases are 12 cm apart.

EXERCISE 9-3   A cement building block with three rectangular spaces in it has measurements as shown in Figure 9-32. Find the volume of the cement in the block. [*Hint:* Don't forget to subtract the holes.]

EXERCISE 9-4   Find the total surface area for the cement block in Figure 9-32.

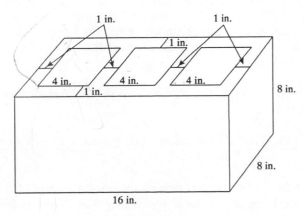

Figure 9-32

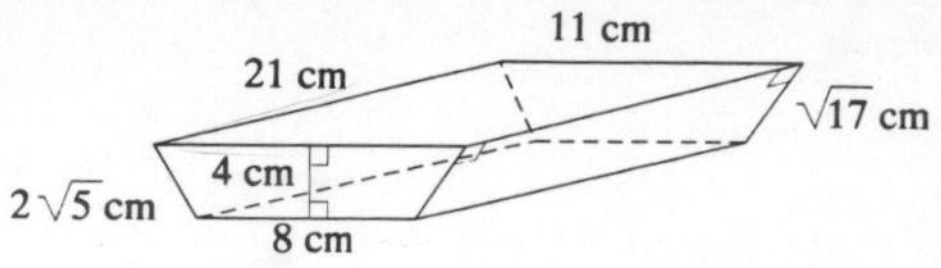

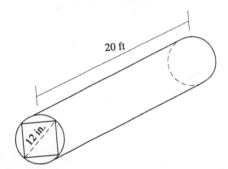

Figure 9-33

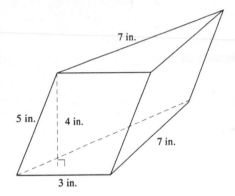

Figure 9-34

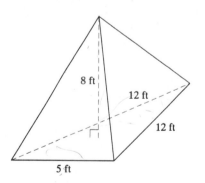

Figure 9-35

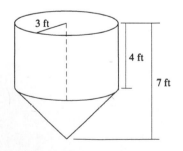

Figure 9-36

EXERCISE 9-5   Find the volume of the solid shown in Figure 9-33.

EXERCISE 9-6   Find the volume of the solid shown in Figure 9-34.

EXERCISE 9-7   Find the volume of the largest beam with a square cross section that can be cut from a 20-foot cylindrical log with diameter 12 inches. See Figure 9-35.

EXERCISE 9-8   A regular pyramid has an hexagonal base whose side is 3 cm. Given that the lateral edge is 8 cm, find (**a**) the altitude h, (**b**) the slant height l, (**c**) the lateral surface area S_L, and (**d**) the volume V of the pyramid.

EXERCISE 9-9   A pyramid with a square base has height of 5 ft and base edge of 3 ft. Find (**a**) the volume V, (**b**) the slant height l, and (**c**) the lateral surface area S_L of the pyramid.

EXERCISE 9-10   Find the volume of the pyramid shown in Figure 9-36.

EXERCISE 9-11   A regular triangular pyramid has congruent equilateral triangles as its base and lateral faces. The base side is 6 inches. Find (**a**) the altitude h, (**b**) the slant height l, (**c**) the total surface area S_T, and (**d**) the volume V of the pyramid.

EXERCISE 9-12   A cylinder has height 6 ft and base radius 4 ft. Find (**a**) the lateral surface area S_L, (**b**) the total surface area S_T, and (**c**) the volume V of the cylinder.

EXERCISE 9-13   Find the height of a cylinder whose lateral surface area is 18π ft^2 and whose base area is 9π ft^2.

EXERCISE 9-14   A right cone has height 6 cm and base radius 4 cm. Find (**a**) the lateral surface area and (**b**) the total surface area of the cone.

EXERCISE 9-15   A storage bin is shaped as a cylinder on top of a cone as shown in Figure 9-37. Find (**a**) the volume and (**b**) the total surface area (including the top).

EXERCISE 9-16   An oblique cone has height 8 cm and base radius 3 cm. A plane parallel to the base cuts through the cone 5 cm above the base. Find (**a**) the area A of the cross section and (**b**) the volume V of the frustrum that is created.

EXERCISE 9-17   Two right cones have the same vertex and concentric bases whose radii are 5 cm and 2 cm. The vertex is 7 cm above the base plane. A plane 1 cm from the vertex cuts through both cones. Find the volume of the space that is inside the larger cone but outside the smaller cone and between the base and the cutting plane.

EXERCISE 9-18   Find (**a**) the volume and (**b**) the surface area of a sphere whose radius is 2 feet.

EXERCISE 9-19   How do the volume and surface area of a sphere change if the radius is (**a**) doubled? (**b**) tripled?

EXERCISE 9-20   A barn silo is formed by placing a hemispherical dome on top of a cylinder. If the silo has total overall height of 26 meters and base radius

Figure 9-37

of 5 meters, what is **(a)** the surface area of the hemispherical dome and **(b)** the volume of the silo?

EXERCISE 9-21   Four stone spheres of radius 4 cm are placed inside a hollow cylinder of height 32 cm and base radius 4 cm. Water is then poured into the cylinder until it is totally full. Find the volume of the water inside the cylinder.

Answers to Review Exercises

9-1   **(a)** 64 in^3 **(b)** 112 in^2

9-2   360 cm^2

9-3   448 in^3

9-4   976 in^2

9-5   798 cm^3

9-6   41.02 in^3

9-7   10 ft^3

9-8   **(a)** $h = 7.42$ cm,   **(c)** $S_L = 70.72$ cm^2,
   (b) $l = 7.86$ cm,   **(d)** $V = 57.80$ cm^3

9-9   **(a)** $V = 15$ ft^3,   **(c)** $S_L = 31.32$ ft^2
   (b) $l = 5.22$ ft,

9-10   78.24 ft^3

9-11   **(a)** $h = 4.90$ ft,   **(c)** $S_T = 62.35$ ft^2,
   (b) $l = 5.20$ ft,   **(d)** $V = 25.46$ ft^3

9-12   **(a)** $S_L = 150.80$ ft^2,   **(c)** $V = 301.59$ ft^3
   (b) $S_T = 251.33$ ft^2,

9-13 $h = 3$ ft

9-14   **(a)** $S_L = 90.62$ cm^2,   **(b)** $S_T = 140.88$ cm^2

9-15   **(a)** $V = 141.37$ ft^3,   **(b)** $S_T = 143.66$ ft^2

9-16   **(a)** $A = 3.98$ cm^2,   **(b)** $V = 71.42$ cm^3

9-17   153.49 cm^3

9-18   **(a)** 33.51 ft^3,   **(b)** 50.27 ft^2

9-19   **(a)** Volume is multiplied by 8; surface area is multiplied by 4.
   (b) Volume is multiplied by 27; surface area is multiplied by 9

9-20   **(a)** 157.08 ft^2,   **(b)** 1911.14 ft^3

9-21   536.17 cm^3

10 INTRODUCTION TO TRIGONOMETRY

THIS CHAPTER IS ABOUT

☑ **Angles Reconsidered**
☑ **Cosine and Sine**
☑ **Tangent, Cotangent, Secant, and Cosecant**
☑ **Squared Formulas**
☑ **Sum Formulas**
☑ **Double- and Half-Angle Formulas**

10-1. Angles Reconsidered

Let's begin reconsidering angles by reconsidering circles. We know that what characterizes a circle is its radius r, which is always the same distance from the center C. So we can think of a circle as the path—or **locus**—a moving point P must take on the plane if P travels a constant distance (r) from a fixed point (C). Thus, from what we know about distance, we cay say that a circle whose center is at the origin $O(0, 0)$ of a Cartesian plane consists of a set of locations, or coordinates (x, y), that P can be in such that $P(x, y)$ satisfies the equation

CIRCLE
[$C = (0, 0)$]
$$x^2 + y^2 = r^2$$

That is, the position of P with respect to the x and y axes must always be a distance $r = \sqrt{x^2 + y^2}$ from O.

Now suppose we have a *unit circle*—a circle whose center is at the origin O of a Cartesian plane and whose radius is 1. If we pick any P on the unit circle and draw a ray OP through it, this ray will contain the radius $r = \overline{OP}$ and the coordinates of P will satisfy

$$x^2 + y^2 = r^2 = 1^2 = 1$$

Then if we pick a fixed point S on the positive x axis, say $(1, 0)$, and draw a ray $\overrightarrow{OS}$ through it, as in Figure 10-1, we see that the intersection of $\overrightarrow{OP}$ and $\overrightarrow{OS}$ forms an angle, to which we assign a value θ. This value θ is a measure of the angle formed when a ray $\overrightarrow{OP}$ containing the moving point P intersects the ray $\overrightarrow{OS}$ containing the fixed point S on the positive x axis. We consider the side of the angle containing the fixed point to be the **initial side** and the side of the angle containing the moving point to be the **terminal side**. If P moves in a clockwise direction, we say that the resulting angle is *negative*; but if P moves in a counterclockwise direction, we say that the resulting angle is *positive* (see Figure 10-1).

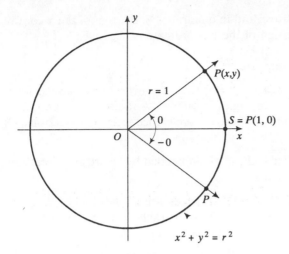

Figure 10-1

note: We always measure θ from the side containing the positive x axis—the initial side—to the side containing the moving point P— the terminal side.

A. Angle measure and the coordinates of P

1. Degree measure

As P moves around the circle, the coordinates of P change. In Figure 10-2a, we see what happens to the degree measure of θ when P moves around the circle. When P is at $(1, 0)$, $\theta = 0°$; when P is at $(\sqrt{3}/2, 1/2)$, $\theta = 30°$; when P is at $(1/\sqrt{2}, 1/\sqrt{2})$, $\theta = 45°$; when P is at $(1/2, \sqrt{3}/2)$, $\theta = 60°$; and when P is at $(0, 1)$, $\theta = 90°$. And P can go on around the circle, continuing to increase the θ value of its angle as it travels counterclockwise or decrease the θ value of its angle as it travels clockwise.

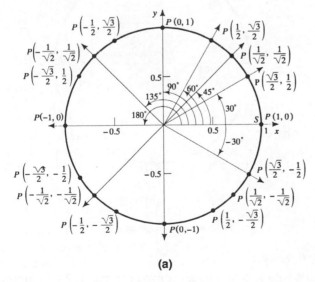

(a)

Figure 10-2

Notice that when P moves out of the first quadrant—quadrant I—of the Cartesian plane, the sign of one or both of P's coordinates changes. In quadrant I, both the x and y coordinates are positive; in quadrant II, the sign of the x coordinate is negative and the sign of the y coordinate is positive; in quadrant III, the signs of both the x and y coordinates are negative; and in quadrant IV, the sign of the x coordinate is positive and the sign of the y coordinate is negative.

2. Radian measure

Degree measure is not the only measure we can use for angles—we can also measure angles in radians. A radian is just another way of dividing up the circle, whose circumference is $2\pi r$ (see Section 7-2). Formally defined,

- A **radian** is the angle subtended by an arc of a circle equal to the radius of the circle.

Thus, if the whole circumference of the circle is $2\pi r$, this circumference can be divided into 2π parts each of which has a length r. We can also say that the whole angle of a circle is 2π radians; that is,

$$360° = 2\pi \text{ radians}$$

so

$$1° = \frac{2\pi}{360} \text{ radians}$$

note: When we use radians, it's customary to leave out the unit designation. Some books, however, do use a unit designation—"rad."

In Figure 10-2b, we see the same angles shown in Figure 10-2a except for the fact that the angles in this figure are measured in radians rather than degrees.

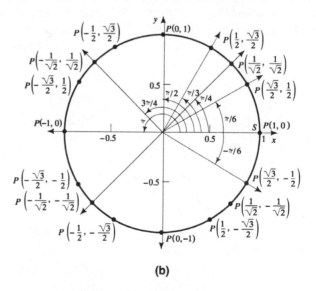

(b)

Figure 10-2. Continued

B. Finding θ values

The trouble with P is that it can continue to move. As P continues to move counterclockwise around the circle, θ keeps increasing without bound; and as P continues to move clockwise around the circle, θ keeps decreasing without bound. But, at any particular instant, we can see that

θ VALUE OF AN ANGLE $\theta = n(90°) + \alpha$ or $\theta = n\left(\dfrac{\pi}{2}\right) + \alpha$

where n is the number of times that P has crossed either an x or y axis since starting at $\overrightarrow{OS}$, 90° (or $\pi/2$) is the measure of a quadrantal angle, and

α is the angle $\overrightarrow{OP}$ forms with the last axis it crossed, as shown in Figure 10-3. Notice in Figure 10-3 that $\theta = 2(90°) + \alpha = 2(\pi/2) + \alpha = \pi + \alpha$.

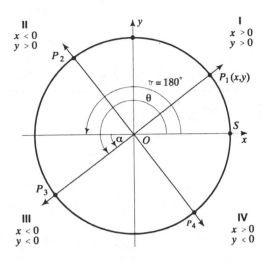

Figure 10-3

note: You can now see that angles are not just "acute" or "obtuse." Rather, angles may have arbitrarily large or arbitrarily small (large absolute value, negative sign) θ values. In fact, finding the value of θ for angles formed when the terminal side contains a moving point P can get tricky because every time P makes one full circuit around the circle in either direction, the θ value changes by $\pm 360°$ (or $\pm 2\pi$). Thus, although you can tell from the θ value exactly where P is in the Cartesian plane, you cannot uniquely determine θ unless you know where P has *been*.

EXAMPLE 10-1: Give the θ value in degrees and radians for P if P is at $(-1/2, \sqrt{3}/2)$ and P has traveled **(a)** counterclockwise, but has not made one full revolution around the circle; **(b)** clockwise, but has not made one full revolution around the circle; **(c)** counterclockwise, and made two full revolutions around the circle; **(d)** clockwise, and made one full revolution around the circle.

Solution: Draw ray $\overrightarrow{OP}$ as shown in Figure 10-4. Notice that $\overrightarrow{OP}$ and the negative x axis have a $60°$ ($\pi/3$) angle between them and $\overrightarrow{OP}$ and the positive y axis have a $30°$ ($\pi/6$) angle between them.

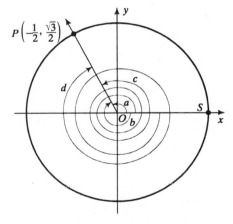

Figure 10-4

(a) As P moves counterclockwise from $(1, 0)$ to $(-1/2, \sqrt{3}/2)$, it crosses one axis (the positive y axis). Thus

$$\theta = 1(90°) + 30° = 120°$$

or

$$\theta = 1(\pi/2) + \pi/6 = 2\pi/3$$

(b) As P moves clockwise from $(1, 0)$ to $(-1/2, \sqrt{3}/2)$, it crosses two axes (the negative y axis and the negative x axis). Thus

$$\theta = -2(90°) - 60° = -240°$$

or

$$\theta = -2(\pi/2) - \pi/3 = -4\pi/3$$

(c) As P moves counterclockwise from $(1, 0)$ to $(-1/2, \sqrt{3}/2)$, it crosses nine axes. Thus

$$\theta = 9(90°) + 30° = 840°$$

or

$$\theta = 9(\pi/2) + \pi/6 = 14\pi/3$$

(**d**) As P moves clockwise from $(1,0)$ to $(-1/2, \sqrt{3}/2)$, it crosses six axes. Thus

$$\theta = -6(90°) - 60° = -600°$$

or

$$\theta = -6(\pi/2) - \pi/3 = 10\pi/3$$

10-2. Cosine and Sine

A. Cosine and sine reconsidered

Let's look at Figure 10-5a, where θ is acute. Here, OP is the hypotenuse of a right triangle whose legs are the projections of OP onto the x and y axes. We know that the length of the hypotenuse OP is 1 and the lengths of the legs are given by the x and y coordinates of P. And, from Section 3-3, we know that

$$\cos\theta = \frac{\text{adjacent side}}{\text{hypotenuse}} = \frac{x}{1} = x \quad \text{and} \quad \sin\theta = \frac{\text{opposite side}}{\text{hypotenuse}} = \frac{y}{1} = y$$

So we can see that the x and y coordinates of P in quadrant I are actually the cosine and sine, respectively, of the acute angle θ.

Then, when θ is obtuse, as in Figure 10-5b, OP is again the hypotenuse of a right triangle whose legs are OP's projections onto the x and y axes. But in this triangle, the lengths of the legs are $|x|$ (since $x < 0$) and y. Now Figure 10-5b also shows us that $|x| = \cos(\pi - \theta)$ [where π is 180° in radian measure], so $x = -\cos(\pi - \theta)$ and $y = \sin(\pi - \theta)$. But in Sections 4-3 and 4-5, we extended the sine and cosine functions to obtuse angles, as

$$\sin\theta = \sin(\pi - \theta) \quad \text{and} \quad \cos\theta = -\cos(\pi - \theta)$$

Thus, again, the x and y coordinates of P in quadrant II are, respectively, the cosine and sine of the obtuse angle θ.

Having made these observations, let's redefine the cosine and sine of *any* angle:

- The cosine and sine of any angle θ are the respective x and y coordinates of a point P on the unit circle for that angle.

B. Rules for finding the sign of the cosine and sine

The definition of cosine and sine just given lets us see several rules that can make trigonometry easier:

- $\sin(-\theta) = -\sin\theta$ • $\sin(\theta + 360°) = \sin\theta$ • $\sin(\theta + 180°) = -\sin\theta$
- $\cos(-\theta) = \cos\theta$ • $\cos(\theta + 360°) = \cos\theta$ • $\cos(\theta + 180°) = -\cos\theta$

And we can also see that

- When the terminal side of θ is in quadrants I and II, $\sin\theta > 0$
 III and IV, $\sin\theta < 0$
 I and IV, $\cos\theta > 0$
 II and III, $\cos\theta < 0$

C. A word about tables and some words about calculators

The word about using tables of trig functions is

DON'T

You'll find that a calculator—with plenty of function keys—will make your life in trigonometry immeasurably easier. So the words about calculators are

BUY ONE OR BORROW ONE

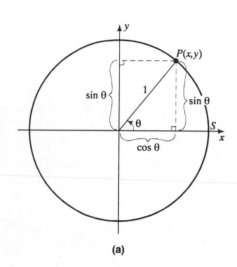

(a)

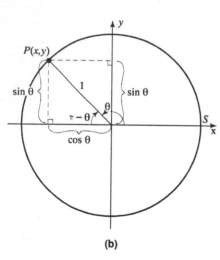

(b)

Figure 10-5

note: If you borrow your calculator, be sure to borrow the instructions, too. Calculators vary in their mode of action.

It's not difficult to use a calculator to find the cosine or sine of any angle θ.

note: The angle returned by your calculator's $\boxed{\sin^{-1}}$ key will always fall between $-90°$ and $90°$ ($-\pi/2$ and $\pi/2$). The $\boxed{\cos^{-1}}$ key will always return an angle between $0°$ and $180°$ (0 and π). These are the so-called **principal value** angles. For these angles $\theta = \cos^{-1}(\cos \theta)$ and $\theta = \sin^{-1}(\sin \theta)$. These relations are not valid outside this range of angles. If you use the $\boxed{\sin^{-1}}$ or $\boxed{\cos^{-1}}$ button for angles in quadrants II, III, or IV, you will also have to use the relations in Section 10-2B to determine the actual angle.

EXAMPLE 10-2: Suppose you have an angle whose measure is (**a**) 72°, (**b**) 138°, (**c**) 215°, (**d**) 330°, (**e**) 400°, (**f**) 542°, (**g**) $-12°$, (**h**) $-163°$. Identify the quadrant in which the terminal side of each angle lies, predict the signs of the sine and cosine of each angle, and find the values of the sine and cosine.

Solution: See Figure 10-6.

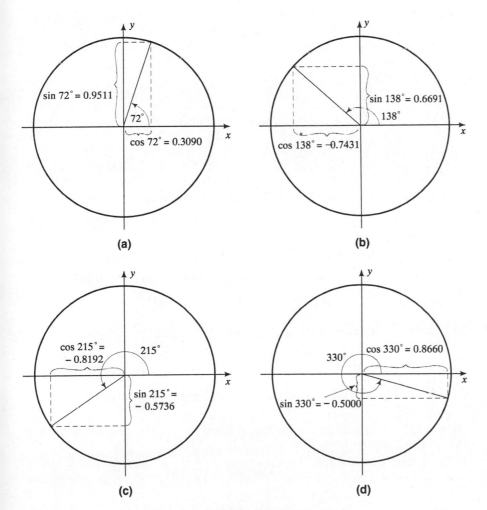

(a)   (b)

(c)   (d)

Figure 10-6

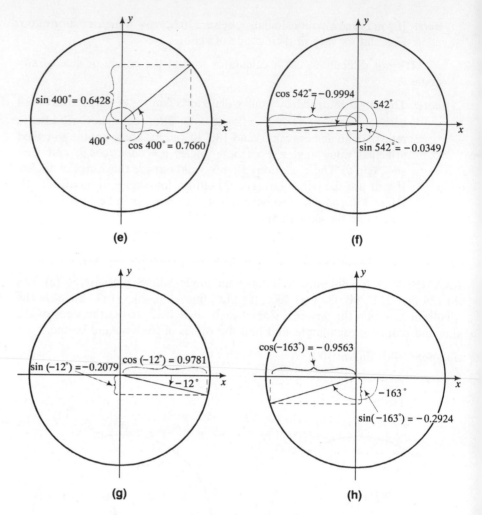

Figure 10-6.   Continued

(**a**) $\theta = 72°$:   Quadrant I; both the sine and cosine are positive.

$$\sin 72° = 0.9551 \qquad \cos 72° = 0.3090$$

(**b**) $\theta = 138°$:   Quadrant II; the sine is positive and the cosine is negative.

$$\sin 138° = \sin(180° - 138°) = \sin 42° = 0.6691$$
$$\cos 138° = -\cos 42° = -0.7431$$

(**c**) $\theta = 215°$:   Quadrant III; both the sine and cosine are negative.

$$\sin 215° = \sin(180° - 215°) = \sin(-35°) = -\sin 35°$$
$$= -0.5736$$
$$\cos 215° = -\cos 35° = -0.8192$$

(**d**) $\theta = 330°$:   Quadrant IV; the sine is negative and the cosine is positive.

$$\sin 330° = \sin(180° - 330°) = \sin(-150°) = -\sin 150°$$
$$= -\sin(180° - 150°) = -\sin 30° = -0.5000$$
$$-\cos 30° = -0.8660$$

(e) $\theta = 400°$:   Quadrant I; both the sine and cosine are positive.

$$\sin 400° = \sin(400° - 360°) = \sin 40° = 0.6428$$

$$\cos 40° = 0.7660$$

(f) $\theta = 542°$:   Quadrant III; both the sine and cosine are negative.

$$\sin 542° = \sin(542° - 360°) = \sin 182°$$
$$= \sin(180° - 182°) = \sin(-2°) = -\sin 2° = -0.0349$$
$$-\cos 2° = -0.9994$$

(g) $\theta = -12°$:   Quadrant IV; the sine is negative and the cosine is positive.

$$\sin(-12°) = -\sin 12° = -0.2079$$

$$\cos(-12°) = \cos 12° = 0.9789$$

(h) $\theta = -163°$: Quadrant III; both the sine and the cosine are negative.

$$\sin(-163°) = -\sin 163° = -\sin(180° - 163°)$$
$$= -\sin 17° = -0.2924$$
$$-\cos 17° = -0.9563$$

> *note:* Example 10-2 illustrates the fact that you can find the sine and cosine of any angle by determining the quadrant that the terminal side OP lies in and then computing the sine and cosine of the acute angle formed by OP and the x axis. This acute angle is often called the auxiliary angle. Notice that for positive angles in quadrants I and III, the auxiliary angle has the same absolute degree measure as α, whereas in quadrants II and IV the auxiliary angle has the absolute degree measure of α's complement. For negative angles in quadrants IV and II, α has the same absolute degree measure as the auxiliary angle, whereas in quadrants III and I α has the absolute degree measure as the auxiliary angle's complement.

EXAMPLE 10-3:   Give the absolute degree measure of α and of the auxiliary angle of all the angles given in Example 10-2.

Solution

(a) $\alpha = 72°$; since the given angle's terminal side is in quadrant I, the auxiliary angle is also $72°$.

(b) $\alpha = 138° - 90° = 48°$; since the given angle's terminal side is in quadrant II, the auxiliary angle is $90° - 48° = 42°$.

(c) $\alpha = 215° - 2(90°) = 35°$; since the given angle's terminal side is in quadrant III, the auxiliary angle is also $35°$.

(d) $\alpha = 330 - 3(90°) = 60°$; since the given angle's terminal side is in quadrant IV, the auxiliary angle is $90° - 60° = 30°$.

(e) $\alpha = 400° - 4(90°) = 40°$; since the given angle's terminal side is in quadrant I, the auxiliary angle is also $40°$.

(f) $\alpha = 542° - 6(90°) = 2°$; since the given angle's terminal side is in quadrant III, the auxiliary angle is also $2°$.

(g) $\alpha = -12°$; since the given angle is negative and has its terminal side in quadrant IV, the auxiliary angle is also $12°$.

(h) $\alpha = -163° - (-90°) = -73°$; since the given angle is negative and has its terminal side in quadrant III, the auxiliary angle is $90° - 73° = 17°$.

See Figure 10-7.

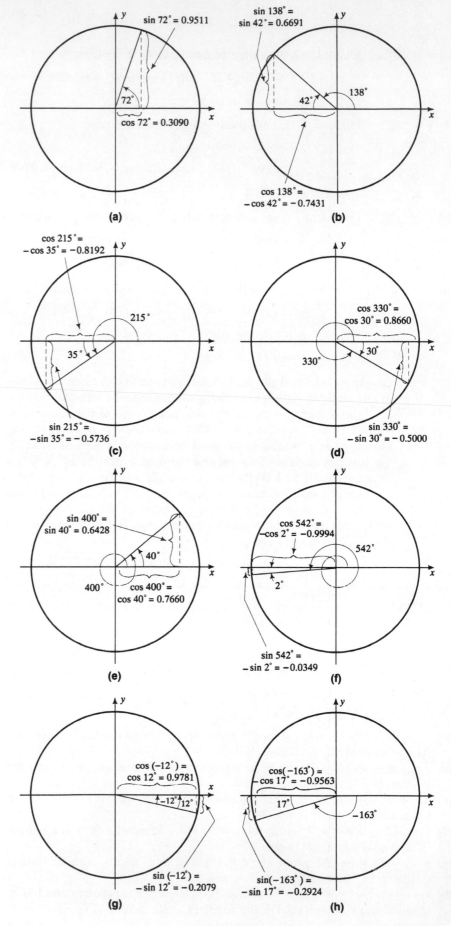

Figure 10-7

10-3. Tangent, Cotangent, Secant, and Cosecant

There are four additional functions of θ, which are frequently studied in their own right—the tangent, cotangent, secant, and cosecant of θ:

TANGENT $\tan \theta = \dfrac{\sin \theta}{\cos \theta}$ **COTANGENT** $\cot \theta = \dfrac{\cos \theta}{\sin \theta}$

SECANT $\sec \theta = \dfrac{1}{\cos \theta}$ **COSECANT** $\csc \theta = \dfrac{1}{\sin \theta}$

It's handy to know these functions because doing algebraic calculations with $\sin \theta$ or $\cos \theta$ in the denominator is a pain in the neck. If we can use a calculator to find these commonly occurring ratios, we can usually keep from having to divide by $\sin \theta$ or $\cos \theta$.

note: If you simply treat these four functions as fancy names for messy expressions of $\sin \theta$ and $\cos \theta$, your life in trigonometry will be a lot easier. You can easily develop all their properties from those of the sine and cosine. For example,

$$\tan(-\theta) = \frac{\sin(-\theta)}{\cos(-\theta)} = \frac{-\sin \theta}{\cos \theta} = -\tan \theta$$

$$\cot(-\theta) = \frac{\cos(-\theta)}{\sin(-\theta)} = \frac{\cos \theta}{-\sin \theta} = \frac{-\cos \theta}{\sin \theta} = -\cot \theta$$

$$\sec(-\theta) = \frac{1}{\cos(-\theta)} = \frac{1}{\cos \theta} = \sec \theta$$

$$\csc(-\theta) = \frac{1}{\sin(-\theta)} = \frac{1}{-\sin \theta} = \frac{-1}{\sin \theta} = -\csc \theta$$

And since $\sin(\theta + 360°) = \sin \theta$ and $\cos(\theta + 360°) = \cos \theta$,

$$\tan(\theta + 360°) = \tan \theta \qquad \cot(\theta + 360°) = \cot \theta$$
$$\sec(\theta + 360°) = \sec \theta \qquad \csc(\theta + 360°) = \csc \theta$$

EXAMPLE 10-4: Find the tangent, cotangent, secant, and cosecant of θ if $\theta = $ **(a)** $72°$, **(b)** $135°$, **(c)** $-30°$.

Solution

(a) $\sin 72° = 0.9511$; $\cos 72° = 0.3090$. So

$$\tan 72° = \frac{\sin 72°}{\cos 72°} = \frac{0.9511}{0.3090} = 3.0777$$

$$\cot 72° = \frac{\cos 72°}{\sin 72°} = \frac{0.3090}{0.9511} = 0.3249$$

$$\sec 72° = \frac{1}{\cos 72°} = \frac{1}{0.3090} = 3.2361$$

$$\csc 72° = \frac{1}{\sin 72°} = \frac{1}{0.9511} = 1.0515$$

(b) $\sin 135° = 0.7071$; $\cos 135° = -0.7071$. So

$$\tan 135° = \frac{\sin 135°}{\cos 135°} = \frac{0.7071}{-0.7071} = -1.0000$$

$$\cot 135° = \frac{\cos 135°}{\sin 135°} = \frac{-0.7071}{0.7071} = -1.0000$$

$$\sec 135° = \frac{1}{\cos 135°} = \frac{1}{-0.7071} = -1.4142$$

$$\csc 135° = \frac{1}{\sin 135°} = \frac{1}{0.7071} = 1.4142$$

(c) $\sin(-30°) = -0.5000$; $\cos(-30°) = 0.8660$. So

$$\tan(-30°) = \frac{\sin(-30°)}{\cos(-30°)} = \frac{-0.5000}{0.8660} = -0.5774$$

$$\cot(-30°) = \frac{\cos(-30°)}{\sin(-30°)} = \frac{0.8660}{-0.5000} = -1.7321$$

$$\sec(-30°) = \frac{1}{\cos(-30°)} = \frac{1}{0.8660} = 1.1547$$

$$\csc(-30°) = \frac{1}{\sin(-30°)} = \frac{1}{-0.5000} = -2.0000$$

note: Tan θ and sec θ are *undefined* when $\theta = \pm 90°$, $\pm 270°$, $\pm 450°$, ...; cot θ and csc θ are undefined when $\theta = 0°$, $\pm 180°$, $\pm 360°$, ... Can you see why?

10-4. Squared Formulas

As P moves around the unit circle ($r = 1$), its x and y coordinates satisfy the equation of the unit circle $x^2 + y^2 = 1$. But these x and y coordinates are, respectively, $\cos \theta$ and $\sin \theta$. Putting these two facts together, we get the most important formula of trigonometry—the **basic trigonometric identity**:

BASIC TRIGONOMETRIC IDENTITY
$$\cos^2\theta + \sin^2\theta = 1 \tag{10-1}$$

for every angle.

note: The notation used in Eq. (10-1) is a kind of shorthand: $\cos^2\theta$ and $\sin^2\theta$ are just short versions of $(\cos \theta)^2$ and $(\sin \theta)^2$. This short notation is used with all trigonometric functions except for $(\sin \theta)^{-1}$, $(\cos \theta)^{-1}$ where the short notation conflicts with the inverse trigonometric function notation.

Many other formulas in trigonometry are really just rewritings of this important equation. For example, if we divide both sides of Eq. (10-1) by $\cos^2\theta$, we get

$$\frac{\cos^2\theta}{\cos^2\theta} + \frac{\sin^2\theta}{\cos^2\theta} = \frac{1}{\cos^2\theta}$$

or

$$\tan^2\theta + 1 = \sec^2\theta \tag{10-2}$$

which is the relation between the tangent and the secant of an angle θ. Although this relation is also quite important, we don't have to memorize it, as long as we remember the definitions of tan θ and sec θ and Eq. (10-1). Equation (10-2) follows easily. Similarly, if we divide both sides of Eq. (10-1) by $\sin^2\theta$, we get

$$\frac{\cos^2\theta}{\sin^2\theta} + \frac{\sin^2\theta}{\sin^2\theta} = \frac{1}{\sin^2\theta}$$

or

$$\cot^2\theta + 1 = \csc^2\theta \qquad \textbf{(10-3)}$$

which is the relation between the cotangent and the cosecant of an angle θ.

EXAMPLE 10-5: Simplify

(a) $\dfrac{1 - \sin^2\theta}{\sin\theta\cos\theta}$ **(b)** $\tan^2\theta - \sec^2\theta$ **(c)** $\sqrt{\dfrac{1 - \sin\theta}{1 + \sin\theta}}$

Solution

(a) You know from Eq. (10-1) that $\cos^2\theta + \sin^2\theta = 1$, so

$$\frac{1 - \sin^2\theta}{\sin\theta\cos\theta} = \frac{\cos^2\theta + \sin^2\theta - \sin^2\theta}{\sin\theta\cos\theta}$$

Now you can cancel equal positive and negative terms in the numerator and divide through by $\cos\theta$:

$$\frac{\cos^2\theta + \sin^2\theta - \sin^2\theta}{\sin\theta\cos\theta} = \frac{\cos^2\theta}{\sin\theta\cos\theta} = \frac{\cos\theta}{\sin\theta}$$

And by the definition of the cotangent of θ,

$$\frac{\cos\theta}{\sin\theta} = \cot\theta$$

(b) You know from Eq. (10-2) that $\sec^2\theta = 1 + \tan^2\theta$, so

$$\tan^2\theta - \sec^2\theta = \tan^2\theta - (1 + \tan^2\theta) = \tan^2\theta - 1 - \tan^2\theta = -1$$

(c) This one's just a little bit tricky. Start by multiplying the numerator and denominator by $1 + \sin\theta$:

$$\sqrt{\frac{1 - \sin\theta}{1 + \sin\theta}} = \sqrt{\frac{(1 - \sin\theta)(1 + \sin\theta)}{(1 + \sin\theta)(1 + \sin\theta)}} = \sqrt{\frac{1 - \sin^2\theta}{(1 + \sin\theta)^2}}$$

Now from Eq. (10-1) you can see that $1 - \sin^2\theta = \cos^2\theta$, so

$$\sqrt{\frac{1 - \sin^2\theta}{(1 + \sin\theta)^2}} = \sqrt{\frac{\cos^2\theta}{(1 + \sin\theta)^2}} = \pm\frac{\cos\theta}{1 + \sin\theta}$$

note: If $\sqrt{}$ denotes a positive number, then

$$\sqrt{\frac{1 - \sin\theta}{1 + \sin\theta}} = \frac{\cos\theta}{1 + \sin\theta} \qquad \text{when } \cos\theta > 0$$

$$= \frac{-\cos\theta}{1 + \sin\theta} \qquad \text{when } \cos\theta < 0$$

10-5. Sum Formulas

Suppose we have two angles α and β. If we already know the sine and cosine of α and β, can we make use of these values to compute $\sin(\alpha + \beta)$ and $\cos(\alpha + \beta)$ and save ourselves some work? How?

Yes ... and here's how. Start with three angles α, β, and $\alpha + \beta$ such that α, β, and $\alpha + \beta$ are all in the first quadrant, as shown in Figure 10-8a. In this figure we have two rays $\overrightarrow{OP_2}$ and $\overrightarrow{OP_1}$ on the unit circle ($r = OP_2 = OP_1 = 1$) such that $\angle SOP_2 = \alpha$ and $\angle P_2OP_1 = \beta$.

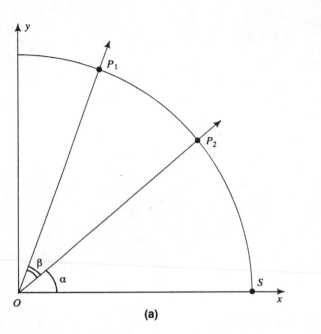

(a)

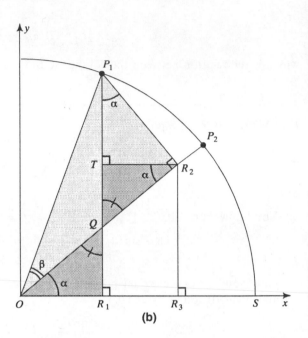

(b)

Figure 10-8

Now we do some constructing to generate some figures we can work with. We drop three perpendiculars:

P_1R_1 from P_1 to the x axis, such that P_1R_1 intersects OP_2 at a point Q
P_1R_2 from P_1 to OP_2
R_2R_3 from the intersection of perpendicular P_1R_2 with OP_2

The result of this construction is a bunch of right triangles, which we know enough about to make some deductions.

We begin with a pair of right triangles $\triangle OR_1Q$ and $\triangle R_2TQ$. We see that the corresponding pair of angles $\angle R_1QO$ and $\angle TQR_2$ are vertical angles formed by intersecting line segments OR_2 and TR_1, so $\angle R_1QO \cong \angle TQR_2$. And since these angles form a congruent pair of corresponding angles in two right triangles, the other pair of corresponding angles must also be congruent; thus $\angle R_1OQ \cong \angle TR_2Q$. Next we see we have another right triangle $\triangle P_1TR_2$, so its two acute angles must be

$$\angle P_1R_2T + \angle TP_1R_2 = 90°$$

And since P_1R_2 is perpendicular to QR_2, we know that

$$\angle P_1R_2T + \angle TR_2Q = 90°$$

A. Sin($\alpha + \beta$)

We can see that $\angle TP_1R_2 \cong \angle TR_2Q$. Then, since $\angle SOP_2 = \alpha = \angle R_1OQ \cong \angle TR_2Q$, $\angle P_1R_2T + \alpha = 90°$ and $\angle TP_1R_2$ must also be equal to α.

note: Recall that AB can indicate both the line segment AB itself and the *measure* of the line segment AB. You figure out whether you're dealing with the line segment itself or its measure by context. You can do the same sort of thing with angles. Angles themselves are *congruent*, while their measures are *equal*—but you have to judge from the context which you're dealing with. In the following discussion you'll notice that we're dealing with measures—numbers upon which operations $(+, -, \times, \div)$ can be performed.

Now we look at another set of right triangles. In $\triangle OR_2P_1$, which includes β, we have

$$\sin \beta = \frac{P_1R_2}{P_1O} = \frac{P_1R_2}{1} = P_1R_2 \qquad \textbf{(a)}$$

$$\cos \beta = \frac{OR_2}{OP_1} = \frac{OR_2}{1} = OR_2 \qquad \textbf{(b)}$$

And in right triangle $\triangle OR_1P_1$, which includes $\alpha + \beta$,

$$\sin(\alpha + \beta) = \frac{P_1R_1}{P_1O} = \frac{P_1R_1}{1} = P_1R_1 \qquad \textbf{(c)}$$

But we know that

$$P_1R_1 = P_1T + TR_1 = P_1T + R_2R_3 \qquad \textbf{(d)}$$

Now in right triangle $\triangle OR_3R_2$

$$\sin \alpha = \frac{R_2R_3}{OR_2}$$

or

$$R_2R_3 = OR_2 \sin \alpha \qquad \text{(that is, } OR_2 \text{ times } \sin \alpha\text{)} \qquad \textbf{(e)}$$

And since we know that $OR_2 = \cos \beta$ [Eq. (b)], we now have

$$R_2R_3 = \cos \beta \sin \alpha \qquad \textbf{(f)}$$

Similarly, in right triangle $\triangle P_1TR_2$

$$\cos \alpha = \frac{P_1T}{P_1R_2} \quad \text{or} \quad P_1T = P_1R_2 \cos \alpha \qquad \textbf{(g)}$$

And since $P_1R_2 = \sin \beta$ [Eq. (a)], we now have

$$P_1T = \sin \beta \cos \alpha \qquad \textbf{(h)}$$

Substituting these last two results [Eqs. (f) and (h)] into Eq. (d), we get the sine of the sum of two angles:

SINE OF THE SUM OF TWO ANGLES $\qquad \sin(\alpha + \beta) = \sin \alpha \cos \beta + \cos \alpha \sin \beta \qquad \textbf{(10-4)}$

note: Although relation (10-4) is for the case in which both angles and their sum are in the first quadrant, this relation holds for any two angles—even if they are not configured as in Figure 10-8.

B. Cos($\alpha + \beta$)

We see from Figure 10-8b that

$$\cos(\alpha + \beta) = \frac{OR_1}{OP_1} = \frac{OR_1}{1} = OR_1 \qquad \textbf{(i)}$$

But

$$OR_1 = OR_3 - R_1R_3 \qquad \textbf{(j)}$$

And in right triangle $\triangle OR_2R_3$

$$\cos \alpha = \frac{OR_3}{OR_2} \quad \text{or} \quad OR_3 = OR_2 \cos \alpha \qquad \textbf{(k)}$$

And since we know that $OR_2 = \cos \beta$, we have

$$OR_3 = \cos \beta \cos \alpha \qquad \textbf{(l)}$$

Now we can see from right triangle $\triangle P_1 TR_2$ that

$$\sin \alpha = \frac{TR_2}{P_1R_2} \qquad \text{or} \qquad TR_2 = P_1R_2 \sin \alpha \qquad \textbf{(m)}$$

Then since $P_1R_2 = \sin \beta$ [Eq. (a)] and $R_1R_3 \cong TR_2$ (by construction), we have

$$TR_2 = R_1R_3 = \sin \beta \sin \alpha \qquad \textbf{(n)}$$

So if we substitute these results [(i), (l), (n)] into Eq. (j), we get the cosine of the sum of two angles:

COSINE OF THE SUM OF TWO ANGLES $\cos(\alpha + \beta) = \cos \alpha \cos \beta - \sin \alpha \sin \beta \qquad \textbf{(10-5)}$

C. Sine and cosine of the difference of two angles

We can easily develop formulas for finding the sine and cosine of the difference between two angles α and β if we remember that $\sin(-\beta) = -\sin \beta$, $\cos(-\beta) = \cos \beta$, and $\alpha - \beta = \alpha + (-\beta)$. Thus

$$\sin(\alpha - \beta) = \sin[\alpha + (-\beta)] = \sin \alpha \cos(-\beta) + \cos \alpha \sin(-\beta)$$

or

SINE OF THE DIFFERENCE BETWEEN TWO ANGLES $\sin(\alpha - \beta) = \sin \alpha \cos \beta - \cos \alpha \sin \beta \qquad \textbf{(10-6)}$

Also

$$\cos(\alpha - \beta) = \cos[\alpha + (-\beta)] = \cos \alpha \cos(-\beta) - \sin \alpha \sin(-\beta)$$

or

COSINE OF THE DIFFERENCE BETWEEN TWO ANGLES $\cos(\alpha - \beta) = \cos \alpha \cos \beta + \sin \alpha \sin \beta \qquad \textbf{(10-7)}$

EXAMPLE 10-6:   Given that α and β are the angles indicated in the triangles shown in Figure 10-9, find **(a)** $\sin(\alpha + \beta)$, **(b)** $\cos(\alpha + \beta)$, **(c)** $\sin(\beta - \alpha)$, and **(d)** $\cos(\beta - \alpha)$.

Solution: You can see that

$$\sin \alpha = \frac{5}{13} \qquad \cos \alpha = \frac{12}{13} \qquad \sin \beta = \frac{4}{5} \qquad \cos \beta = \frac{3}{5}$$

(a) By the sum formula (10-4)

$$\sin(\alpha + \beta) = \sin \alpha \cos \beta + \cos \alpha \sin \beta$$

$$= \left(\frac{5}{13}\right)\left(\frac{3}{5}\right) + \left(\frac{12}{13}\right)\left(\frac{4}{5}\right)$$

$$= \frac{63}{65}$$

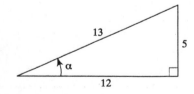
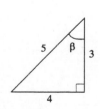

Figure 10-9

(b) By the sum formula (10-5)

$$\cos(\alpha + \beta) = \cos \alpha \cos \beta - \sin \alpha \sin \beta$$

$$= \left(\frac{12}{13}\right)\left(\frac{3}{5}\right) - \left(\frac{5}{13}\right)\left(\frac{4}{5}\right)$$

$$= \frac{16}{65}$$

(c) By the difference formula (10-6)

$$\sin(\alpha - \beta) = \sin \alpha \cos \beta - \cos \alpha \sin \beta$$

so

$$\sin(\beta - \alpha) = \sin \beta \cos \alpha - \cos \beta \sin \alpha$$

$$= \left(\frac{4}{5}\right)\left(\frac{12}{13}\right) - \left(\frac{3}{5}\right)\left(\frac{5}{13}\right)$$

$$= \frac{33}{65}$$

(d) By the difference formula (10-7)

$$\cos(\alpha - \beta) = \cos \alpha \cos \beta + \sin \alpha \sin \beta$$

so

$$\cos(\beta - \alpha) = \cos \beta \cos \alpha + \sin \beta \sin \alpha$$

$$= \left(\frac{3}{5}\right)\left(\frac{12}{13}\right) + \left(\frac{4}{5}\right)\left(\frac{5}{13}\right)$$

$$= \frac{56}{65}$$

EXAMPLE 10-7: Show that the formula for the sine of the sum of two angles (10-4) holds true when the angles α, β, and $\alpha + \beta$ are configured as in Figure 10-10a.

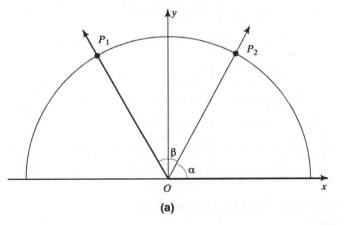

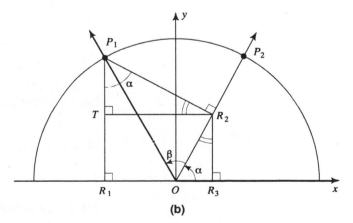

Figure 10-10

Solution: Begin by dropping some perpendiculars: P_1R_1 perpendicular to the x axis, P_1R_2 perpendicular to $\overrightarrow{OP_2}$, R_2R_3 perpendicular to the x axis, and TR_2 perpendicular to P_1R_1. The result of this construction is shown in Figure 10-10b. Notice that P_1R_1 must be parallel to R_2R_3, since both these line segments are constructed perpendicular to the x axis. Hence TR_2, which is constructed perpendicular to P_1R_1, must also be perpendicular to R_2R_3.

Next, looking at angles, you have

$$\angle OR_2R_3 + \angle OR_2T = 90° \qquad \text{(since } TR_2 \text{ is perpendicular to } R_2R_3\text{)}$$

and

$$\angle P_1R_2T + \angle OR_2T = 90° \qquad \text{(since } P_1R_2 \text{ is perpendicular to } \overrightarrow{OP_2}\text{)}$$

so

$$\angle OR_2R_3 \cong \angle P_1R_2T$$

Then in right triangle $\triangle OR_2R_3$, you have

$$\angle R_2OR_3 + \angle OR_2R_3 = \alpha + \angle OR_2R_3 = 90°$$

And, in right triangle $\triangle P_1TR_2$

$$\angle TP_1R_2 + \angle P_1R_2T = 90°$$

but $\angle P_1R_2T \cong \angle OR_2R_3$, so

$$\angle TP_1R_2 + \angle OR_2R_3 = 90°$$

and, since $\alpha + \angle OR_2R_3 = 90°$,

$$\angle TP_1R_2 = \alpha$$

From the definition of the sine, you know that

$$\sin(\alpha + \beta) = \frac{P_1R_1}{P_1O} = \frac{P_1R_1}{1} = P_1R_1$$

And you can see that

$$P_1R_1 = P_1T + TR_1$$

Then in right triangle $\triangle P_1TR_2$, you have $\angle TP_1R_2 = \alpha$ and $\cos\alpha = P_1T/P_1R_2$ or

$$P_1T = P_1R_2 \cos\alpha$$

And in right triangle $\triangle P_1R_2O$, you have $\sin\beta = P_1R_2/P_1O = P_1R_2/1 = P_1R_2$. Thus

$$P_1T = P_1R_2 \cos\alpha = \cos\alpha \sin\beta$$

Next, observe that $TR_1 \cong R_2R_3$. And in right triangle $\triangle R_2R_3O$, $\sin\alpha = R_2R_3/R_2O$ or

$$R_2R_3 = R_2O \sin\alpha$$

But in right triangle $\triangle P_1R_2O$, $\cos\beta = R_2O/1 = R_2O$, so

$$R_2R_3 = R_2O \sin\alpha = \sin\alpha \cos\beta$$

Therefore, if $\sin(\alpha + \beta) = P_1R_1 = P_1T + TR_1 = P_1T + R_2O_3$, then

$$\sin(\alpha + \beta) = \sin\alpha \cos\beta + \cos\alpha \sin\beta \qquad \text{QED}$$

10-6. Double- and Half-Angle Formulas

A. Double-angle formulas

If we let $\alpha = \beta$ in the sum formulas (10-4) and (10-5), we get, respectively,

$$\sin(\alpha + \alpha) = \sin\alpha \cos\alpha + \cos\alpha \sin\alpha$$

or

**SINE OF A
DOUBLE ANGLE**
$$\sin 2\alpha = 2 \sin\alpha \cos\alpha \qquad \text{(10-8)}$$

and

$$\cos(\alpha + \alpha) = \cos\alpha\cos\alpha - \sin\alpha\sin\alpha$$

or

COSINE OF A DOUBLE ANGLE $\qquad \cos 2\alpha = \cos^2\alpha - \sin^2\alpha \qquad$ **(10-9)**

These important formulas are called the **double-angle formulas**. They allow us to calculate the sine and cosine of twice the angle when we know only the sine and cosine of the original angle.

From the basic trig formula (10-1) we know that $\cos^2\alpha + \sin^2\alpha = 1$; thus

$$\cos^2\alpha = 1 - \sin^2\alpha$$

If we substitute this into the double-angle formula (10-9), we get an alternative form of the double-angle formula for cosine:

$$\cos 2\alpha = \cos^2\alpha - \sin^2\alpha = (1 - \sin^2\alpha) - \sin^2\alpha$$

or

$$\cos 2\alpha = 1 - 2\sin^2\alpha \qquad \textbf{(10-10)}$$

Similarly,

$$\sin^2\alpha = 1 - \cos^2\alpha$$

If we put this into formula (10-9), we get yet another common form of the cosine double-angle formula:

$$\cos 2\alpha = \cos^2\alpha - \sin^2\alpha = \cos^2\alpha - (1 - \cos^2\alpha)$$

or

$$\cos 2\alpha = 2\cos^2\alpha - 1 \qquad \textbf{(10-11)}$$

B. Half-angle formulas

Alternative formulas (10-10) and (10-11) are particularly useful for deriving half-angle formulas for sine and cosine. If we let $\alpha = \theta/2$ in formula (10-10), we get

$$\cos\frac{2\theta}{2} = 1 - 2\sin^2\frac{\theta}{2}$$

or

$$\cos\theta = 1 - 2\sin^2\frac{\theta}{2}$$

so

$$\sin^2\frac{\theta}{2} = \frac{1 - \cos\theta}{2} \qquad \textbf{(10-12)}$$

and

$$\sin\frac{\theta}{2} = \pm\sqrt{\frac{1 - \cos\theta}{2}} \qquad \textbf{(10-12a)}$$

note: If you use formula (10-12a), you must pay particular attention to what quadrant $\theta/2$ lies in, and then choose the $+$ to $-$ sign accordingly.

Similarly,

$$\cos\frac{2\theta}{2} = 2\cos^2\frac{\theta}{2} - 1$$

or

$$\cos \theta = 2 \cos^2 \frac{\theta}{2} - 1$$

so

$$\cos^2 \frac{\theta}{2} = \frac{1 + \cos \theta}{2} \qquad \textbf{(10-13)}$$

and

$$\cos \frac{\theta}{2} = \pm \sqrt{\frac{1 + \cos \theta}{2}} \qquad \textbf{(10-13a)}$$

EXAMPLE 10-8:   Compute the sine and cosine for 2α where α is the angle from Figure 10-9.

Solution: You know that $\sin \alpha = 5/13$ and $\cos \alpha = 12/13$. So, by formula (10-8),

$$\sin 2\alpha = 2 \sin \alpha \cos \alpha$$

$$= 2 \left(\frac{5}{13} \right) \left(\frac{12}{13} \right) = \frac{120}{169}$$

$$= 0.7101$$

and by formula (10-9),

$$\cos 2\alpha = \cos^2\alpha - \sin^2\alpha$$

$$= \left(\frac{12}{13} \right)^2 - \left(\frac{5}{13} \right)^2 = \frac{144 - 25}{169} = \frac{119}{169}$$

$$= 0.7041$$

EXAMPLE 10-9:   Use the ideas of Section 10-6 to develop formulas for **(a)** $\sin 3\theta$ and **(b)** $\cos 3\theta$ in terms of $\sin \theta$ and $\cos \theta$.

Solution: Note that $3\theta = 2\theta + \theta$.

(a) $\sin 3\theta = \sin 2\theta \cos \theta + \cos 2\theta \sin \theta$ By Eq. (10-4)

$\quad\quad = (2 \sin \theta \cos \theta) \cos \theta + (\cos^2\theta - \sin^2\theta) \sin \theta$ By Eqs. (10-8) and (10-9)

$\quad\quad = 2 \sin \theta \cos^2\theta + \sin \theta \cos^2\theta - \sin^3\theta$ Multiply

$\quad\quad = \sin \theta (2 \cos^2\theta + \cos^2\theta - \sin^2\theta)$ Factor

$\quad\quad = \sin \theta (3 \cos^2\theta - \sin^2\theta)$ Simplify

(b) $\cos 3\theta = \cos 2\theta \cos \theta - \sin 2\theta \sin \theta$ By Eq. (10-5)

$\quad\quad = (\cos^2\theta - \sin^2\theta) \cos \theta - 2 \sin \theta \cos \theta \sin \theta$ By Eqs. (10-8) and (10-9)

$\quad\quad = (\cos^3\theta - \sin^2\theta \cos \theta - 2 \sin^2\theta \cos \theta$ Multiply

$\quad\quad = \cos \theta (\cos^2\theta - \sin^2\theta - 2 \sin^2\theta)$ Factor

$\quad\quad = \cos \theta (\cos^2\theta - 3 \sin^2\theta)$ Simplify

note: You can also see that

$$\sin 3\theta = \sin \theta (4 \cos^2\theta - 1)$$

and

$$\cos 3\theta = \cos \theta (1 - 4 \sin^2\theta)$$

SUMMARY

1. The sine of an angle θ (sin θ) is the projection on the y axis of the radius r of a unit circle ($r = 1$) that forms an angle θ with the positive x axis.
2. The cosine of an angle θ (cos θ) is the projection on the x axis of the radius r of a unit circle ($r = 1$) that forms an angle θ with the positive x axis.
3. $\sin(-\theta) = -\sin\theta$ and $\cos(-\theta) = \cos\theta$
4. All the other trigonometric functions—tangent, cotangent, secant, and cosecant—are formed sin θ and cos θ:

$$\tan\theta = \frac{\sin\theta}{\cos\theta} \qquad \cot\theta = \frac{\cos\theta}{\sin\theta}$$

$$\sec\theta = \frac{1}{\cos\theta} \qquad \csc\theta = \frac{1}{\sin\theta}$$

5. All six trigonometric functions repeat after every 360°; that is, $\sin\theta + 360° = \sin\theta$, $\cos\theta + 360° = \cos\theta$, etc.
6. For θ in quadrant I, $\sin\theta > 0$ and $\cos\theta > 0$.

   For θ in quadrant II, $\sin\theta > 0$ and $\cos\theta < 0$.

   For θ in quadrant III, $\sin\theta < 0$ and $\cos\theta < 0$.

   For θ in quadrant IV, $\sin\theta < 0$ and $\cos\theta > 0$.

7. The sine and cosine of any angle θ may be found by using the auxiliary angle, which is the acute angle formed by θ's terminal side and the x axis.
8. The fundamental formula of trigonometry is

$$\sin^2\theta + \cos^2\theta = 1$$

9. The relations between the tangent and secant and between the cotangent and the cosecant of an angle θ are described by the squared formulas

$$\tan^2\theta + 1 = \sec^2\theta$$

and

$$1 + \cot^2\theta = \csc^2\theta$$

10. The sine and cosine of the sum of two angles α and β are given by the sum formulas

$$\sin(\alpha + \beta) = \sin\alpha\cos\beta + \cos\alpha\sin\beta$$

and

$$\cos(\alpha + \beta) = \cos\alpha\cos\beta - \sin\alpha\sin\beta$$

11. The sine and cosine of the difference between two angles are given by the difference formulas

$$\sin(\alpha - \beta) = \sin\alpha\cos\beta - \cos\alpha\sin\beta$$

and

$$\cos(\alpha - \beta) = \cos\alpha\cos\beta + \sin\alpha\sin\beta$$

12. The double-angle formulas are

$$\sin 2\theta = 2\sin\theta\cos\theta$$
$$\cos 2\theta = \cos^2\theta - \sin^2\theta$$
$$\cos 2\theta = 1 - 2\sin^2\theta$$
$$\cos 2\theta = 2\cos^2\theta - 1$$

13. The half-angle formulas are

$$\sin^2\frac{\theta}{2} = \frac{1 - \cos\theta}{2} \qquad \text{or} \qquad \sin\frac{\theta}{2} = \pm\sqrt{\frac{1 - \cos\theta}{2}}$$

and

$$\cos^2 \frac{\theta}{2} = \frac{1 + \cos \theta}{2} \qquad \text{or} \qquad \cos \frac{\theta}{2} = \pm\sqrt{\frac{1 + \cos \theta}{2}}$$

RAISE YOUR GRADES

Can you ...?

☑ find the sine and cosine of any angle, given its measure in degrees or radians
☑ find the tangent, cotangent, secant, and cosecant, given the sine and cosine of an angle
☑ find the sine and cosine of an angle in the second, third, or fourth quadrant
☑ determine the sign of a trigonometric function, given the quadrant the angle is in
☑ rewrite any trigonometric expression in terms of the sine and cosine
☑ use the identity $\sin^2\theta + \cos^2\theta = 1$ to simplify involved trigonometric expressions
☑ find $\sin(\alpha + \beta)$ and $\cos(\alpha + \beta)$, given $\sin \alpha$, $\cos \alpha$, $\sin \beta$, and $\cos \beta$
☑ derive the double- and half-angle formulas
☑ develop a formula for $\tan(\alpha + \beta)$ in terms of $\tan \alpha$ and $\tan \beta$

SOLVED PROBLEMS

PROBLEM 10-1   Determine the quadrants in which the terminal sides of the following angles lie:

(a) 115°   (b) 210°   (c) 323°   (d) 410°   (e) −128°   (f) −220°

Solution   For positive angles θ, first subtract multiples of 360° until the resulting difference α is between 0° and 360°. For negative angles θ, add multiples of 360° until the resulting sum α is between 0° and 360°. The angle is in quadrant I if $0 < \alpha \le 90°$, quadrant II if $90° < \alpha \le 180°$, quadrant III if $180° < \alpha \le 270°$, and quadrant IV if $270° < \alpha \le 360°$.

(a) $\theta = 115° \Rightarrow$ quadrant II
(b) $\theta = 210° \Rightarrow$ quadrant III
(c) $\theta = 323° \Rightarrow$ quadrant IV
(d) $\theta = 410°$, $\alpha = 410° - 360° = 50° \Rightarrow$ quadrant I
(e) $\theta = -128°$, $\alpha = -128° + 360° = 232° \Rightarrow$ quadrant III
(f) $\theta = -220°$, $\alpha = -220° + 360° = 140° \Rightarrow$ quadrant II

PROBLEM 10-2   Find the auxiliary angles as well as the sine and cosine for:

(a) 115°   (b) 210°   (c) 323°   (d) 410°   (e) −128°   (f) −220°

Solution   Remember that the absolute value of the sine and cosine for any angle is equal to that of the auxiliary angle—and that the signs of these functions are determined by the quadrant in which the terminal side lies.

(a) auxiliary angle = $180° - 115° = 65°$; $\sin 115° = \sin 65° = 0.9063$; $\cos 115° = -\cos 65° = -0.4226$
(b) auxiliary angle = $210° - 180° = 30°$; $\sin 210° = -\sin 30° = -0.5000$; $\cos 210° = -\cos 30° = -0.8660$

(c) auxiliary angle $= 360° - 323° = 37°$; $\sin 323° = -\sin 37° = -0.6018$; $\cos 323° = \cos 37° = 0.7986$

(d) auxiliary angle $= 410° - 360° = 50°$; $\sin 410° = \sin 50° = 0.7660$; $\cos 410° = \cos 50° = 0.6428$

(e) auxiliary angle $= 180° - 128° = 52°$; $\sin -128° = -\sin 52° = -0.7880$; $\cos -128° = -\cos 52° = -0.6157$

(f) auxiliary angle $= 220° - 180° = 40°$; $\sin -220° = -\sin 40° = -0.6428$; $\cos -220° = -\cos 40° = -0.7660$

PROBLEM 10-3 If θ is in the second quadrant and $\sin \theta = 1/5$, find the other five trigonometric functions.

Solution Start with the basic trigonometric identity:

$$\sin^2\theta + \cos^2\theta = \frac{1}{25} + \cos^2\theta = 1$$

or

$$\cos^2\theta = 1 - \frac{1}{25} = \frac{24}{25}$$

Since θ is in the second quadrant, $\cos \theta < 0$. Thus

$$\cos \theta = -\sqrt{\frac{24}{25}} = -\sqrt{\frac{2^2 \cdot 6}{5^2}} = -\frac{2\sqrt{6}}{5} = -0.9798$$

Now

$$\tan \theta = \frac{\sin \theta}{\cos \theta} = \frac{-1/5}{2\sqrt{6}/5} = \frac{-1}{2\sqrt{6}} = -0.2041$$

$$\cot \theta = \frac{1}{\tan \theta} = -4.8990$$

$$\sec \theta = \frac{1}{\cos \theta} = \frac{1}{-0.9798} = -1.0206$$

$$\csc \theta = \frac{1}{\sin \theta} = \frac{1}{1/5} = 5$$

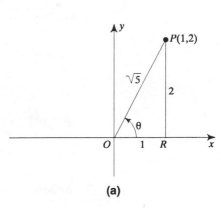

(a)

PROBLEM 10-4 A ray is drawn in the Cartesian plane from the origin O to a point P. Angle θ is measured from the x axis to OP. Find $\sin \theta$ and $\cos \theta$ for P if it has the following coordinates:

(a) $(1, 2)$ (b) $(-1, 3)$ (c) $(-2, -1)$ (d) $(2, -5)$

Solution

(a) θ is in the first quadrant and is the acute angle of right triangle $\triangle ORP$ as shown in Figure 10-11a. This triangle has legs of 1 (the x coordinate) and 2 (the y coordinate), so its hypotenuse is $\sqrt{1^2 + 2^2} = \sqrt{5}$. Thus,

$$\sin \theta = \frac{2}{\sqrt{5}}, \qquad \cos \theta = \frac{1}{\sqrt{5}}$$

(b) θ is in the second quadrant. The auxiliary angle α is the acute angle of right triangle $\triangle OSP$ as shown in Figure 10-11b. This triangle has legs 3 and 1 and hypotenuse $\sqrt{3^2 + 1^2} = \sqrt{10}$, so

$$\sin \theta = \sin \alpha = \frac{3}{\sqrt{10}}, \qquad \cos \theta = -\cos \alpha = -\frac{1}{\sqrt{10}}$$

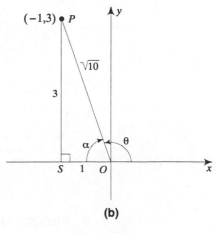

(b)

Figure 10-11

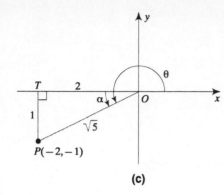

(c)

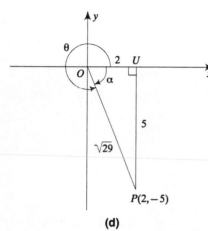

(d)

Figure 10-11. Continued

(c) θ is in the third quadrant. The auxiliary angle α is the acute angle of right triangle $\triangle OTP$ as shown in Figure 10-11c. This triangle has legs 1 and 2 and hypotenuse $\sqrt{5}$, so

$$\sin \theta = -\sin \alpha = -\frac{1}{\sqrt{5}} \qquad \cos \theta = -\cos \alpha = -\frac{2}{\sqrt{5}}$$

(d) θ is in the fourth quadrant. The auxiliary angle α is the acute angle of right triangle $\triangle OUP$ in Figure 10-11d. This triangle has legs 2 and 5 and hypotenuse $\sqrt{2^2 + 5^2} = \sqrt{29}$, so

$$\sin \theta = -\sin \alpha = -\frac{5}{\sqrt{29}}, \qquad \cos \theta = \cos \alpha = \frac{2}{\sqrt{29}}$$

PROBLEM 10-5   Simplify $\cos^4\theta - \sin^4\theta$.

Solution

$$\begin{aligned}
\cos^4\theta - \sin^4\theta &= (\cos^2\theta + \sin^2\theta)(\cos^2\theta - \sin^2\theta) \\
&= 1\cos 2\theta \\
&= \cos 2\theta
\end{aligned}$$

PROBLEM 10-6   Express $\tan(\alpha + \beta)$ in terms of $\tan \alpha$ and $\tan \beta$.

Solution

$$\tan(\alpha + \beta) = \frac{\sin(\alpha + \beta)}{\cos(\alpha + \beta)} = \frac{\sin \alpha \cos \beta + \cos \alpha \sin \beta}{\cos \alpha \cos \beta - \sin \alpha \sin \beta}$$

$$= \frac{\dfrac{\sin \alpha \cos \beta}{\cos \alpha \cos \beta} + \dfrac{\cos \alpha \sin \beta}{\cos \alpha \cos \beta}}{\dfrac{\cos \alpha \cos \beta}{\cos \alpha \cos \beta} - \dfrac{\sin \alpha \sin \beta}{\cos \alpha \cos \beta}}$$

$$= \frac{\tan \alpha + \tan \beta}{1 - \tan \alpha \tan \beta}$$

PROBLEM 10-7   Show that

$$\frac{1 + \sec \theta}{\sec \theta} = 2\cos^2\left(\frac{\theta}{2}\right)$$

Solution

$$\frac{1 + \sec \theta}{\sec \theta} = \frac{1 + \dfrac{1}{\cos \theta}}{\dfrac{1}{\cos \theta}} = \frac{\cos \theta \left(1 + \dfrac{1}{\cos \theta}\right)}{1}$$

$$= \cos \theta + 1 = \cos\left(2\frac{\theta}{2}\right) + 1$$

$$= 2\cos^2\frac{\theta}{2} - 1 + 1$$

$$= 2\cos^2\frac{\theta}{2}$$

PROBLEM 10-8   You know that $\sin 45° = \cos 45° = 1/\sqrt{2}$, $\sin 30° = 1/2$, and $\cos 30° = \sqrt{3}/2$. Use these values to find $\sin 15°$ and $\cos 15°$.

Solution

$$\sin 15° = \sin(45° - 30°)$$
$$= \sin 45° \cos 30° - \cos 45° \sin 30°$$
$$= (1/\sqrt{2})(\sqrt{3}/2) - \left(\frac{1}{\sqrt{2}}\right)\left(\frac{1}{2}\right) = 0.2588$$

$$\cos 15° = \cos(45° - 30°)$$
$$= \cos 45° \cos 30° + \sin 45° \sin 30°$$
$$= \left(\frac{1}{\sqrt{2}}\right)\left(\frac{\sqrt{3}}{2}\right) + \left(\frac{1}{\sqrt{2}}\right)\left(\frac{1}{2}\right) = 0.9659$$

PROBLEM 10-9   Given that $\sin \theta = 12/13$ and $\cos \theta = 5/13$; find $\cos(\theta - 30°)$.

Solution

$$\sin 30° = \frac{1}{2}$$

$$\cos 30° = \frac{\sqrt{3}}{2}$$

$$\cos(\theta - 30°) = \cos \theta \cos 30° + \sin \theta \sin 30°$$
$$= \left(\frac{5}{13}\right)\left(\frac{\sqrt{3}}{2}\right) + \left(\frac{12}{13}\right)\left(\frac{1}{2}\right)$$
$$= 0.7946$$

Review Exercises

EXERCISE 10-1   A ray is drawn in the Cartesian plane from the origin to a point P. Angle θ is measured from the x axis to OP. Determine what quadrant θ lies in if P has the following coordinates:

(a) $(1, -2)$ (c) $(-1, -1)$ (e) $(-1, -2)$ (g) $(2, -1)$
(b) $(2, 3)$ (d) $(-1, 2)$ (f) $(1, 1)$ (h) $(-4, -1)$

EXERCISE 10-2   Find the sine and cosine for each of the angles in Exercise 10-1.

EXERCISE 10-3   Find $\sin \theta$, $\cos \theta$, $\tan \theta$, $\cot \theta$, $\sec \theta$, and $\csc \theta$ for $\theta =$

(a) $10°$ (b) $-78°$ (c) $163°$ (d) $212°$

EXERCISE 10-4   Determine what quadrant θ lies in if (a) $\sin \theta > 0$ and $\tan \theta < 0$, (b) $\sec \theta > 0$ and $\sin \theta < 0$, (c) $\sin \theta < 0$ and $\tan \theta > 0$.

EXERCISE 10-5   Suppose $\sin \theta = 3/7$ and $\cos \theta < 0$. Draw a careful figure showing which quadrant θ is in. Then find the other five trigonometric functions.

EXERCISE 10-6   Suppose $\tan \theta = 5$ and $\cos \theta < 0$. Draw a careful figure showing which quadrant θ is in. Then find the other five trigonometric functions.

EXERCISE 10-7   Simplify $(\sin \theta + \cos \theta)^2 + (\sin \theta - \cos \theta)^2$.

EXERCISE 10-8 Simplify $\sin^3\theta \cos\theta + \sin\theta \cos^3\theta$.

EXERCISE 10-9 Simplify

$$\frac{\tan\theta + \sin\theta}{1 + \cos\theta}$$

EXERCISE 10-10 Show that

$$\frac{1 - \sin\theta}{1 + \sin\theta} = (\sec\theta - \tan\theta)^2$$

EXERCISE 10-11 Show that

$$\frac{1 + \sin\theta}{\cos\theta} + \frac{\cos\theta}{1 + \sin\theta} = 2\sec\theta$$

EXERCISE 10-12 Show that

$$\sec^4\theta(1 - \sin^4\theta) = \sec^2\theta + \tan^2\theta$$

EXERCISE 10-13 Show that

$$1 - \frac{\cos^2\theta}{1 + \sin\theta} = \sin\theta$$

EXERCISE 10-14 Given that $\cos\alpha = 0.9563$, $\sin\alpha = 0.2924$, $\cos\beta = 0.9135$, and $\sin\beta = 0.4067$, find **(a)** $\cos(\alpha + \beta)$, **(b)** $\cos(\alpha - \beta)$, **(c)** $\sin(\alpha + \beta)$, **(d)** $\sin(\alpha - \beta)$.

EXERCISE 10-15 Given that $\cos\alpha = 0.0698$ and $\sin\alpha = 0.9976$, find **(a)** $\cos 2\alpha$ and **(b)** $\sin 2\alpha$.

EXERCISE 10-16 Given that $\cos\alpha = 0.8480$ and $\sin\alpha = 0.5299$, find $\tan 2\alpha$.

EXERCISE 10-17 Given that $\cos\alpha = 4/5$ and $\sin\alpha = 3/5$, find **(a)** $\sin(\alpha + 45°)$ and **(b)** $\cos(30° - \alpha)$.

EXERCISE 10-18 Given that $\tan\alpha = 12/5$ and $\cos\alpha = 5/13$, find **(a)** $\sin(\alpha + 30°)$ and **(b)** $\cos(\alpha - 45°)$.

EXERCISE 10-19 Given that $\sin\alpha = 24/25$ and $\cos\alpha = 7/25$, find $\sin(\alpha/2)$ and $\cos(\alpha/2)$.

EXERCISE 10-20 Given that $\sin\alpha = 15/17$ and $\cos\alpha = 8/17$, find $\tan(\alpha/2)$.

Answers to Review Exercises

10-1 **(a)** IV **(d)** II **(g)** = IV
(b) I **(e)** III **(h)** III
(c) III **(f)** I

10-2 **(a)** $\sin\theta = -2/\sqrt{5} = -0.8944$,
$\cos\theta = 1/\sqrt{5} = 0.4472$
(b) $\sin\theta = 3/\sqrt{13} = 0.8321$,
$\cos\theta = 2/\sqrt{13} = 0.5547$
(c) $\sin\theta = -1/\sqrt{2} = -0.7071$,
$\cos\theta = -1/\sqrt{2} = -0.7071$

(d) $\sin\theta = 2/\sqrt{5} = 0.8944$,
$\cos\theta = -1/\sqrt{5} = -0.4472$
(e) $\sin\theta = -2/\sqrt{5} = -0.8944$,
$\cos\theta = -1/\sqrt{5} = -0.4472$
(f) $\sin\theta = 1/\sqrt{2} = 0.7071$,
$\cos\theta = 1/\sqrt{2} = 0.7071$
(g) $\sin\theta = -1/\sqrt{5} = -0.4472$,
$\cos\theta = 2/\sqrt{5} = 0.8944$
(h) $\sin\theta = -1/\sqrt{17} = -0.2425$,
$\cos\theta = -4/\sqrt{17} = -0.9701$

	θ	sin	cos	tan	cot	sec	csc
(a)	10°	0.1736	0.9848	0.1763	5.6713	1.0154	5.7603
(b)	−78°	−0.9781	0.2079	−4.7046	−0.2125	4.8097	−1.0223
(c)	163°	0.2924	−0.9563	−0.3057	−3.2709	−1.0457	3.4203
(d)	212°	−0.5299	−0.8480	0.6249	1.6003	−1.1792	−1.8871

10-3

10-4 (a) II   (b) IV   (c) III

10-5 See Figure 10.12: $\cos \theta = -\sqrt{40}/7 = -0.9035$, $\tan \theta = 3/-\sqrt{40} = -0.4743$, $\cot \theta = -\sqrt{40}/3 = -2.1082$, $\sec \theta = 7/-\sqrt{40} = -1.1068$; $\csc \theta = 7/3 = 2.3333$

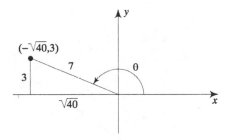

Figure 10-12

10-6 See Figure 10-13: $\sin \theta = -5/\sqrt{26} = -0.9806$, $\cos \theta = -1/\sqrt{26} = -0.1961$, $\cot \theta = -1/-5 = 0.2000$, $\sec \theta = -\sqrt{26}/1 = -5.0990$, $\csc \theta = -1.0198$

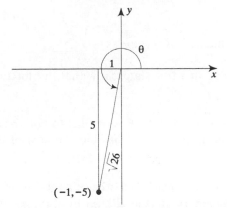

Figure 10-13

10-7 2

10-8 $\sin \theta \cos \theta$

10-9 $\tan \theta$

10-14 (a) 0.7547   (c) 0.6560
         (b) 0.9925   (d) −0.1218

10-15 (a) −0.9903   (b) 0.1393

10-16 2.0504

10-17 (a) 0.9899   (b) 0.9928

10-18 (a) 0.9917   (b) 0.9247

10-19 (a) 0.6000   (b) 0.8000

10-20 0.6000

TEST 1

1. Three angles are specified in Figure T-1. Find all of the other angles indicated. (Assume that *AB* is parallel to *CD*.)

2. Line segment *AB* in Figure T-2 joints point *A* whose coordinates are (1, 2) to point *B* whose coordinates are (7, 8). Point *P* lies on this line segment, and its distance to *A* is one-half its distance to *B*. Find *P*'s coordinates.

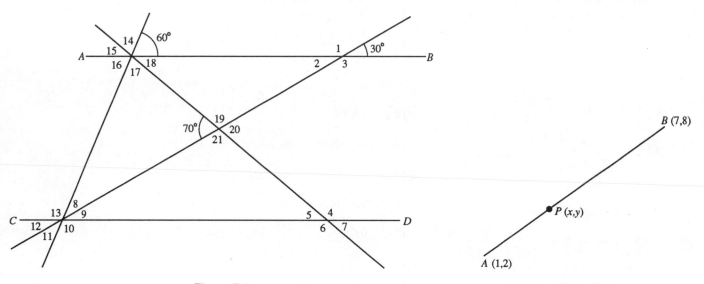

Figure T-1   **Figure T-2**

3. Use a truth table to show that $q \wedge (r \vee s)$ is the same as $(q \wedge r) \vee (q \wedge s)$.

4. If the logical expressions $q \to p$, $s \vee q$, and $\sim s$ are all in a true state, show that this forces p to be true.

5. Quadrilateral *ABCD* in Figure T-3 is a trapezoid. Line segment *EF* is parallel to both bases and lies a distance 3 from base *BC*. (a) Find *x*, the length of *EF*. (b) Find the area of the trapezoid *BCFE*.

6. The rhombus of Figure T-4 has sides of length 4 inches each. Two adjacent sides meet at an angle of 70°. (a) Find the lengths of the two diagonals *AC* and *BD*. (b) Find the area of the rhombus.

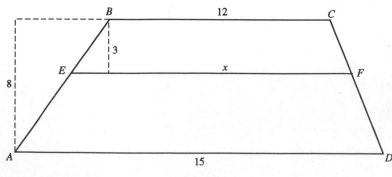

Figure T-3

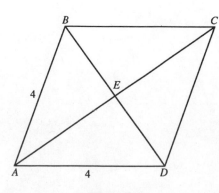

Figure T-4

7. In △*ABC* of Figure T-5, ∠*A* ≅ ∠*C* and *AD* ≅ *CE*. Show that ∠*BDE* ≅ ∠*BED*.

8. The triangles in Figure T-6 are all congruent right triangles. Point *B* is the center of a diameter for a semicircle of radius 4, and points *A* and *C* lies on the diameter. Find the area of the three triangles combined.

9. In △*ABC* of Figure T-7, ∠*A* = 110°, *AB* measures 6 inches, and *AC* measures 8 inches. **(a)** Find the area of △*ABC*. **(b)** Find the length of the side *BC*. **(c)** Find the length of the altitude *AD* drawn from *A* to side *BC*.

10. In Figure T-8, points *D*, *E*, and *F* are the midpoints of line segments *AB*, *BC*, and *CA*, respectively. Points *G*, *H*, and *I* are the midpoints of the line segments *DE*, *DF*, and *FE*, respectively. The area of △*GHI* is 12 square inches. Find the area of △*ABC*.

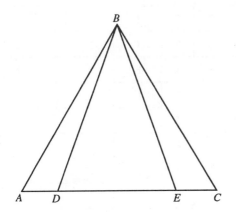

Figure T-5

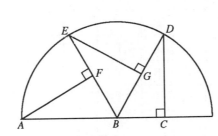

Figure T-6

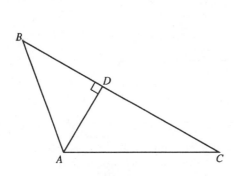

Figure T-7

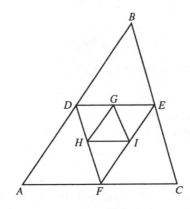

Figure T-8

Answers to Test 1

---

1. ∠1 = 150°, ∠2 = 30°, ∠3 = 150°, ∠4 = 140°, ∠5 = 40°, ∠6 = 140°, ∠7 = 40°, ∠8 = 30°, ∠9 = 30°, ∠10 = 120°, ∠11 = 30°, ∠12 = 30°, ∠13 = 120°, ∠14 = 80°, ∠15 = 40°, ∠16 = 60°, ∠17 = 80°, ∠18 = 40°, ∠19 = 110°, ∠20 = 70°, ∠21 = 110°

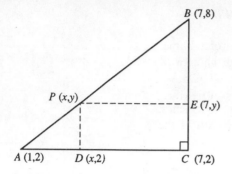

Figure T-9

2. (a) *Step 1:* Draw a line from A parallel to the x axis and a line from B parallel to the y axis as in Figure T-9. They meet at point C with coordinates $(7, 2)$.

Step 2: Draw a perpendicular PD from P to AC and a perpendicular PE from P to BC. D has coordinates $(x, 2)$ and E has coordinates $(7, y)$.

Step 3: $\triangle ABC$, $\triangle APD$, and $\triangle PBE$ are similar right triangles. Thus

$$\frac{AD}{AC} = \frac{AP}{AB}, \qquad \text{so} \quad \frac{x-1}{7-1} = \frac{1}{3} \qquad \text{and}$$

$$\frac{BE}{BC} = \frac{PB}{AB}, \qquad \text{so} \quad \frac{8-y}{8-2} = \frac{2}{3}$$

Step 4: $x = 1 + \frac{6}{3} = 3$ and $y = 8 - \frac{12}{3} = 4$.

3.

q	r	s	r∨s	q∧(r∨s)
T	T	T	T	T
T	T	F	T	T
T	F	T	T	T
T	F	F	F	F
F	T	T	T	F
F	T	F	T	F
F	F	T	T	F
F	F	F	F	F

q	r	s	q∧r	q∧s	(q∧r)∨(q∧s)
T	T	T	T	T	T
T	T	F	T	F	T
T	F	T	F	T	T
T	F	F	F	F	F
F	T	T	F	F	F
F	T	F	F	F	F
F	F	T	F	F	F
F	F	F	F	F	F

4. ∼s true means s is false. Since s is false, and s∨q is true, q must be true. Since the implication q → p is true and the premise q is true, p must also be true.

5. (a) *Step 1:* Draw BH parallel to CD as in Figure T-10. Quadrilateral $BGFC$ is also a parallelogram and GF has length 12.

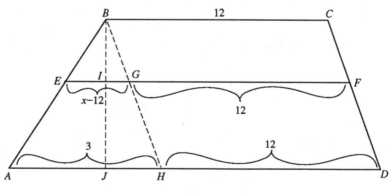

Figure T-10

Step 2: In $\triangle BEG$ and $\triangle BAH$ draw respective altitudes BI and BJ.

Step 3: $\triangle BEG$ is similar to $\triangle BAH$. Thus $\dfrac{BI}{BJ} = \dfrac{EG}{AH}$ or $\dfrac{x-12}{3} = \dfrac{3}{8}$.

Step 4: Solve this equation for $x \cdot x = 12 + \frac{9}{8}$ or $x = 13.125$.

(b) Area $= \frac{1}{2} \cdot 3[12 + 13.125] = 37.69$.

6. (a) *Step 1:* BD and AC are perpendicular bisectors of each other at E. AC bisects $\angle A$ and $\angle C$, and BD bisects $\angle B$ and $\angle D$.

Step 2: $\triangle ABE$ is a right triangle with hypotenuse AB of length 4 and acute angle $70°/2 = 35°$.

Step 3: $BE = 4\sin(35°) = 2.29$ and $AE = 4\cos(35°) = 3.28$.

Step 4: $BD = 2(2.29) = 4.58$ and $AC = 2(3.28) = 6.55$.

(b) The area of $\triangle ABE = (\frac{1}{2})(2.29)(3.28) = 3.76$. The area of $ABCD = 4(3.76) = 15.02$.

7. *Step 1:* $\angle A \cong \angle C$ is given. This implies $\triangle ABC$ is an isosceles triangle with legs BA and BC of equal length. Thus $BA \cong BC$.

Step 2: $AD \cong CE$ is given.

Step 3: $\triangle ABD \cong \triangle CBE$ by SAS.

Step 4: $BD \cong BE$ as corresponding sides of congruent triangles

Step 5: $\triangle DBE$ is isosceles.

Step 6: $\angle BDE$ and $\angle BED$ are base angles of an isosceles triangle and so are equal.

8. *Step 1:* $\angle ABF + \angle FBG + \angle GBC = 180°$.

Step 2: $\angle ABF \cong \angle FBG \cong \angle GBC$. Let x be the common measure.

Step 3: $3x = 180°$ and thus $x = 60°$.

Step 4: $\angle FAB = 30°$, and $\triangle FAB$ is a 30°-60°-90° triangle.

Step 5: $FB = 4(\sin 30°) = \frac{4}{2} = 2$, and $FA = 4(\cos 30°) = \frac{4\sqrt{3}}{2} = 2\sqrt{3}$.

Step 6: The area of the combined triangles is 3 times the area of $\triangle AFB = \frac{1}{2}(2)(2\sqrt{3}) = 2\sqrt{3}$, so the total area is $6\sqrt{3}$.

9. **(a)** From the law of sines, area $\triangle ABC = (\frac{1}{2})(6)(8)\sin(110°) = 22.55$ in^2.

(b) From the law of cosines, $BC^2 = 6^2 + 8^2 - 2(6)(8)\cos(110°)$. So $BC^2 = 132.83$ and $BC = \sqrt{132.83} = 11.53$.

(c) The area of $\triangle ABC = \frac{1}{2}(AD)(BC)$. Since the area of $\triangle ABC = 22.55$, $AD = \dfrac{2(22.55)}{11.53} = 3.91$.

10. $\triangle GHI \sim \triangle FED \sim \triangle BAC$. The corresponding parts of $\triangle GHI$ stand in 1:2 ratio with the sides of $\triangle FED$, and the corresponding parts of $\triangle FED$ stand in 1:2 ratio with the sides of $\triangle BAC$. Consequently, $AB = 2EF = 4GH$, $AC = 2DE = 4HI$, and $\angle BAC \cong \angle DEF \cong \angle GHI$. Using the law of sines, the area of

$$\triangle ABC = \frac{1}{2}(AB)(AC)\sin(\angle BAC)$$
$$= \frac{1}{2}(4GH)(4HI)\sin(\angle GHI)$$
$$= (16)\frac{1}{2}(GH)(HI)\sin(\angle GHI)$$

which is 16 times the area of $\triangle GHI$, or $16(12$ in$^2) = 192$ in^2.

TEST 2

1. In $\triangle ABC$ of Figure T-11, AB has length 12 cm. $\angle A$ and $\angle B$ are angles of 70° and 50°, respectively. Find x and y, the lengths of AC and BC.

2. Quadrilateral $ABCD$ of Figure T-12 is a parallelogram. E, F, G, and H are the midpoints of AB, BC, CD, and DA, respectively. Show that quadrilateral $EFGH$ also must be a parallelogram.

3. Quadrilateral $ABCD$ of Figure T-13 is an isosceles trapezoid with height 4 and bases 8 and 12. **(a)** Find the length of side AB. **(b)** Find the length of diagonal AC.

4. $\triangle ABC$ in Figure T-14 is an equilateral triangle that circumscribes a circle. If a side of the triangle measures 7 inches, **(a)** find the area of the shaded region that lies outside the circle but inside the triangle, and **(b)** find the perimeter of this shaded region.

5. A six-pointed star is formed when all of the diagonals of a regular hexagon are drawn, as is shown in Figure T-15. If the radius of the circle that circumscribes the hexagon is 5 inches, **(a)** find the perimeter of the star, and **(b)** find the area of the star.

6. The regular octagon shown in Figure T-16 has side 4 inches. **(a)** Find the area of the circle that circumscribes the octagon. **(b)** Find the area of the circle that is inscribed in the octagon. **(c)** Find the area of the region between these two circles that is also outside the octagon.

7. In Figure T-17, O is the center of a circle of radius 4 inches. **(a)** Find the measure of arc x in both degrees and inches. **(b)** Find the measure of angle y in degrees. **(c)** Find the shaded area of the circle, if secant AB measures 7 inches.

8. The larger circle in Figure T-18 has B as its center and is inscribed in a square of side 6 cm. The smaller circle is tangent to the larger circle and the square at point A, and passes through B. Find the perimeter of the shaded region.

9. In Figure T-19 chords AD and BC meet at point E in a circle. Minor arc $\overset{\frown}{AB}$ measures 20° and minor arc $\overset{\frown}{DC}$ measures 14.5 cm. $\angle CED$ measures 40°. **(a)** Find the radius of the circle. **(b)** Find the lengths of chords AB and CD.

10. In Figure T-18, if point B has coordinates $(1, 3)$ and point A has coordinates $(-2, 3)$, find the equations of both circles.

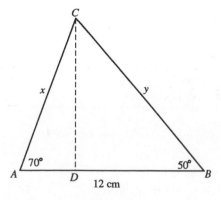

Figure T-11

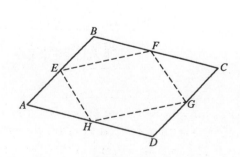

Figure T-12

234

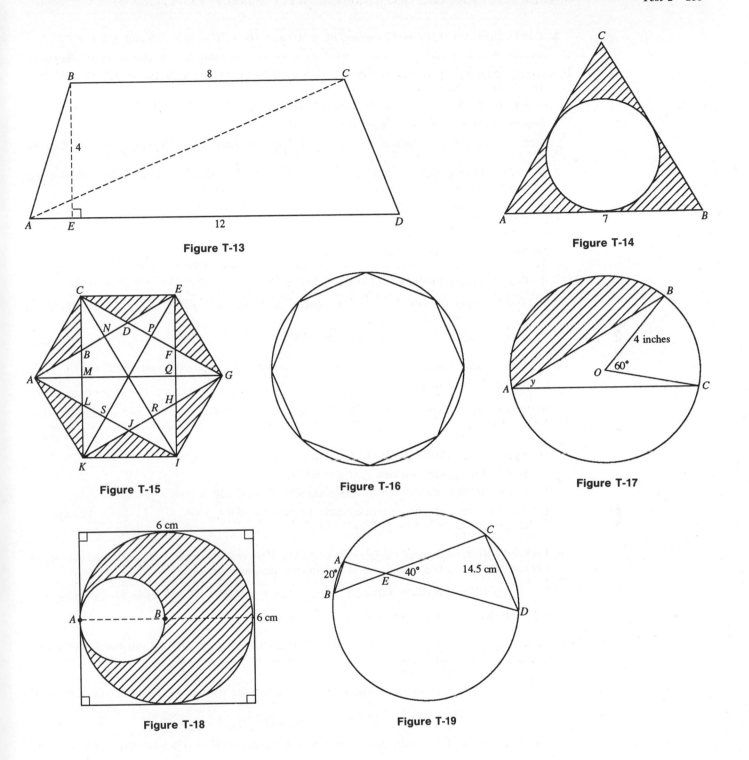

Figure T-13

Figure T-14

Figure T-15

Figure T-16

Figure T-17

Figure T-18

Figure T-19

Answers to Test 2

1. *Step 1:* $CD = x \sin(70°)$ and $CD = y \sin(50°)$. Thus $x \sin(70°) = y \cos(50°)$ or $x = 0.684y$.

Step 2: $AD = x \cos(70°)$ and $DB = y \cos(50°)$.

Step 3: $AD + DB = 12$. Thus $x \sin(70°) + y \cos(50°) = 12$ or $x(0.342) + y(0.643) = 12$.

Step 4: Substitute for x from step 1 and get $(0.685)(0.342)y + (0.643)y = 12$ or $(0.877)y = 12$.

Step 5: $y = 13.684$ cm and $x = (0.684)(13.684) = 9.36$ cm.

2. *Step 1:* Since $ABCD$ is a parallelogram, $BC \cong AD$ and $AB \cong DC$, $\angle B \cong \angle D$ and $\angle A \cong \angle C$.

Step 2: Since F and H are the midpoints of congruent line segments BC and AD, $BF \cong HD$. Similarly $EB \cong GD$.

Step 3: Thus $\triangle BEF \cong \triangle DGH$ by SAS.

Step 4: $EF \cong GH$ as corresponding parts of congruent triangles.

Step 5: Similarly, $\triangle HEA \cong \triangle FGC$ and $EH \cong GF$.

Step 6: Since $EFGH$ is a quadrilateral with pairs of opposite sides of equal length, it is a parallelogram.

3. (a) *Step 1:* Draw perpendicular BE to base AD. This cuts off segment AE. Because $AD - BC = 4$ and the two sides of an isosceles trapezoid are symmetric, AE has length 2.

Step 2: $\triangle ABE$ is a right triangle with legs 2 and 4. The hypotenuse is $\sqrt{2^2 + 4^2} = \sqrt{20} = 2\sqrt{5} = 4.47$.

(b) *Step 1:* Let x be the measure of $\angle ABE$. Since $\sin(x) = \dfrac{2}{\sqrt{20}} = \dfrac{1}{\sqrt{5}}$, x is approximately $27°$.

Step 2: Since $\angle EBC$ is a right angle, $\angle ABC$ is $27° + 90° = 117°$.

Step 3: Apply the law of cosines to $\triangle ABC$ and find $AC^2 = 20 + 64 - 2\sqrt{20}\,8\cos(117°)$ or $AC^2 = 116.49$.

Step 4: Diagonal AC measures $\sqrt{116.49} = 10.78$.

4. (a) *Step 1:* Draw a perpendicular from the circles center O to the point of tangency D on AB.

Step 2: Draw $\triangle OAD$.

Step 3: $\angle OAD = 30°$ and $AD = 3.5$. Thus in 30-60-90 $\triangle AOD$ $OD = \dfrac{3.5}{\sqrt{3}} = 2.02$ and the circle has a radius 2.02.

Step 4: Area $\triangle AOD = (\tfrac{1}{2})(2.02)(3.5) = 3.34$.

Step 5: Area $\triangle ABC = 6$ area $\triangle AOD = 6(3.34) = 21.21$.

Step 6: Shaded area $= \triangle ABC -$ area circle $= 21.21 - \pi(2.02)^2 = 8.39$.

(b) The perimeter of the shaded area is merely the perimeter of the $\triangle ABC$ ($3 \times 7 = 21$) + the perimeter of the circle ($2\pi \times 2 \cdot 20 = 12.69$) $= 33.69$.

5. Let's concentrate on a typical triangle such as $\triangle COE$. If we can find the area of $\triangle CDE$ and the lengths CD and DE, we can find our answer by multiplying these numbers by 6.

Step 1: $\triangle COE$ is an equilateral triangle. Thus, $\angle OCE = \angle OEC = 60°$ and $CE = 5$ inches. Also, area $\triangle COE = \tfrac{1}{2}(5)(5)\sin 60° = \dfrac{25}{2}\dfrac{\sqrt{3}}{2} = 25\sqrt{3}/4$ in^2.

Step 2: $\angle ACK$, $\angle KCI$, $\angle ICG$, and $\angle GCE$ all are congruent since they cut out equal chords, and hence also equal arcs of the circumscribed circle. They all sum to $120°$. Thus, $\angle GCE$ (or $\angle DCE$) $= 30°$. Similarly, $\angle DEC$ is $30°$.

Step 3: $\triangle CDE$ is an isosceles triangle with base 5 inches and base angles $30°$. Thus, the altitude of $\triangle CDE = \left(\dfrac{5}{2}\right)\tan(30°) = \dfrac{5}{2\sqrt{3}}$ and $CD = \dfrac{5/2}{\cos(30°)} = \dfrac{5}{2}\dfrac{2}{\sqrt{3}} = \dfrac{5}{\sqrt{3}}$ inches.

Step 4: The area of the entire shaded region is $6 \times$ area $\triangle CDE = 6(\tfrac{1}{2})(5/2\sqrt{3})(5) = 21.65$ in^2. The perimeter of the star is $12 \times \dfrac{5}{\sqrt{3}} = 43.64$ inches.

Step 5: The area of the star is the area of the entire hexagon $\left(6 \times \dfrac{25\sqrt{3}}{4} = 64.95\right)$ minus the area of the shaded region (21.65) $= 43.30$ in^2.

6. (a) *Step 1:* The radius of the octagon is $r = \dfrac{4/2}{\sin(180°/8)} = 5.23$ inches. This is the radius of the circumscribing circle.

Step 2: The area of the larger circle is $\pi(5.23)^2 = 85.81$.

(b) *Step 1:* The apothem of the octagon is $\dfrac{4}{2} \tan\left(\dfrac{135°}{2}\right) = 4.83$. This is the radius of the inscribed circle.

  Step 2: The area of the smaller circle is $(4.83)^2\pi = 73.24$.

(c) *Step 1:* The area of the octagon is $\dfrac{8(2^2)}{\tan(180°/8)} = 77.25$.

  Step 2: The area outside the octagon but between the two circles is $85.81 - 77.25 = 8.56$.

7. (a) $x = 60°$ (the same measure as the central angle)

(b) $y = \frac{1}{2}(x \text{ in degrees}) = 30°$

(c) *Step 1:* Draw radius OA and consider $\triangle AOB$. This is an isosceles triangle with sides 4 and base 7.

  Step 2: Let $x = \angle AOB$. From the law of cosines, $7^2 = 4^2 + 4^2 - 2(4)(4)\cos(x)$ or $49 = 32 - 32\cos(x) \cdot \cos(x) = \dfrac{49 - 32}{-32} = -0.53$.

  Step 3: $x = 122°$.

  Step 4: Area of pie-shaped sector $AOB = \dfrac{122°}{360°}\pi(4)^2 = 17.03 \text{ in}^2$.

  Step 5: From the law of sines, area of $\triangle AOB = \frac{1}{2} \cdot 4^2 \sin(122°) = 6.78 \text{ in}^2$.

  Step 6: Shaded area $= 17.03 - 6.78 = 10.25 \text{ in}^2$.

8. *Step 1:* The radius of the larger circle is 3 cm. Its perimeter is $2\pi3 = 18.85$ cm.

  Step 2: The radius of the smaller circle is 1.5 cm. Its perimeter is $2\pi(1.5) = 9.42$ cm.

  Step 3: The perimeter of the shaded region is $18.85 + 9.42 = 28.27$ cm.

9. (a) *Step 1:* Let $x =$ degree measure of arc CD. By the chord-chord angle theorem, $\dfrac{20° + x}{2} = 40°$.

  Step 2: $x = 60°$.

  Step 3: $\dfrac{14.5}{2\pi r} = \dfrac{60°}{360°}$.

  Step 4: $r = \dfrac{(6)(14.5)}{2\pi} = 13.85$ cm.

(b) *Step 1:* $\dfrac{\text{arc } AB}{2\pi(13.85)} = \dfrac{20°}{360°}$, so arc $\widehat{AB} = \dfrac{2\pi(13.85)}{18} = 4.83$ cm.

  Step 2: Let O be the center of the circle, and consider the triangles $\triangle OCD$ and $\triangle OAB$. From the law of cosines $CD^2 = 13.85^2 + 13.85^2 - 2 \times 13.85^2 \cos(60°) = 13.85^2$ and $AB^2 = 13.85^2 + 13.85^2 - 2 \times 13.85^2 \cos(20°) = 13.85^2 \times 0.12 = 23.14$.

  Step 3: Thus $CD = 13.85$ cm and $AB = 4.8$ cm.

  note: $\triangle ABE \sim \triangle CDE$ and the corresponding sides stand in a ratio of $\dfrac{4.8}{13.85} = 0.35$.

10. The equation of the larger circle is $(x - 1)^2 + (y - 3)^2 = 3^2$. The center of the smaller circle is $\left(\dfrac{1-2}{2}, \dfrac{3+3}{2}\right)$ or $\left(-\dfrac{1}{2}, 3\right)$. The equation of the smaller circle is $\left(x + \dfrac{1}{2}\right)^2 + (y - 3)^2 = \left(\dfrac{3}{2}\right)^2$.

TEST 3

1. Find the distance in 3-space between the points whose coordinates are $(1, 3, 2)$ and $(2, -1, 7)$.

2. The line segment joining the points whose coordinates are $(1, 2, 0)$ and $(3, 4, 1)$ is the diagonal of a rectangular solid (a box) whose sides are parallel to the coordinate planes in Cartesian 3-space. (a) Find the coordinates of the other corners of the rectangular solid. (b) Find the angle this diagonal makes with an edge of the box that is parallel to the z axis. (c) Find the angle that this diagonal makes with a side of the box that is parallel to the xy plane.

3. Find the coordinates of the midpoint of the line segment joining points whose coordinates are $(0, 4, 5)$ and $(-1, 2, 3)$.

4. A sphere of radius 3 is centered at $(1, 2, 1)$. (a) Find the 3-space Cartesian equation of this sphere. (b) This sphere intersects the xy plane in a circle. Find the equation of this circle in the xy plane. What are the center and the radius of this circle?

5. An oblique circular cone has a vertex with coordinates $(2, 3, 5)$. Its base is a circle in the xy plane with radius 3 and center at $(0, 0)$. (a) Find the volume of this cone. (b) A plane parallel to the base and 2 units above it passes through the cone and cuts out an oblique frustrum and a smaller oblique cone. Find the volume of each of these solids

6. A pyramid is formed by 4 intersecting planes in space. The planes meet (three at a time) at the points $(0, 0, 2)$, $(1, 0, 0)$, $(0, 3, 0)$, and $(1, 4\ 6)$. The edges of the pyramid are the line segments joining these points. Find the surface area of this pyramid.

7. An ice cream cone has a height of 4 inches and a circular opening of radius 1.5 inches. The cone is filled completely with ice cream and a hemispherical ball of ice cream is placed on the top of the cone so as to fit perfectly onto the cone's circular opening. (a) Find the volume of ice cream used. (b) If the ice cream end of the cone is dipped into chocolate and submerged until only 3 inches of the cone sticks above the chocolate surface, find the surface area of the chocolate coating.

8. An oblique prism is formed as in Figure T-20 with base and top an isosceles trapezoid whose bases are 6 and 8 cm long, and whose sides have length $\sqrt{2}$ cm. The edges of the sides of the prism measure 4 cm and make an angle of $40°$ with the plane of the base. Find the volume of the prism.

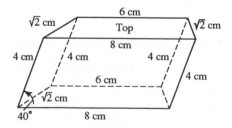

Figure T-20

9. Find $\tan(\alpha)$, $\cot(\alpha)$, $\sec(\alpha)$, $\csc(\alpha)$, $\sin(2\alpha)$, and $\cos(2\alpha)$ without a calculator if $\sin(\alpha) = \dfrac{1}{3}$ and $\cos(\alpha) = \dfrac{2\sqrt{2}}{3}$.

10. Find $\tan(\alpha + \beta)$ if $\sin(\alpha) = \dfrac{1}{3}$, $\cos(\alpha) = \dfrac{2\sqrt{2}}{3}$, $\sin(\beta) = \dfrac{2}{5}$, and $\cos(\beta) = \dfrac{\sqrt{21}}{5}$.

Answers to Test 3

1. Distance $= \sqrt{(2-1)^2 + (-1-3)^2 + (7-2)^2} = \sqrt{(1)^2 + (-4)^2 + (5)^2} = \sqrt{1+16+25} = \sqrt{25} = 7.21$.

2. (a) Draw the rectangular solid as in Figure T-21 with vertices A, B, C, D, E, F, G, H. We are given points $A(1,2,0)$ and $G(3,4,1)$. Thus, the base of the solid is in the $z=0$ plane and the top is in the $z=1$ plane. The point E (above A) is in the $z=1$ plane and projects onto A. E has coordinates $(1,2,1)$. Point G projects onto point C, which thus has coordinates $(3,4,0)$. The remaining sides are in the parallel pairs of planes $x=1$ and $x=3$ and the planes $y=2$ and $y=4$. B's coordinates are $(3,2,0)$, F's coordinates are $(3,2,1)$, H's coordinates are $(1,4,1)$, and D's coordinates are $(1,4,0)$.

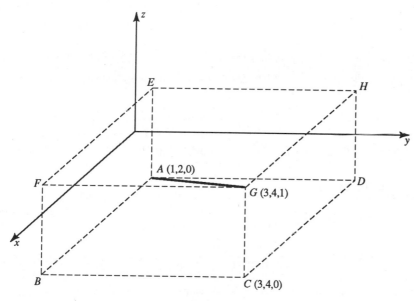

Figure T-21

(b) Points A, E, G, and C lie on two parallel lines and hence are coplanar. Draw $\triangle AEG$ in the plane that they define. The angle you want is $\angle A$ between AE and AG. Now $\triangle AEG$ is a right triangle with hypotenuse AG whose length is $\sqrt{(3-1)^2 + (4-2)^2 + (1-0)^2} = 3$. Leg AE measures 1. Thus, the cosine of $\angle A$ is $\frac{1}{3}$ and $\angle A$ is approximately $71°$.

(c) note: Since AE is parallel to the z axis, it is perpendicular to the xy plane, and the plane of $AEGC$ is perpendicular to the xy plane. Thus, the angle between the diagonal AG and base $ABCD$ is merely $90° - \angle EAG = 19°$.

3. Let $P = (x, y, z)$ be the midpoint of line segment AB, where $A = (-1, 2, 3)$ and $B = (0, 4, 5)$. Then $x = \dfrac{-1+0}{2} \simeq -\dfrac{1}{2}$, $y = \dfrac{2+4}{2} = 3$, and $z = \dfrac{3+5}{2} = 4$. So $P = \left(-\dfrac{1}{2}, 3, 4\right)$.

4. (a) $(x_0, y_0, z_0) = (1, 2, 1)$ and $r = 3$. Thus, the equation of the sphere is $(x-1)^2 + (y-2)^2 + (z-1)^2 = 3^2$.

(b) There are two ways to do this problem.

Method 1: When the sphere cuts the xy plane, $z = 0$. Thus, the equation becomes $(x-1)^2 + (y-2)^2 + (0-1)^2 = 3^2$ or $(x-1)^2 + (y-2)^2 = 8$. This is the equation of a circle in the xy plane with center at $(1, 2)$ and radius $\sqrt{8} = 2\sqrt{2}$.

Method 2: Project the center of the sphere onto the xy plane. This gives $(1, 2, 0)$ as the center of the circle. From Figure T-22 we see that the radius of the planar circle is the leg of a right triangle with hypotenuse 3 and the other leg 1. From the Pythagorean theorem, $r = \sqrt{3^2 - 1^2} = \sqrt{8} = 2\sqrt{2}$. The equation of the circle is $(x-1)^2 + (y-2)^2 = 8$.

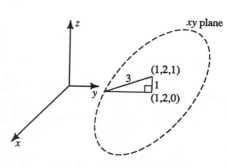

Figure T-22

5. (a) The vertex of the cone is 5 units above the xy plane. Thus, $h = 5$ and $r = 3$. The volume of the cone is $V = \frac{1}{3}\pi r^2 h = \frac{1}{3}\pi 3^2 5 = 47.12$.

(b) The plane cuts out a circle of radius s from the cone. This is the base of the small oblique cone whose height is $5 - 2 = 3$. Now

$$\frac{r_{smaller\ circle}}{r_{larger\ circle}} = \frac{h_{small\ cone}}{h_{large\ cone}}$$

or $\frac{s}{3} = \frac{3}{5}$, which gives $s = \frac{9}{5} = 1.8$. Thus, the volume of the smaller cone is $\frac{1}{3}\pi(1.8)^2 3 = 3.24\pi = 10.18$. The volume of the frustrum $= 47.12 - 10.18 = 36.94$.

6. Let the vertices be $A = (0, 0, 2)$, $B = (1, 0, 0)$, $C = (0, 3, 0)$, and $D = (1, 4, 6)$.

Step 1: Lengths of $AB = \sqrt{1^2 + 0^2 + 2^2} = \sqrt{5}$, $AC = \sqrt{0^2 + 3^2 + 2^2} = \sqrt{13}$, $AD = \sqrt{1^2 + 4^2 + 4^2} = \sqrt{33}$, $BC = \sqrt{1^2 + 3^2 + 0^2} = \sqrt{10}$, $BD = \sqrt{0^2 + 4^2 + 6^2} = \sqrt{52}$, and $CD = \sqrt{1^2 + 1^2 + 6^2} = \sqrt{38}$.

Step 2: The surface area of the pyramid is the sum of the surface areas of each of the triangular faces.

Step 3: Find the surface area of each triangle by the law of sines after you've found one angle of the triangle by the law of cosines.

Step 4: For $\triangle ABC$, $\cos(A) = \dfrac{AB^2 + AC^2 - BC^2}{2(AC)(AB)} = \dfrac{5 + 13 - 10}{2\sqrt{5}\sqrt{13}} = \dfrac{8}{16.12} = 0.50$, so $\angle A = 60°$. Area

$$\triangle ABC = \frac{1}{2}(AB)(AC)\sin(A) = \frac{\sqrt{5}\sqrt{13}}{2}\sin(60°) = 3.49.$$

Step 5: For $\triangle BCD$, $\cos(C) = \dfrac{CB^2 + CD^2 - BD^2}{2(CB)(CD)} = \dfrac{10 + 38 - 52}{2\sqrt{10}\sqrt{38}} = \dfrac{-4}{38.99} = -0.10$, so $\angle C = 96°$.

$$\text{Area } \triangle BCD = \frac{1}{2}(CB)(CD)\sin(C) = \frac{\sqrt{10}\sqrt{38}}{2}\sin(96°) = 9.69.$$

Step 6: For $\triangle ABD$, $\cos(A) = \dfrac{AB^2 + AD^2 - BD^2}{2(AB)(AD)} = \dfrac{5 + 33 - 52}{2\sqrt{5}\sqrt{33}} = \dfrac{-14}{25.69} = -0.54$, so $\angle A = 123°$.

$$\text{Area } \triangle ABD = \frac{1}{2}(AB)(AD)\sin(A) = \frac{\sqrt{5}\sqrt{33}}{2}\sin(123°) = 5.39.$$

Step 7: For $\triangle ACD$, $\cos(A) = \dfrac{AC^2 + AD^2 - CD^2}{2(AC)(AD)} = \dfrac{13 + 33 - 38}{2\sqrt{13}\sqrt{33}} = \dfrac{8}{41.42} = 0.19$, so $\angle A = 79°$. Area

$$\triangle ACD = \frac{1}{2}(AC)(AD)\sin(A) = \frac{\sqrt{13}\sqrt{33}}{2}\sin(79°) = 10.17.$$

Step 8: The entire surface area is $= 3.49 + 9.69 + 5.39 + 10.17 = 28.74$.

7. (a) The volume of the cone is $\frac{1}{3}\pi r^2 h = \frac{1}{3}\pi(1.5)^2 4 = 3\pi = 9.42$ in^3. The volume of the hemisphere $= \frac{1}{2}\left(\dfrac{4\pi r^3}{3}\right) = 2.25\pi = 7.07$ in^3. The cone uses 16.49 in^3 of ice cream.

(b) *Step 1:* The surface area of the hemisphere is $2\pi r^2 = 4.5\pi = 14.14$ in^2.

Step 2: The surface area of the frustrum $=$ the area of the entire cone $-$ the area of the cone above the chocolate.

Step 3: Since the smaller cone is similar to the larger cone, $\dfrac{r_{small}}{1.5} = \dfrac{h_{small}}{4}$ or $r_{small} = \dfrac{(1.5)(3)}{4} = 1.13$.

Step 4: Thus, the area of the frustrum $= \sqrt{1.5^2 + 4^2}\,\pi(1.5) - \sqrt{1.13^2 + 3^2}\,\pi(1.13) = (6.41 - 3.62)\pi = 8.77$ in^2.

Step 5: The area of the chocolate is $14.14 + 8.77 = 22.91$ in^2.

8. *Step 1:* The prism's height (distance between the two trapezoidal bases) is $4\sin(40°) = 2.57$ cm.

Step 2: The height of the isosceles trapezoidal base is $\sqrt{(\sqrt{2})^2 - \left(\dfrac{8 - 6}{2}\right)^2} = \sqrt{1} = 1$.

Step 3: The base area is $\frac{1}{2}(1)(6 + 8) = 7$ cm^2.

Step 4: The volume of the prism is $h \times$ base area $= (2.57)(7) = 18.0$ cm^3.

9. $\tan(\alpha) = \dfrac{\sin(\alpha)}{\cos(\alpha)} = \dfrac{\dfrac{1}{3}}{\dfrac{2\sqrt{2}}{3}} = \dfrac{1}{2\sqrt{2}}$

$\cot(\alpha) = \dfrac{1}{\tan(\alpha)} = 2\sqrt{2}$

$\sec(\alpha) = \dfrac{1}{\cos(\alpha)} = \dfrac{3}{2\sqrt{2}}$

$\csc(\alpha) = \dfrac{1}{\sin(\alpha)} = 3$

$\sin(2\alpha) = 2\sin(\alpha)\cos(\alpha) = 2\left(\dfrac{1}{3}\right)\left(\dfrac{2\sqrt{2}}{3}\right) = \dfrac{4\sqrt{2}}{9}$

$\cos(2\alpha) = \cos^2(\alpha) - \sin^2(\alpha) = \dfrac{8}{9} - \dfrac{1}{9} = \dfrac{7}{9}.$

10. $\tan(\alpha + \beta) = \dfrac{\sin(\alpha + \beta)}{\cos(\alpha + \beta)} = \dfrac{\sin(\alpha)\cos(\beta) + \cos(\alpha)\sin(\beta)}{\cos(\alpha)\cos(\beta) - \sin(\alpha)\sin(\beta)}$

$= \dfrac{\left(\dfrac{1}{3}\right)\left(\dfrac{\sqrt{21}}{5}\right) + \left(\dfrac{2\sqrt{2}}{3}\right)\left(\dfrac{2}{5}\right)}{\left(\dfrac{2\sqrt{2}}{3}\right)\left(\dfrac{\sqrt{21}}{5}\right) - \left(\dfrac{1}{3}\right)\left(\dfrac{2}{5}\right)} = \dfrac{\sqrt{21} + 4\sqrt{2}}{2\sqrt{42} - 2} = \dfrac{4.58 + 5.66}{12.96 - 2} = \dfrac{10.24}{10.96} = 0.93$

INDEX